ANATOMY AND PHYSIOLOGY OF FARM ANIMALS

Anatomy and Physiology of Farm Animals

Editor:

Stephen Rego, Ph.D.

Delve Publishing
www.delvepublishing.com

Anatomy and Physiology of Farm Animals
Editor: **Stephen Rego, Ph.D.**

708 3rd Avenue, 6th Floor
New York, NY 10017
United States of America
www.delvepublishing.com

ISBN: 978-1-68095-859-1

Library of Congress Control Number: 2016951612

About the Editor

Stephen Rego, Ph.D.

Stephen obtained his PhD from the University of North Carolina at Charlotte in 2013 where his graduate research focused on cancer immunology and the tumor microenvironment. He received postdoctoral training in regenerative and translational medicine, specifically gastrointestinal tissue engineering, at the Wake Forest Institute of Regenerative Medicine. Currently, Stephen is an instructor for anatomy and physiology and biology at Forsyth Technical Community College.

Contents

List of Contributors

A. M. Scholz
Livestock Center Oberschleißheim, Ludwig-Maximilians-University Munich, Sankt-Hubertus-strasse 12, 85764 Oberschleißheim, Germany;

L. Bünger
SRUC, Animal and Veterinary Sciences, Roslin Institute Building, Easter Bush, Midlothian, Scotland EH25 9RG, UK

J. Kongsro
Norsvin, Department of Animal and Aquacultural Sciences, c/o Norwegian University of Life Sciences, PO Box 5003, N-1432 Ås, Norway

U. Baulain
Institute of Farm Animal Genetics, Friedrich-Loeffler-Institut, Hoeltystr.10, 31535 Neustadt, Germany;

A. D. Mitchell
Agricultural Research Service (Retired), US Department of Agriculture, 10300 Baltimore Avenue, BARC-West, Beltsville, MD 20705, USA

C. Egger-Danne
ZuchtData EDV-Dienstleistungen GmbH, Dresdner Str. 89/19, A-1200 Vienna, Austria;

J. B. Cole
Animal Genomics and Improvement Laboratory, ARS, USDA, 10300 Baltimore Avenue, Beltsville, MD 20705-2350, USA

J. E. Pryce
Department of Environment and Primary Industries, La Trobe University, Agribio, 5 Ring Road, Bundoora, Victoria 3083, Australia;

N. Gengler
University of Liège, Gembloux Agro-Bio Tech (GxABT), Animal Science Unit, Passage des Déportés 2, B-5030 Gembloux, Belgium;

B. Heringstad
Department of Animal and Aquacultural Sciences, Norwegian University of Life Sciences, PO Box 5003, N-1432 Ås, Norway;

A. Bradley
Quality Milk Management Services Ltd, Cedar Barn, Easton Hill, Easton, Wells, Somerset, BA5 1EY, UK; University of Nottingham, School of Veterinary Medicine and Science, Sutton Bonington Campus, Sutton Bonington, Leicestershire, LE12 5RD, UK;

K. F. Stock
Vereinigte Informationssysteme Tierhaltung w.V. (vit), Heideweg 1, D-27283 Verden, Germany

G. J. Stefánsdóttir
Department of Equine Science, Hólar University College, 551 Sauðárkrókur, Iceland;
Department of Animal Nutrition and Management, Swedish University of Agricultural Sciences, 750 07 Uppsala, Sweden

S. Ragnarsson
Department of Equine Science, Hólar University College, 551 Sauðárkrókur, Iceland;

V. Gunnarsson
Department of Equine Science, Hólar University College, 551 Sauðárkrókur, Iceland;

A. Jansson
Department of Equine Science, Hólar University College, 551 Sauðárkrókur, Iceland;
Department of Animal Nutrition and Management, Swedish University of Agricultural Sciences, 750 07 Uppsala, Sweden

R. Jha
Department of Human Nutrition, Food and Animal Sciences, College of Tropical Agriculture and Human Resources, University of Hawaii at Manoa, Honolulu, HI 96822, US

J D. Berrocoso
Department of Human Nutrition, Food and Animal Sciences, College of Tropical Agriculture and Human Resources, University of Hawaii at Manoa, Honolulu, HI 96822, US

D. I. Batonon-Alavo
INRA, UR83 Recherches Avicoles, Nouzilly, France;
Research Centre for Animal Nutrition and Health, DSM Nutritional Products France, Saint Louis, France;

D. Bastianelli
CIRAD, UMR SELMET, Systèmes d'élevage méditerranéens et tropicaux, Baillarguet TA C-112/A, Montpellier Cedex 05, France;

P. Lescoat
AgroParisTech, UMR 1048 SADAPT, Paris, France;

G. M. Weber
Nutrition Innovation Center, DSM Nutritional Products Ltd, Basel, Switzerland

M. Umar Faruk
Research Centre for Animal Nutrition and Health, DSM Nutritional Products France, Saint Louis, France;

K. M. Hemmingsa
School of Biosciences, Division of Nutritional Sciences, The University of Nottingham, Sutton Bonington Campus, Leicestershire, LE12 5RD, UK

Z. C. T. R. Daniel
School of Biosciences, Division of Nutritional Sciences, The University of Nottingham, Sutton Bonington Campus, Leicestershire, LE12 5RD, UK

P. J. Buttery
School of Biosciences, Division of Nutritional Sciences, The University of Nottingham, Sutton Bonington Campus, Leicestershire, LE12 5RD, UK

T. Parr
School of Biosciences, Division of Nutritional Sciences, The University of Nottingham, Sutton Bonington Campus, Leicestershire, LE12 5RD, UK

J. M. Brameld
School of Biosciences, Division of Nutritional Sciences, The University of Nottingham, Sutton Bonington Campus, Leicestershire, LE12 5RD, UK

L. M. Pluym
Technology and Food Sciences Unit, Institute of Agricultural and Fisheries Research, Burgemeester Van Gansberghelaan 115 bus 1, 9820 Merelbeke, Belgium
Department of Reproduction, Obstetrics and Herd Health, Ghent University, Salisburylaan 133, 9820 Merelbeke, Belgium

A. Van Nuffel
Technology and Food Sciences Unit, Institute of Agricultural and Fisheries Research, Burgemeester Van Gansberghelaan 115 bus 1, 9820 Merelbeke, Belgium

S. Van Weyenberg
Technology and Food Sciences Unit, Institute of Agricultural and Fisheries Research, Burgemeester Van Gansberghelaan 115 bus 1, 9820 Merelbeke, Belgium

D. Maes
Department of Reproduction, Obstetrics and Herd Health, Ghent University, Salisburylaan 133, 9820 Merelbeke, Belgium

A. O'Connor
Department of Veterinary Diagnostic and Production Animal Medicine, College of Veterinary Medicine, Iowa State University, Ames, 50011 IA, USA;

R. Anthony
Department of Philosophy, University of Alaska Anchorage, Anchorage, 99508 AK, USA;

L. Bergamasco
Department of Animal and Poultry Sciences, College of Agriculture and Life Science, Virginia Tech, Blacksburg, VA 24060, USA;

J. F. Coetzee
Department of Veterinary Diagnostic and Production Animal Medicine, College of Veterinary Medicine, Iowa State University, Ames, 50011 IA, USA;

R. S. Dzikamunhenga
Department of Veterinary Diagnostic and Production Animal Medicine, College of Veterinary Medicine, Iowa State University, Ames, 50011 IA, USA;

A. K. Johnson
Department of Animal Science, College of Agriculture and Life Sciences, Iowa State University, Ames, 50011 IA, USA;

L. A. Karriker
American Association of Swine Veterinarians (AASV), Perry, 50220 IA, USA;

J. N. Marchant-Forde
USDA-ARS, Livestock Behavior Research Unit, West Lafayette, 47907 IN, USA;

G. P. Martineau
Department of Animal Production, National Veterinary School, Swine Clinic, 31076 Toulouse, France; 7

S. T. Millman
Department of Veterinary Diagnostic and Production Animal Medicine, College of Veterinary Medicine, Iowa State University, Ames, 50011 IA, USA;
Department of Biomedical Science, College of Veterinary Medicine, Iowa State University, Ames, 50011 IA, USA;

E. A. Pajor
Department of Production Animal Health, Faculty of Veterinary Medicine, University of Calgary, AB T2N 1N4 Calgary, Canada

K. Rutherford
Scotland's Rural College (SRUC), EH9 3JG Edinburgh, United Kingdom;

M. Sprague
American Association of Swine Veterinarians (AASV), Perry, 50220 IA, USA;

M. A. Sutherland
AgResearch Ltd, Ruakura Research Centre, 3240 Hamilton, New Zealand;

E. von Borell
Department of Animal Husbandry and Ecology, Institute of Agricultural and Nutritional Sciences, Martin Luther-University HalleWittenberg, 06108 Halle (Saale), Germany;

S. R. Webb
National Pork Board, Des Moines, 50325 IA, USA;

Aurora Villarroel
Department of Clinical Sciences, College of Veterinary Medicine, Oregon State University, 158 Magruder Hall, Corvallis, OR 97331, USA

Taylor B Miller
Department of Clinical Sciences, College of Veterinary Medicine, Oregon State University, 158 Magruder Hall, Corvallis, OR 97331, USA

Justin K Ward
Department of Clinical Sciences, College of Veterinary Medicine, Oregon State University, 158 Magruder Hall, Corvallis, OR 97331, USA

Elisha D Johnson
Department of Clinical Sciences, College of Veterinary Medicine, Oregon State University, 158 Magruder Hall, Corvallis, OR 97331, USA

Kimberly R Noyes
Department of Clinical Sciences, College of Veterinary Medicine, Oregon State University, 158 Magruder Hall, Corvallis, OR 97331, USA

Afshin Hosseini
Department of Animal Sciences and Division of Nutritional Sciences, University of Illinois, Urbana, IL 61801, USA

Rekha Sharma
National Bureau of Animal Genetics Resources (NBAGR), Karnal, India

Massimo Bionaz
Department of Animal and Rangeland Sciences, Oregon State University, Corvallis, OR 97330, USA

Juan J Loor
Department of Animal Sciences and Division of Nutritional Sciences, University of Illinois, Urbana, IL 61801, USA

Basil Alexander
Department of Biomedical Sciences, Ontario Veterinary College (OVC), University of Guelph, Ontario, Canada
Department of Farm Animal Production and Health, Faculty of Veterinary Medicine and Animal Science, University of Peradeniya, Sri Lanka

Gabriela Mastromonaco
Department of Biomedical Sciences, Ontario Veterinary College (OVC), University of Guelph, Ontario, Canada
Reproductive Physiology, Toronto Zoo, 361A Old Finch Avenue, Scarborough, ON, M1B 5K7

W. Allan King
Department of Biomedical Sciences, Ontario Veterinary College (OVC), University of Guelph, Ontario, Canada

Córdova-Izquierdo Alejandro
Department of Agricultural and Animal Production, Metropolitan Autonomous University, Xochimilco, Academic Division of Agricultural Sciences, Juárez Autonomous University of Tabasco, Mexico

Villa-Mancera Abel
Laboratory of Genetics and Reproduction, Faculty of Veterinary Medicine, Autonomous University of Puebla, Mexico

Olivares Pérez Jaime
Autonomous University of Guerrero-Unit Academic Veterinary Medicine and Animal Science, Mexico

Sánchez-Aparicio Pedro
Faculty of Veterinary Medicine, Autonomous University of the State of Mexico, Bioterio Veterinary Hospital for small species UAEM, Toluca, Mexico

Massimo Bionaz
Department of Animal and Rangeland Sciences, Oregon State University, USA

M. T. W. Verhoeven
Livestock Research, Wageningen University and Research Centre, P.O. Box 338, 6700 AH Wageningen, The Netherlands;
Adaptation Physiology Group, Department of Animal Sciences, Wageningen University, P.O. Box 338, 6700 AH Wageningen, The Netherlands;

M. A. Gerritzen
Livestock Research, Wageningen University and Research Centre, P.O. Box 338, 6700 AH Wageningen, The Netherlands;

L. J. Hellebrekers
Faculty of Veterinary Medicine, Utrecht Unversity, P.O. Box 80154, 3508 TD Utrecht, The Netherlands

B. Kemp
Adaptation Physiology Group, Department of Animal Sciences, Wageningen University, P.O. Box 338, 6700 AH Wageningen, The Netherlands;

Preface

The anatomy and physiology of farm animals is an important topic that has implications ranging from human health to the economy. With the broad scope of topics associated with farm animal anatomy and physiology this book covers various animals including, cows (which are discussed in detail), horses, sheep, pig, and goats. The specific topics covered in this book include imaging modalities to study anatomy, physiologic responses to environment and diet, pharmacokinetics, reproductive physiology and developmental physiology. This book may interest both lay and professional individuals interested in furthering their understanding of a wide array of farm animal anatomy and physiology topics. The final section of this book introduces the consciousness of farm animals being slaughtered to highlight the importance of anatomy and physiology in the humane treatment of commercial animals. The text is comprised of fifteen chapters. First chapter provides a summary of the recent developments in the application of dual-energy X-ray absorptiometry, X-ray computed tomography, magnetic resonance imaging and ultrasound in livestock. Second chapter provides an overview of new traits and phenotyping strategies in dairy cattle with a focus on functional traits. Succedingly, the aim of third chapter was to describe the response in terms of heart and respiratory rate (RR), Htc, rectal temperature (RT), and some plasma variables in stallions and mares of different ages performing a BEFT. Fourth chapter provides relevant information on the utilization of dietary fiber (DF) by pigs and its effects on gut physiological functions, microbiota and health. It delivers in-depth insight on both negative and positive effects of different fibers inclusion in swine diets. Additional knowledge to the use of sorghum, millet and cottonseed meal in commercial poultry production has been provided in fifth chapter and sixth chapter provides mechanistic information on the anabolic responses to growth hormone and β-adrenergic agonist in sheep muscle. Seventh chapter highlights on the economic impact of claw lesions and lameness helps to understand how both the diseases influence economic losses, and indicates how these problems should best be addressed. Furthermore, it implicates that focus on housing in the insemination stable is important to improve the prevention strategy. In eighth chapter, we discuss approaches for comprehensive reporting in animal welfare studies. Ninth chapter aims to determine if TP concentrations differ in fresh and frozen/thawed samples; and to determine the correlation between serum and plasma samples. Tenth chapter focuses on factors affecting serum total protein and immunoglobulin G concentration in replacement dairy calves. Transcriptomics comparisons of Mac-T cells versus mammary tissue during late pregnancy, and recent advances in reproductive biotechnologies in sheep and goat are discussed in chapter eleven and twelve respectively. Thirteenth chapter reviews on environmental stress effect on animal reproduction. At present, it has been concluded that stress is one of the environmental factors that affects handling and decreases the production of livestock, it is considered that the main factor which should be controlled in units Animal Production, since it is closely related to pathogens and infectious agents that may try to the health of animals. Fourteenth chapter presents nutrigenomics approaches to fine-tune metabolism and milk production. Last chapter shows validation of indicators used to assess unconsciousness in veal calves at slaughter.

Editor

Stephen Rego, Ph.D.

INTRODUCTION

The term anatomy has come to refer to the science that deals with the form and structure of all organisms. Literally, the word means to cut apart; it was used by early anatomists when speaking of complete dissection of a cadaver. In contrast to anatomy, which deals primarily with structure, physiology is the study of the integrated functions of the body and the functions of all its parts (systems, organs, tissues, cells, and cell components), including biophysical and biochemical processes. When anatomy and physiology courses are taught separately, the approach to the laboratory portion of each course is considerably different. Study in a typical gross anatomy laboratory is based primarily on dissection of animal cadavers.

These usually have been preserved by embalming, and one or more parts of the vascular system have been injected with a colored material to facilitate identification of the vessels. Careful dissection coupled with close observation gives the student a concept of the shape, texture, location, and relations of structures visible to the unaided eye that can be gained in no other way. Similarly, the use of the microscope with properly prepared tissue sections on slides is essential for understanding structures that are so small they cannot be seen without optical or electron microscopic assistance. In the physiology laboratory, the student studies the response of whole animals, isolated organs, or individual cells to changes in their environment (both internal and external). Changes may be induced by almost any agent or manipulation, for example, drugs, changes in temperature or altitude, surgical modifications (such as neutering), and changes in diet.

Monitoring of the responses may be as simple as monitoring changes in body weight or as complex as measuring the electrical potential across the cell membrane of a single cell. Anatomists and physiologists working in research use some of the same techniques that are used in teaching laboratories but with considerable refinement. Both types of scientists use equipment and methods developed in the physical sciences, particularly chemistry and physics. The anatomist applies the principles of physics to the use of microscopes and applies knowledge of chemistry in the staining of various parts of cells and tissues.

The combination of chemistry and microscopic anatomy is known as histochemistry. Although anatomy and physiology are commonly pursued as more or less independent disciplines, they are both facets of the study of the animal body. A thorough knowledge of structure imparts much information about its function. However, a mere description of structure without describing function would be of little practical value. Conversely, it is impossible to gain a thorough understanding of function without a basic knowledge of structure.

The science of anatomy has become so extensive that it is now divided into many specialized branches. In fact, Dorland's Medical Dictionary defines 30 subdivisions of anatomy. This text chiefly describes gross (macroscopic) anatomy. This is the study of the form and relations (relative positions) of the structures of the body that can be seen with the unaided eye. Comparative anatomy is a study of the structures of various species of animals, with particular emphasis on those characteristics that aid in classification. Embryology is the study of developmental anatomy, covering the period from conception (fertilization of the egg) to birth. Another large branch of anatomy consists of the study of tissues and cells that can be seen only with the aid of a microscope. This is known as microscopic anatomy, or histology. The most recent development in the study of anatomy is ultra-structural cytology, which deals with portions of cells and tissues as they are visualized with the aid of the electron microscope. The term fine structure is used frequently in reference to structures seen in electron micrographs (photographs made with the electron microscope). Our approach to the study of anatomy will be chiefly by systems—systematic anatomy. To name the study, the suffix-ology, which means branch of knowledge or science, is added to the root word referring to the system. Table 1 indicates the commonly accepted systems, the name of the study of those systems, and the chief structures involved in each system.

Table 1: Nomenclature for Systematic Anatomy

System	*Name of Study*	*Chief Structures*
Skeletal system	Osteology	Bones
Articular system	Arthrology	Joints
Muscular system	Myology	Muscles
Digestive system	Splanchnology	Stomach and intestines
Respiratory system	Splanchnology	Lungs and airways
Urinary system	Splanchnology	Kidneys and urinary bladder
Reproductive system	Splanchnology	Ovaries and testes
Endocrine system	Endocrinology	Ductless glands
Nervous system	Neurology	Brain, spinal cord, and nerves
Circulatory system	Cardiology	Heart and vessels
Sensory system	Esthesiology	Eye and ear

Physiology has also become so extensive in scope that many areas of specialization are recognized. Like anatomy, these may be based on body systems (e.g., neurophysiology, gastrointestinal physiology, cardiovascular physiology, respiratory physiology, endocrine physiology, and reproductive physiology) or the level of biological organization (cell physiology and organismal physiology). All of these subdivisions become the parts of such overall areas of study as applied physiology, comparative physiology, pathophysiology, medical physiology, and mammalian physiology. We will be concerned with these systems and studies as they relate specifically to farm animals.

Descriptive Terms Useful in the Study of Anatomy

When giving geographic locations, we make use of certain arbitrary frames of reference known as meridians of latitude and longitude. However, since an animal is rarely oriented exactly with a line on the earth's surface, our frames of reference must be in relation to the animal itself and must apply regardless of the position or direction of the animal (Figure 1). Many terms of direction differ significantly between human and domestic animal anatomy because of the orientation of bipedal versus quadrupedal stance. Although use of human anatomical nomenclature in quadrupeds usually leads to confusion, the terms anterior, posterior, superior, and inferior are frequently used to describe the eye and aspects of dental anatomy of both human beings and domestic animals.

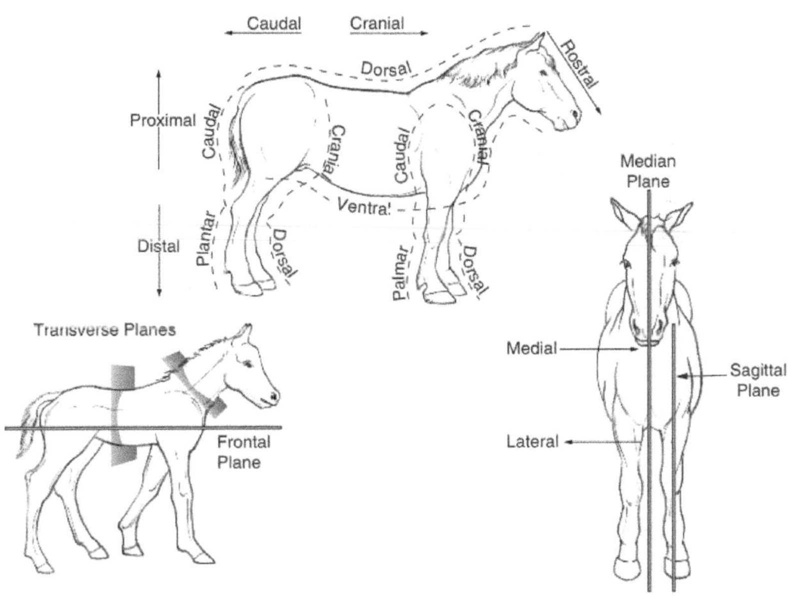

Figure 1: Directional terms and planes of the animal body.

- *Cranial* is a directional term meaning toward the head. The shoulder is cranial to the hip; it is closer to the head than is the hip.
- *Caudal* means toward the tail. The rump is caudal to the loin.
- *Rostral and caudal* are directional terms used in reference to features of the head to mean toward the nose (rostral) or toward the tail (caudal).
- *The median plane* is an imaginary plane passing through the body so as to divide

the body into equal right and left halves. A beef carcass is split into two halves on the median plane.

- *A sagittal plane* is any plane parallel to the median plane. The median plane is sometimes called the midsagittal plane.
- *A transverse plane* is at right angles to the median plane and divides the body into cranial and caudal segments. A cross-section of the body would be made on a transverse plane. The cinch of a saddle defines a transverse plane through the thorax of a horse.
- *A horizontal plane* is at right angles to both the median plane and transverse planes. The horizontal plane divides the body into dorsal (upper) and ventral (lower) segments. If a cow walks into a lake until the water comes above the chest, the surface of the water is in a horizontal plane in relation to the cow. In addition to the planes of reference, other descriptive terms are valuable in defining an area we wish to discuss.
- *Medial* is an adjective meaning close to or toward the median plane. The heart is medial to the lungs; it is closer to the median plane than are the lungs. The chestnut is on the medial aspect (inside) of a horse's limb; it is on the side closest to the median plane.
- *Lateral* is the antonym of medial; it means away from the median plane. The ribs are lateral to the lungs, that is, farther from the median plane.
- *Dorsal* means toward or beyond the backbone or vertebral column. The kidneys are dorsal to the intestines; they are closer to the vertebral column.
- *Dorsum* is the noun referring to the dorsal portion or back. A saddle is placed on the dorsum of a horse.
- *Ventral* means away from the vertebral column or toward the midabdominal wall. The udder is the most ventral part of the body of a cow, the part of the body farthest from the vertebral column.
- *Deep and internal* indicate proximity to the center of an anatomical structure. The humerus (arm bone) is deep in relation to all other structures in the arm.
- *Superficial and external* refer to proximity to the surface of the body. Hair is superficial to all other structures of the body.
- *Proximal* means relatively close to a given part, usually the vertebral column, body, or center of gravity. Proximal is generally used in reference to an extremity or limb. The carpus or knee is proximal to the foot.
- *Distal* means farther from the vertebral column, and like proximal, it is generally used in reference to portions of an extremity. The hoof is distal to the carpus or knee.
- *The suffix-ad* is used to form an adverb from any of the above-named directional terms, indicating movement in the direction of or toward, as in dorsad, ventrad, caudad, and craniad, that is, respectively, toward the dorsum, toward the belly, toward the tail, and toward the head. For example, the superficial digital flexor tendon inserts on the distal limb (the adjective distal describes noun limb), but it

passes distad as it runs along the palmar aspect of the manus (the adverb distad describes the verb passes).

- The thoracic limb (forelimb) distal to the carpus, palmar refers to the flexor or caudal surface. Dorsal is used in this region to refer to the opposite (cranial) side. In describing the pelvic limb (hindlimb) distal to the hock, plantar refers to the caudal surface, and dorsal here, too, refers to the side directly opposite (the cranial side).
- *Prone* refers to a position in which the dorsal aspect of the body or any extremity is uppermost.
- *Pronation* refers to the act of turning toward a prone position.
- *Supine* refers to the position in which the ventral aspect of the body or palmar or plantar aspect of an extremity is uppermost. Supination refers to the act of turning toward a supine position.

The term median is often confused with medial. Both words are used as adjectives when describing anatomical structures. Median means on the midline (as in the median plane, or the median artery). Medial is subtly different, as it means toward the midline and is a term of relativity (as it implies that there is a lateral).

Importance of Anatomy and Physiology in Animal Production

Animals are complex organisms, and understanding anatomy and physiology is important in knowing how they carry out life processes. Anatomy is the study of the form, shape, and appearance of an animal. Since mammals are among the most common animals, most of the information on anatomy will focus on these animals. Scientists divide anatomy into the following branches:

- Gross anatomy is the study of body features that can be seen with the naked eye. As an example, gross anatomy includes studying feet, tails, tongues, and teeth.
- Microscopic anatomy, or histology, is the study of tissue and tissue features that can be seen only with magnification. Microscopic anatomy includes the study of features such as cells and sperm.
- Comparative anatomy is the study of similarities and differences in the structure of different species.
- Embryology is the study of the body as it develops in utero in mammals and in the eggs of poultry.

Physiology is the study of the functions of the cells, tissues, organs, and organ systems of a living organism. Physiology includes relationships between functions by different systems of an organism.

People who care for animals need to understand the fundamentals of anatomy and physiology to promote animal well-being. Animal well-being is the state in which animals' needs are met and the animals do not suffer. Conditions for raising and keeping animals must be considered for their well-being. Species have different environmental requirements. Animal producers are more effective in meeting these requirements when they know the unique anatomy and physiology of a species. For example, some breeds of cattle are more resistant to extreme temperatures than others. Producing a breed outside its preferred temperature range means that steps must be taken to provide shade to protect from the heat or housing to protect from the cold. The design of facilities can accommodate the unique anatomy needs of organisms. The size, shape, and form influence facility arrangement and design. For example, keeping dairy cattle housing clean requires a way to handle animal wastes, including feces and urine. Facility design can help collect and remove wastes from the area. Young animals require different care than older animals. Feed for young animals should be appropriate to their digestive system and nutrient needs. For example, young animals typically require feed with a higher percentage of protein than older animals. Animal productivity is based on animal capacity. Meat animals are required to have muscling in areas used to make the higher-priced cuts. Examples include the loin and hams of hogs. Dairy animals need to have the capacity for high milk production. For example, a dairy cow needs a well-developed mammary system. Animals used for other products are required to have the capacity to produce those products, including egg-laying capacity of chickens and wool quality of sheep. Knowledge of the reproductive system of animals helps a producer provide conditions that promote reproduction.

Cell Specialization

Animals are very complex organisms, yet the structural basis of all animals begins with cells. A cell is the most basic structure of an animal and is considered the building block from which an animal's body is made. All cells of an embryo have the same number and kinds of genes because they all descended from the same zygote. In the development process, cell differentiation occurs. In cell differentiation, or cell specialization, cells become specialized in structure and function by activating and suppressing their genes in selective and unique ways. Specialization is different in each cell, allowing the cell to perform unique activities for an organism. Cell differentiation is very important because organisms could not exist if all cells were alike. Tissues, organs, and organ systems would not exist, and life processes in multicellular organisms would not occur. Specialized groups of cells that are organized to perform a specific function are called tissues. A tissue could also be defined as a group of cells that are alike in activity and structure. The primary types of tissues are epithelial, connective, muscle, and nervous. An organ is a group of tissues that perform a specific function. The tissues work cooperatively so that the organ can carry out its purpose. Examples of organs are the liver, lungs, heart, and brain. An organ system is a group of organs that work together to carry out a specific activity. The following are major organ systems: circulatory, digestive, integumentary, endocrine, urinary, muscular, nervous, reproductive, respiratory, and skeletal.

External Parts of Animals

Producers must be able to describe animals and use the information in selecting and examining animals and in providing health care. The descriptions are based on the external parts of the animals. The presence of various qualities in the external parts indicates the value, health, and condition of an animal. This means that animal producers not only know the names of the parts but they also know the qualities that should be evident upon visual examination of the parts. Qualities vary with the species and the way the species is used. For example, cattle raised for beef have qualities that vary from those raised for dairy production. Considerable study is needed to learn the qualities that indicate the desired characteristics of animals.

General Plan of the Animal Body

All farm animals are vertebrates, and as such they have a vertebral column. The body (with the exception of some of the internal organs) exhibits bilateral symmetry. This means that the right and left sides of the body are mirror images of each other. Similar right and left structures are called paired structures, such as a pair of gloves that are similar but not interchangeable. Most unpaired structures are on or near the median plane, and of course, only one of each unpaired structure exists in any given animal. The tongue, trachea, vertebral column, and heart are examples of unpaired structures. The ribs, limbs, eyes, and most muscles are paired structures. Wherever organs are expected to be in more or-less constant motion and must glide past one another without friction (e.g., the beating heart and moving gut), a serosal cavity is present. The simple squamous epithelium lining various body cavities is also called mesothelium, and the cavities have within them only a scant amount of fluid to facilitate free movement of the tissues. The diaphragm divides the embryonic body cavity into a thoracic cavity and the abdominopelvic cavity. Each of these are further subdivided.

The thoracic cavity contains the pericardial sac, which surrounds the heart, and two pleural sacs, which surround the two lungs. These sacs are formed by a serous membrane, the pleura, a layer of simple squamous epithelium with underlying connective tissue, moistened with the small amount of fluid within the cavity of the sac. The abdominopelvic cavity is somewhat arbitrarily divided into the abdominal and pelvic cavities. The abdominal cavity contains the kidneys, most of the digestive organs, and a variable amount of the internal reproductive organs in both sexes. The pelvic cavity contains the terminal part of the digestive system (the rectum) and all of the internal portions of the urogenital system not found in the abdominal cavity. The abdominal and pelvic cavities are continuous with one another, and the brim of the pelvis marks the transition. The serous membrane that surrounds the abdominal viscera and part of the pelvic viscera is called peritoneum.

A transverse section through the abdominal cavity illustrates the general plan of the body as a tube (the digestive tract and its derivatives) within a tube (the body wall) (Figure 2). Normally there are few air-filled spaces in the animal body except in the respiratory system and the ear. However, for the sake of clarity, many illustrations show a considerable separation between structures that in the animal body are actually in contact. The layers of the body wall and the layers of the digestive tract show a striking similarity, although in reverse order. Layers of the body wall from outside inward are (1) epithelium (epidermis of the skin), (2) connective tissue (dermis and fascia), (3) muscle (striated), (4) connective tissue (transverse fascia), and (5) mesothelium (parietal peritoneum).

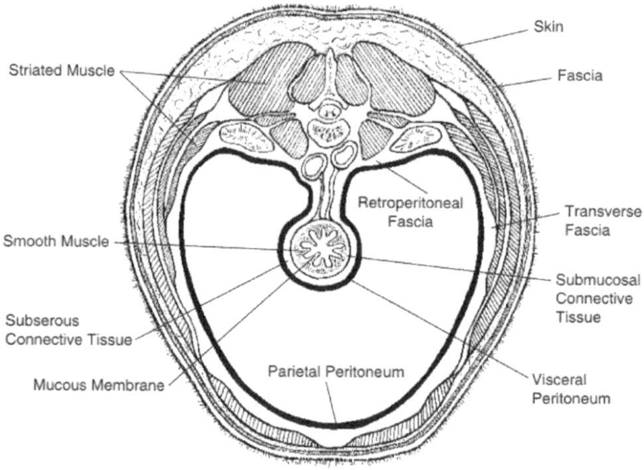

Figure 2: Cross-section of the body wall and digestive tract.

The layers of the gut wall from outside inward are (1) mesothelium (visceral peritoneum), (2) connective tissue (subserous connective tissue), (3) muscle (smooth), (4) connective tissue (submucosa), and (5) epithelium (mucous membrane) (Figure 2).

The serous membranes mentioned previously (pericardium, pleura, and peritoneum) are all derivatives of the lining of the celomic cavity of the embryo. Each serous membrane forms a continuous sac that is usually empty except for a small amount of serous (watery) fluid. In other words, no viscera are found inside any of the serous sacs, although most viscera are covered by at least one layer of a serous membrane. A simple analogy is that of pushing one's first into a partially inflated balloon. The first is never actually within the balloon proper, but still it is surrounded by a portion of the balloon (Figure 3).

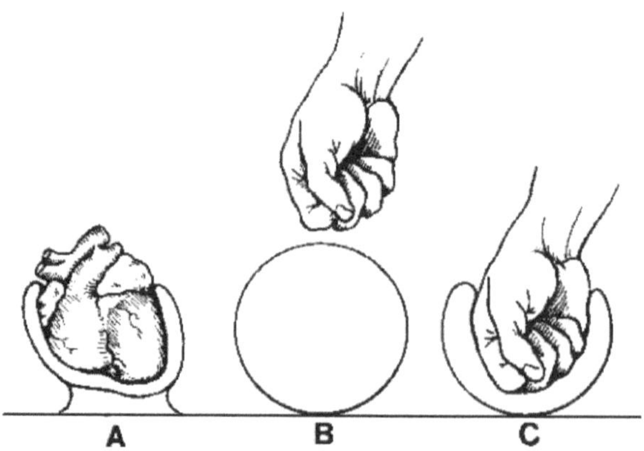

Figure 3: A) Invagination of serous membrane to form outer (parietal) and inner (visceral) layers. B and C) This is similar to a first pushed into a balloon.

The part of the serous membrane covering a viscus is called the visceral serous membrane (visceral pericardium, visceral pleura, and visceral peritoneum). The serous membrane lining a body cavity is called the parietal serous membrane (parietal pericardium, parietal pleura, and parietal peritoneum). The continuity of each serous sac is maintained by connecting layers of serous membrane that extend from the visceral layer of each serous membrane to the parietal layer of the same serous membrane.

1

CHAPTER

NON-INVASIVE METHODS FOR THE DETERMINATION OF BODY AND CARCASS COMPOSITION IN LIVESTOCK: DUAL-ENERGY X-RAY ABSORPTIOMETRY, COMPUTED TOMOGRAPHY, MAGNETIC RESONANCE IMAGING AND ULTRASOUND

A. M. Scholz, L. Bünger, J. Kongsro, U. Baulain and A. D. Mitchell

KEYWORDS: X-ray attenuation, animal, body composition, magnetic resonance imaging, ultrasound

IMPLICATIONS

The ability to accurately and precisely measure body or carcass composition is important for performance testing, grading and finally the selection or payment of meat-producing animals. Advances especially in non-invasive techniques are mainly based on the development of electronic and computer-driven methods in order to provide objective phenotypic data. This review provides a summary of the recent developments in the application of dual-energy X-ray absorptiometry, X-ray computed tomography, magnetic resonance imaging and ultrasound.

INTRODUCTION

Although meat consumption in Europe and worldwide is not increasing at the same rate, meat from farm animals will continue to be the major source of protein for human

Citation: Scholz AM, Bünger L, Kongsro J, Baulain U, Mitchell AD. Non-invasive methods for the determination of body and carcass composition in livestock: dual-energy X-ray absorptiometry, computed tomography, magnetic resonance imaging and ultrasound. Animal. 2015;9(7):1250-1264. doi:10.1017/S1751731115000336.

nutrition throughout the world (OECD/Food and Agriculture Organization of the United Nations, 2014). In order to provide a fair and comparable payment for farmers, it is necessary to base the classification of carcasses or meat-producing animals on harmonized procedures with the least impact on the quality and quantity of the products for human consumption. Non-invasive grading or classification procedures have preference over invasive procedures like dissection or chemical analysis. The ability to accurately and precisely measure body composition or carcass composition is also important for applications related to *PHENOTYPING* in performance testing and breeding programs or scientific studies focusing on growth, nutrition, genetics, housing and behavior or farm animal well-being.

BACKGROUND

Humans started to classify or select animals with the domestication of wild animals thousands of years ago. This classification and selection process was (and still is) mainly based on visual and tactile appraisal with a preference for less-aggressive and easily fed animals. Apart from selection procedures based on the form, size, weight, speed, behavior, fertility, antlers, horns or coat color, it took thousands of years until 'more' scientifically based procedures for farm animal evaluation or classification were invented. In almost all cases, farm or wild animals had or have to be sacrificed in order to be able to process the products for human consumption or utilization – with the exception of milk, egg, wool, work, company or manure. Tissue dissection, however, is still the main European reference standard for the approval of carcass classification procedures or formulas (Nissen *et al.*, 2006 [1]).

First 'non-invasive imaging' methods on farm animals were tested by Kronacher and Hogreve (1936) and Hogreve (1938) using X-radiography in order to study the pelvis shape of different pig breeds and the adipose tissue deposition of fattening pigs, respectively. The first studies using the specific velocity of ultrasound (US; >20 kHz) in different body tissues were started in meat-producing farm animals by Temple*et al.* (1956). More than 30 years ago, the first papers dealing with X-ray-based computed tomography (CT; Skjervold *et al.*, 1981) and nuclear magnetic resonance imaging (tomography – MRI: Groeneveld *et al.*, 1983; even earlier spectroscopy, Casey and Miles, 1974) for the evaluation of meat or carcass and body composition of farm animals were published. Early attempts to cope automatically with MRI inhomogeneity were made by Scholz *et al.* (1993) using a cluster analysis for the segmentation into fat and muscle tissue of pigs *in vivo* only after defining a region of interest of the body part MR scanned. The first dual-energy X-ray absorptiometry (DXA) studies – especially regarding farm animal body composition – started with Mitchell*et al.* (1996).

Since then, technical progress continued providing 'bigger, quicker and smarter' non-invasive imaging or scanning devices for the determination of body and/or carcass composition measurements in farm animal selection programs. Besides Australia, New Zealand, Norway and the United Kingdom, quite a few countries like, for example, Austria, Canada, Denmark, France, Germany, Hungary, Ireland, Spain, Sweden and the United States of America use(d) CT (e.g. Junkuszew and Ringdorfer, 2005; Romvari *et al.,* 2006; Font-i -Furnols and Gispert, 2009; Vester-Christensen *et al.,* 2009; Picouet *et al.*, 2010) or MRI (e.g. Mitchell *et al.*, 2001; Collewet *et al.*, 2005; Monziols *et al.*, 2006; Margeta *et al.*, 2007; Baulain, 2013) as the reference technology for carcass grading in abattoirs (e.g. Branscheid *et al.*, 2011; Daumas *et al.*, 2013) or for performance testing in farm animal breeding programs (e.g. sheep: von Korn *et al.*, 2005; Baulain *et al.*,2011; rabbits: Nagy *et al.*, 2010; Gyovai *et al.*, 2012; Szendrő *et al.*, 2012; pigs: McEvoy *et al.*, 2009; Kremer *et al.*, 2012, 2013; broiler: Davenel *et al.*, 2000; Milisits *et al.*, 2013; laying hens: Szentirmai *et al.*,2013; and turkeys: Andrássy-Baka *et al.*, 2003).

Owing to changes in carcass confirmation caused by breeding progress in various farm animal populations, differences among breeds/species themselves or gender-specific carcass composition, there is a steady need for newly derived or adapted formulas for the (S)EUROP classification in carcass grading (Baulain *et al.*, 2003; Branscheid *et al.*, 2011; Monziols *et al.*, 2013) or for performance testing (Tholen *et al.*, 2003; Bernau*et al.*, 2013 and 2015). This is necessary as long as no whole-body or whole-carcass information would be used (Kongsro *et al.*, 2008).

Non-invasive techniques for body/carcass composition measurements

A common feature of non-invasive techniques for body or carcass composition measurements is that they work with electromagnetic or mechanical energies, which are able to pass completely or partially through body or carcass tissues such as muscle (lean meat=protein+water), adipose tissue (fat, lipids) and bone. Figure 1 summarizes the different (electromagnetic) energy levels that are being used for a number of non-invasive measurement techniques.

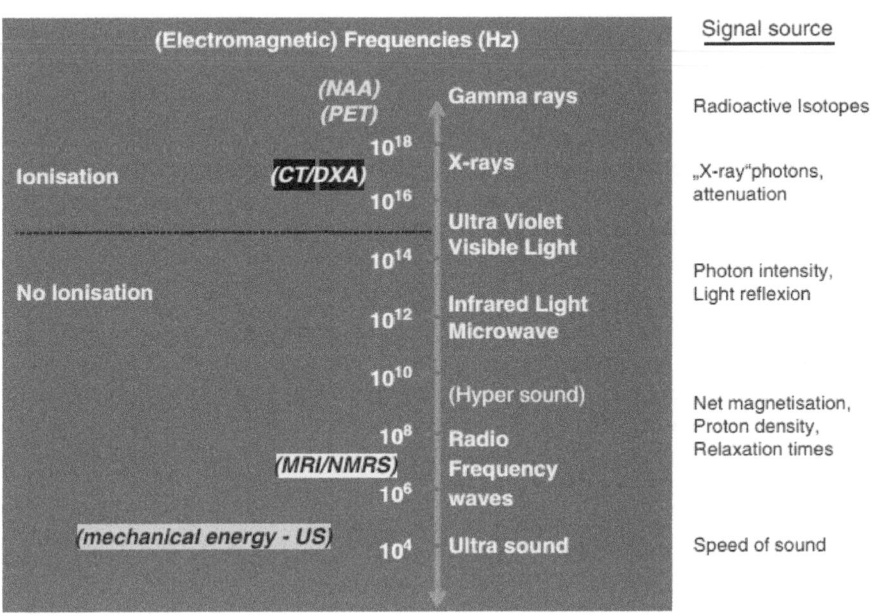

Figure 1: Overview of imaging methods.

All techniques shown in Figure 1 rely on specific device-driven signals, which interact with tissues in the body or carcass at the atomic or molecular level, resulting in secondary or attenuated signals detected by the instruments and analyzed quantitatively. The signal (energy) produced by the instrument may be in the form of sound waves (US), X-radiation (e.g. CT and DXA) or radio frequency (RF) waves (MRI). The signals detected by these instruments are processed to measure, for example, tissue depths, areas, volumes or distributions of fat, muscle (water, protein) and partly bone or bone mineral (Table 1).

Table 1: Traits determined by non-invasive techniques

Non-invasive technique	Trait
Dual-energy X-ray absorptiometry (DXA)	Photon number passage (tissue level)
	X-ray attenuation coefficient from two energy levels
	Bone mineral content
	Bone mineral area
	Bone mineral density
	Soft tissue mass
	Lean tissue mass
	Fat tissue mass
	Whole-body and regional data
Computed tomography (CT)	Photon number passage (tissue level)
	X-ray attenuation in Hounsfield units (HU)
	Tissue areas or volumes depending on anatomical position and HU
	Regional and whole-body data
Magnetic resonance imaging (MRI)	Nuclear magnetic resonance pattern (atomic level)
	Energy level (net magnetization) of nuclei with uneven proton and neutron number
	Longitudinal and transversal relaxation times
	Proton density
	Tissue areas or volumes depending on anatomical position and (arbitrary) signal intensities
	Regional and whole-body data
Ultrasound imaging (US)	Speed of (ultra) sound (tissue level)
	Mechanical energy level *v.* electrical energy level
	Signal amplitude, signal brightness
	Regional distances, areas, volumes

In many cases, these metrology results have to be compared with or validated against a (SI) reference standard directly derived from the carcass. The difference between the arithmetic mean from the non-invasive technique and the arithmetic mean of the reference provides a measure of BIAS or trueness. In addition, the (adjusted) coefficient of determination or regression coefficient (R^2) combined with an error term or term of uncertainty serves as a statistically defined term of accuracy (i.e. precision). For example, partly depending on the statistical modeling procedure, the root mean squared error of estimation and/or prediction or (cross) validation sometimes standardized as residual standard deviation (root mean squared error divided by the standard deviation of the reference mean) provide information about the uncertainty if compared with a reference technique (Johansen *et al.*, 2007, e.g. Table 2). A standard description is provided by ISO 5725 (https://www.iso.org/obp/ui/#iso:std:iso:5725:-1:ed-1:v1:en). If two techniques are being compared without a 'gold standard' or 'true' reference, an improved Bland–Altman analysis might be the first choice as the statistical procedure (Liao and Capen, 2011).

Table 2: Relationship between carcass composition from dissection and DXA carcass or *in vivo* body composition, depending on species (pig, sheep, cattle) studied (all whole-body DXA data from the same GE Lunar DPX-IQ scanner[1])

	Pig ($n = 61$) R^2 ($\sqrt{}$m.s.e./s.d.)			Lamb ($n = 93$) R^2 ($\sqrt{}$m.s.e./s.d.)			Calf ($n = 34$) R^2 ($\sqrt{}$m.s.e./s.d.)		
Dissection *v.* DXA		Carcass	*In vivo*		Carcass	*In vivo*		Carcass	*In vivo*
Reference	CV			CV			CV		
FAT (%)	0.19	0.80 (0.46)	0.74 (0.46)	0.22	0.73 (0.52)	0.51 (0.69)	0.14	0.28 (0.86)	0.003 (1.02)
FAT (g)	0.28	0.90 (0.32)	0.89 (0.43)	0.31	0.83 (0.42)	0.71 (0.27)	0.23	0.64 (0.78)	0.42 (0.39)
Meat (%)/soft lean (%)	0.05	0.70 (0.57)	0.65 (0.64)	0.04	0.57 (0.66)	0.50 (0.70)	0.04	0.53 (0.69)	0.09 (0.97)
Meat (g)/soft lean (g)	0.14	0.94 (0.39)	0.82 (0.55)	0.10	0.88 (0.35)	0.57 (0.33)	0.23	0.98 (0.13)	0.94 (0.12)

DXA = dual-energy X-ray absorptiometry.
[1]This is the only known comparison for the three livestock species using the same DXA device always to compare carcass and *in vivo* data with reference dissection, modified from Scholz *et al.* (2013).

Several publications summarize further invasive or non-invasive methods that are not considered in this review. Simeonova *et al.* (2012), for example, provide a recent review focusing on methods for determining pig body composition, especially on protein deposition during growth. All 'radiology' applications reviewed in this paper are based on the inventions and findings of a number of scientists and engineers summarized in the book Classic Papers in Modern Diagnostic Radiology.

DXA

Both the techniques, CT and DXA, are based on the measurement of the attenuation of X-rays (photons) passing through a body (*in vivo*) or a carcass (*postmortem*). Tissues or elements in the body or carcass are characterized by specific mass attenuation coefficients, depending on the photon energy level being applied for the measurement. DXA and a combination of DXA and CT (dual-energy X-ray computed tomography – DECT; see Johnson, 2009; Magnusson *et al.*, 2011) are based on the application of two different X-ray photon energy levels (high and low), whereas CT works (simplified) only with one (monochromatic) X-ray photon energy level (Kalender, 1988). The ratio (by using the natural logarithm=ln) of the attenuated (I) and the initial X-ray photon number (I_0) for the low (L) and the high (H) energy levels provides the so-called *R*value (X-ray attenuation coefficient). This R value is – depending on the energy levels used – a unique trait for a certain element or compound tissues, such as bone mineral, soft, lean or fat tissues (Crabtree *et al.*, 2007; Wang *et al.*, 2010). Different generations of DXA (or CT) machines use either pencil or fan-beam technology. The fan-beam technology has been extended to a so-called cone-beam or a flash-beam technique. A whole-body scan with a rather slow but very accurate pencil-beam scanner could take up to 35 min, whereas a whole-body scan with a high-speed cone-beam scanner takes <3 min. Different manufacturers of DXA (or DECT) scanners use different approaches to create a high and a low X-ray energy levels (Ulzheimer and Flohr, 2009). Therefore, DXA needs cross-validation for transferring composition results among devices and software modes (Plank, 2005; Scholz *et al.*, 2007 and 2013; Hull *et al.*, 2009; Lösel *et al.*, 2010). In addition, DXA as an indirect tool (Dunshea *et al.*, 2007; Scholz and Mitchell, 2010; Hunter *et al.*, 2011) does not provide a measure of the lean meat percentage. It is still necessary to determine the accuracy of DXA by reference dissection or chemical analysis. The whole-body/carcass composition estimate is available immediately after the scan is completed. Alone, a regional analysis in order to quantify the 2D tissue distribution requires manual manipulation time, depending on the number and anatomical specification of the regions of interest.

DXA studies have been performed on a large variety of farm animal species such as pigs (or pork): Mitchell*et al.* (2000 and 2003); Scholz *et al.* (2002); Suster *et al.* (2004); Marcoux *et al.* (2005); Scholz and Förster (2006); Latorre *et al.* (2008); Kremer *et al.* (2012, 2013); Kogelman *et al.* (2013); chicken/broiler/eggs: Swennen *et al.* (2004); Schreiweis *et al.* (2005); England *et al.* (2012); Salas *et al.* (2012); turkeys: Schöllhorn and Scholz (2007); sheep (or lamb carcasses): Mercier *et al.* (2006); Hopkins *et al.* (2008), Ponnampalam *et al.* (2008); Pearce *et al.* (2009); Hunter *et al.* (2011); Scholz *et al.* (2013); calves/calf carcasses: Bascom (2002); Scholz *et al.* (2003); Hampe *et al.* (2005); fish: Wood (2004), or beef: Ribeiro *et al.* (2011); as well as in the wool and meat industry: Kröger *et al.* (2006) and Ho *et al.* (2013).

The accuracy of DXA measurements comparing pigs, lambs, calves and turkeys has been summarized recently by Scholz *et al.* (2013). To our knowledge, this is the only comparison performed always with the same GE Lunar DPX-IQ (GE Healthcare, Oskar-Schlemmer-Strasse 11, D-80807 München) machine (Table 2). Accuracies for turkeys (n=100) measured with the same GE Lunar DPX IQ pencil-beam scanner comparing DXA carcass with chemical analysis data resulted in the following coefficients of determination as are for fat (%): R^2=0.74 (√m.s.e.=2.11), fat (g): R^2=0.86 (√m.s.e.=254), protein+water *v.* soft lean (%): R^2=0.69 (√m.s.e.=2.33) and protein+water *v.* soft lean (g): R^2=0.99 (√m.s.e.=178) (data from Kreuzer, 2008). The accuracy (low R^2, high r.s.d.) for lean meat percentage in calves (Table 2) is rather low due to the relatively low variability in the lean meat percentage of the young calves in comparison with the relatively high variability of the lean tissue weight especially *in vivo*. The error (inaccuracy) is even inflated during reference dissection, especially by the individual butcher effect (Nissen *et al.*, 2006). The relatively low absolute amount of fat leads to relatively large errors in percentage values for lean and fat tissues. The main difference among calves originates from different BWs causing variations in lean tissue weight. Bascom (2002) concluded that DXA is not suitable for the prediction of the percentage of carcass fat or carcass CP in Jersey calves (adjusted R^2<0.1), although it is unclear what was done during that study in terms of the DXA analysis. Dunshea *et al.* (2007) found higher prediction accuracies for chemically determined reference carcass composition in sheep with R^2 of 0.98 for lean weight and of 0.94 for lean percentage. In addition, Hunter *et al.* (2011) stated that DXA-derived estimates of total and individual tissue masses are highly related to, and can be used to predict, chemical composition *in vivo* or of whole carcasses and carcass halves (in sheep). An adjustment of the prediction equations, however, depends in all cases on the manufacturer (General Electrics, Hologic, Norland, Diagnostic Medical Systems), species, age or weight, software mode and animal positioning on the scan table

CT

Contrary to DXA and DECT, CT works with only one (monochromatic) X-ray level (Kalender, 2006). The mass attenuation coefficient of the object (tissue) of interest is transformed into the so-called Hounsfield units (HU) or CT values by taking the mass attenuation for water and air into account. The almost-fixed range of HU for a given tissue could be used for (fully) automated image segmentation, distinguishing among the body tissue fat, muscle (water) and bone (Glasbey *et al.*, 1999; Johansen *et al.*, 2007; Bünger *et al.*, 2011; Gjerlaug-Enger *et al.*, 2012; Font-i-Furnols *et al.*, 2013; Jay *et al.*, 2013; Judas and Petzet, 2013; Monziols *et al.*, 2013). There is, however, for *in vivo* studies, some overlap between fat and mammary tissue or fat and lung tissue on one side of the HU scale, and bone or muscle, as well as internal organs such as liver, tumor tissue and blood on the upper side of the HU scale. It has to be considered, additionally, that differences in CT protocols may lead to variations of up to 20% in the HU values, especially for (bone containing) tissues with densities >1.1 g/cm^3 (Zurl *et al.*, 2014). As described above, tissue segmentation, for example, by threshold setting is based on assumptions of specific mass attenuation coefficients for different body or carcass tissues, which are calculated as HU. It is, however, not always given – not alone depending on the tissue temperature (Szabó and Babinszky, 2008) – that muscle tissue is detected automatically as muscle tissue (or meat≠meat, fat≠fat) when comparing different CT machines using the same individual(s) (Büngeret *al.*, 2011). Besides small variations for non-adipose tissue (HU=+49 to +52), there is variation in CT values or HU of the adipose tissue within growing pigs. The mean adipose tissue HUs for all pigs (n=9) in a study by McEvoy *et al.* (2008) were −90, −98 and −101 at mean BWs of 51.4, 93.8 and 124.1 kg, respectively. Owing to the anatomical structure of farm animals (or fish: Nanton *et al.*, 2007; Kolstad *et al.*, 2008), however, CT, like DXA, is very well-suited for the discrimination between bone and soft tissues in sheep, chicken, rabbits, beef including buffalo and goose liver *in vivo* (sheep: Johansen *et al.*, 2007; Kvame and Vangen, 2007; Navajas *et al.*, 2007; Macfarlane *et al.*, 2009; Bünger *et al.*, 2011; Ho *et al.*, 2013; chicken: Milisits *et al.*, 2013; Szentirmai *et al.*, 2013; rabbits: Nagy *et al.*, 2010; beef: Hollo *et al.*, 2008; Navajas *et al.*, 2010; buffalo carcass: Holló *et al.*, 2014; goose liver *in vivo*: Locsmandi *et al.*, 2005). Milisitset *al.* (2013) and Szentirmai *et al.* (2013), for example, provided a so-called fat index for determining the body fat content in broiler chicken and laying hens, respectively, by calculating the ratio of the number of fat pixels within the HU range from −20 to −200 to the total number of pixels with HU values for muscle, water and fat between −200 and +200. The muscle index provided additionally by Milisits *et al.* (2013) uses the number of muscle pixels within the HU range from +20 to +200, instead of the fat pixel HU range. The variation found by Chang *et al.* (2011) for various points of visceral and subcutaneous fat in minipigs lies in the range of the above-defined HU threshold values for 'chicken' fat (−20 and −200), with −108.80±5.77 as the lowest mean value (±s.d.) for subcutaneous fat and with −119.41±6.90 as the highest HU value for visceral fat. Johansen *et al.* (2007) provided the following HU thresholds for tissue segmentation in lambs: bone *v.* soft tissue 'kC'=296; soft tissue *v.* background noise (air) 'kA'=−156 and fat *v.* muscle 'kB'=10. The sum of pixels within these thresholds served as estimates of fat and muscle tissue, although according to the thresholds mentioned water was included into the fat tissue.

The latest machines are now the so-called multi-slice spiral (or helical) CTs based on a rotating X-ray source and an array of X-ray photon sensors on the opposite side of the CT gantry (Ulzheimer and Flohr, 2009). Especially for CT, the development of technology occurs at a breathtaking speed. It took only about 10 years from single-slice to multi-slice machines to be developed (Kopp *et al.*, 2000), with now more than 100 slices for one rotation. The body region covered increased from about 1 cm to more than 10 cm/s, whereas the minimal slice thickness decreased from 5 mm to <0.5 mm at the same time (Kalender, 2006). In addition, the gantry size now reaches up to 90 cm providing space for bigger (heavier) farm animals.

After semi-automatic image analysis using OsiriX (Rosset *et al.*, 2004) or ATAR (Animal Tomogram Analysis Routines) software (Haynes *et al.*, 2010; Bünger *et al.*, 2011; Jay *et al.*, 2013), fat, muscle and bone areas can be calculated within the slices of interest. The traits (phenotypes) calculated serve as the basis for the prediction of carcass and tissue weights or volumes and proportions of muscle, fat and bone in combination with additional linear measurements for 2D gigot muscularity, loin eye muscle area and 2D loin eye muscularity and finally as basis for breeding value estimation (Bünger *et al.*, 2011). Present developments aim at whole-body spiral scanning in order to measure the above traits instead of having to predict them, leading additionally to 3D gigot muscularity and 3D loin eye muscularity. The 3D information can even help to include the retailer into the development of new products by applying 'PorkCAD', a new 'design butcher' (Virtual Slaughterhouse) system based on a virtual pig created from CT scanning as suggested by Laursen *et al.* (2013).

Somewhat different approaches and assumptions among the European colleagues from Denmark, France, Hungaria, Ireland, Norway, Sweden, Spain and United Kingdom led to different solutions for carcass grading, and especially formulas for the prediction of the lean meat percentage, which is or should be the basis for the payment of producers (Szabó and Babinszky, 2009). There is, for example, a discussion going on whether meat yield should be determined on the basis of scale weight or on the basis of CT volume (Olsen and Christensen, 2013 *v.* Daumas *et al.,* 2013). Scale weight would require assumptions or knowledge about the 'true' CT density of lean meat (Daumas *et al.*, 2013). Differences in the calculation of CT densities for lean meat result in different lean meat weights for similar lean meat volumes, making the harmonization among different countries or among various CT scanners more complicated (Daumas *et al.*,2013; Olsen and Christensen, 2013).

Correspondingly, CT studies *post mortem* are also aiming to determine the salt content in the dry-cured ham, because the changing NaCl and H_2O proportions lead to modified X-ray attenuations (Fulladosa *et al.*,2010); whereas Frisullo *et al.* (2010) used micro-CT for the rapid estimation of intramuscular fat (IMF) in beef and for the description of the fat microstructure. Anton *et al.* (2013) compared chemical analysis or dissection with CT in order to determine the IMF and carcass fat content of beef in a study focusing on the thyroglobulin (TG) polymorphism. They calculated correlations between IMF (% from Soxhlet analysis) and CT fat (%) in *musculus longissimus dorsi*, and between dissected fat (%) of the right carcass-half and CT fat (%) between the 11th and 13th rib joint of 0.71 and 0.96 ($P<0.001$), respectively. In this context, Jose *et al.*(2009) stated that CT scanning does not negatively affect the quality of (beef or lamb) meat, especially in terms of color. Kongsro and Gjerlaug-Enger (2013) – in pigs – and Clelland *et al.* (2013) – in sheep – started using CT for the measurement of meat quality (IMF content) *in vivo*. The regression coefficients between CT IMF *in vivo* (+further variables) and IMF in the carcass loin eye reached values of adjusted R^2□0.71 (r.m.s.e. □0.36) for Texel lambs, whereas a significantly lower relationship between IMF and CT intensity values was found (R^2=0.18; RMSEP=0.48) according to Kongsro and Gjerlaug-Enger (2013) in pigs. In contrast to the US study by Jiao *et al.* (2014), the relatively low level of IMF and small variation in the Duroc boars studied in comparison with ordinary slaughtered pigs may have led to low prediction accuracies based on CT signal intensities. Font-i-Furnols *et al.* (2013) describe a further method to determine IMF in pork loins using CT. The best prediction of IMF resulted from ordinary linear regression analysis when data from two tomograms were used (R^2=0.83 and RMSEPCV=0.46%). However, genomic selection for IMF improvement based on NIR derived IMF might be a more promising approach according to Gjerlaug-Enger *et al.* (2014).

Deeper insights into the physiological role of IMF in comparison with intermuscular fat (adipose tissue) are provided by Hausman *et al.* (2014).

MRI

The principle of MRI relies on the net magnetization of spinning nuclei with uneven proton and neutron numbers and RF-induced 3D-coded voltage readings with tissue-specific relaxation times depending on spin-lattice (T1) and spin–spin (T2) interactions combined with the proton density (Laurent *et al.*, 2000; Baulain and Henning, 2001; Mitchell *et al.*, 2001). In addition, T1 and T2 depend on the magnetic field strength (Kato *et al.*, 2005). Furthermore, the effect of dehydration plays a crucial role not alone in (dry) cured ham and can be used for MRI applications by taking advantage of changing T1 and T2 relaxation times, which depend on the salt content in the ham (Fantazzini *et al.*, 2009).

A combination of magnetic field produced either by a ferromagnetic, electromagnetic or superconducting system with a field strength between 0.1 and 7 T and so-called gradient coils with a corresponding RF frequency (Larmor frequency) sequence creates a number of cross-sectional images with a 3D voxel definition for the *x*-, *y*- and *z*-axis direction. A Fourier transformation helps in recalculating the signal information from the spectral domain into pixel (or voxel)-wise signal intensity values in a 'gray scale domain' visible on the 'computer' screen. For a T1-weighted sequence with a TR (time between two consecutive RF pulse signals or between successive excitations) of 300 ms and a TE (time between echoes=between middle of exciting RF pulse signal and middle of spin echo production) of 17 ms, the fat tissue containing pixels have rather high signal intensities, whereas the non-fat pixels show lower signal intensities. This pattern, however, changes on chilled objects (Monziols *et al.*, 2005 and 2006). As shown in Figure 2, a T1-weighted sequence would show dark pixels (low signal intensity) for fat tissue and brighter pixels for lean meat tissue (relatively higher signal intensity).

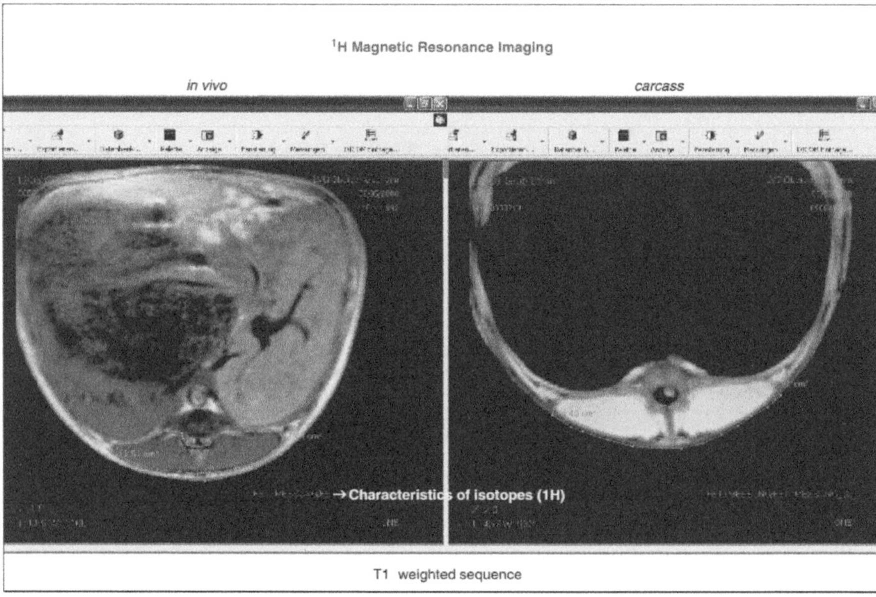

Figure 2: Differences in NMR proton characteristics depending on body temperature (left: lamb *in vivo* ~37°C, right: lamb carcass chilled <8°C, free software DicomWorks, ©Philippe PUECH).

The above-mentioned Larmor (resonance) frequency differs depending on the isotope of interest and the magnetic field strength. Because the isotope ^{1}H has the largest relative sensitivity and highest natural frequency compared with ^{2}H and ^{3}H, proton or ^{1}H nuclear MRI is the most often used method, and is even applied for the study of pork pie (Gaunt *et al.*, 2013).

Usually, an MRI or also a CT (DECT) scan starts with a so-called scout or localizer image sequence in order to be able to define the 'slice' positions and directions as targeted. After successful image acquisition and data storage, a quantitative image analysis is required in order to measure – either in the most simple way – the regions of interest (distances or areas) or calculate – after a more challenging segmentation procedure – the volumes of interest relevant for body or carcass composition measurements. Based on T1 mapping, Kullberg*et al.* (2006 and 2007), for example, described a fully automated protocol for MR image analysis, focusing on the segmentation of visceral and subcutaneous fat in humans, whereas Addeman *et al.* (2015) suggested a so-called fat fraction mapping for the automatic determination of subcutaneous adipose and intra-abdominal adipose tissue within the total adipose tissue.

Various free or commercial software packages are available in order to automate image segmentation into muscle/lean meat, fat, bone and, if necessary, gastrointestinal tract/abdominal content (Figure 3). This procedure can be standardized more easily for CT images than for MRI images, because of the 'unique' application of HU for tissues like bone, muscle (water) and fat. Signals within MR images depend on the tissue-specific relaxation times T1 and T2, including proton density, and on various technical conditions and sequence settings such as the magnetic field strength, the RF pulse sequence(s), slice thickness, distance between slices, number of acquisitions and the specification of (body) coil used.

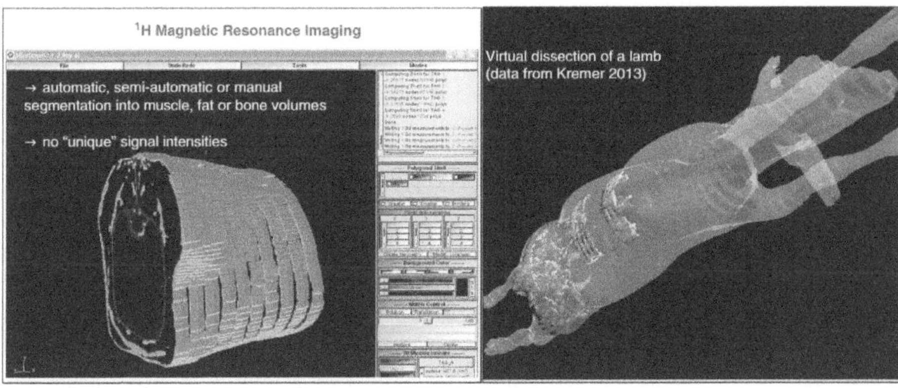

Figure 3: Examples for image analysis and 3D re-calculation (left software used: sliceOmatic, Tomovision Inc.; right software used: 3D DOCTOR, Able Inc., data from Kremer, 2013).

A relatively new non-invasive (but non-imaging) method 'QMR' – quantitative magnetic resonance – is still in the evaluation phase for farm animals (Mitchell *et al.*, 2012).

US Scanning/Imaging

Information retrieval by measuring the velocity of sound is the only method among the methods described in this study that depends on mechanical energy fluctuations. The general principle is based on the partial reflection of (longitudinal) US waves from the interface between different media and/or body tissues (>20 kHz, Scholz and Baulain, 2009; Halliwell, 2010; Scholz and Mitchell, 2010; Pathak *et al.*, 2011).

Different tissues have different (sound) attenuation coefficients depending on the frequency for the creation of the US (waves), whereas the speed of longitudinal sound waves increases with the density of the material the sound wave is travelling through (Halliwell, 2010; Culjat *et al.*, 2012). Because the density of (body) tissues is also temperature-dependent, it makes a difference if a (chilled) carcass or a living animal is 'ultrasonographed'. The speed of US is 1403 m/s in water of a temperature of 0°C and 1472 m/s in water of a temperature of 17°C (Vogt *et al.*, 2008). Van de Sompel *et al.* (2012) obtained a calculated speed of US of 1524 m/s in water with a temperature of 37°C for a salinity of 0% at 0 meter below water.

Because of the accelerated technical improvement of real-time linear-phased array ultrasonic transducers and scanners, this technique has become the most common technology for (farm) animal body and carcass composition assessment (Starck *et al.*, 2001; Mitchell and Scholz, 2009, Scholz and Baulain, 2009).

Two-dimensional US images from so-called B-mode (brightness) devices provide information about adipose tissue depots and cross-sectional areas of muscles, whereas A-mode devices (amplitude) can be used for simple distance measurements of fat or muscle (meat) layers. Real-time B-mode information (2D or 3D images) result from rapid electronic switching or phased array transducers (a number of piezoelectric elements) of different shapes (Starck *et al.*, 2001). At present, most of the US devices for performance testing use (linear) phased array transducers to convert electronic energy to high-frequency ultrasonic (mechanical) energy that travels through the animal body in short pulses. As soon as ultrasonic waves meet at an interface between two tissues that differ in acoustical properties, a part of the (longitudinal) ultrasonic waves are reflected back to the receiver probe (the phased array transducer). Variations in fat, muscle or bone tissue depths or in the distribution of, for example, intermuscular and especially IMF result in time differences in reflected ultrasonic wave signals affected additionally by absorption and refraction (scatter) of the mechanical energy (Starck *et al.*, 2001). These effects combined with variations caused by the expertise of the testing person, age/weight of the animal and the behavior of the animal tested lead in some cases to a challenging interpretation of the US imaging or scanning results, as can be seen from Figure 4.

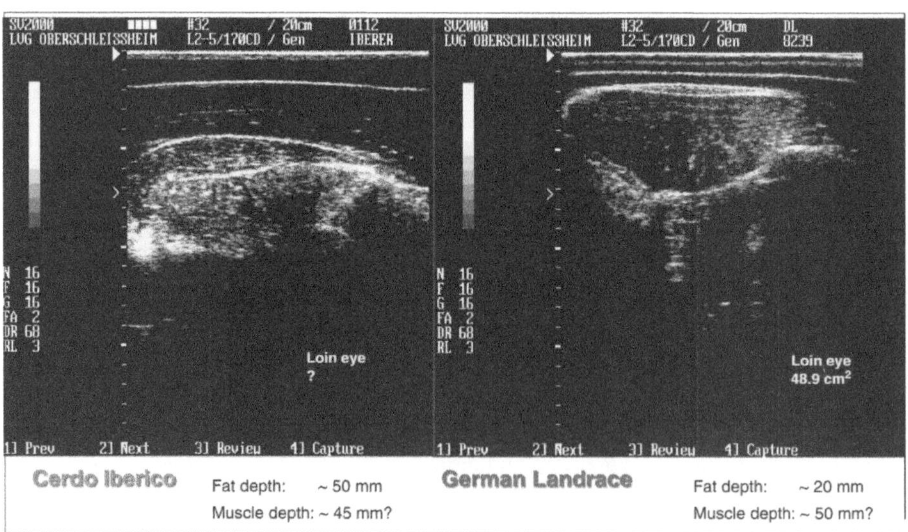

Figure 4: Comparison of 'obese' and 'standard' pigs (using a variable 2.5 to 5 MHz 'backfat' 17-cm transducer).

They make the measurement of areas or even volumes (weights) less accurate in comparison with MRI or CT. Depending on the transducer and on the scan settings in terms of frequency, it might be the case that the measurement of, for example, the loin eye area becomes almost impossible and requires a lot of educated 'anatomical' guessing in order to provide reasonable data (Figure 4). Related to the above measurement site on the (beef) animal, Harangi (2013) stated for Charolais bulls that the relationship between ultrasound rib eye area (UREA) and 'planimeter' carcass rib eye area (CREA) was higher when measured between the 12th and 13th rib instead of between the 11th and 12th rib with R^2 of 0.91 and 0.84 (CV 2.16% *v.* 5.3%), respectively. Török *et al.* (2009) found slightly modified relationships between UREA and CREA for four different beef breeds (Limousin R^2=0.92, Charolais R^2=0.64, Angus and Simmental R^2=0.55).

It must always be considered that the attenuation of ultrasonic energy increases with a rising frequency, whereas the tissue penetration depth of the ultrasonic energy waves decreases with increasing attenuation. Therefore, probes with a frequency between 2 and 5 MHz will be used for measurements, including muscle depth or muscle area (volume), whereas probes with frequency >5 MHz (up to 7.5 MHz) will be used for 'subcutaneous' scanning where deeper muscle regions are not of interest (Schröder and Staufenbiel, 2006; Pillen and van Alfen, 2011).

As muscle tissue has a higher US attenuation than fat tissue, US technology is widely used in farm animal performance testing (Stouffer, 2004; Pathak *et al.*, 2011; Ayuso *et al.*, 2013), obesity diagnostics (Barbero *et al.*, 2013), body condition scoring (Schröder and Staufenbiel, 2006) or for carcass grading (Branscheid *et al.*,2011).

APPLICATION IN EXISTING BREEDING PROGRAMS

The above-mentioned practical applications of US measurements of live animals and carcasses are being extended to genetic selection programs (Müller and Polten, 2004; Kleczek *et al.*, 2009; Emenheiser *et al.*,2010; Case *et al.*, 2012; Maximini *et al.*, 2012; Harangi, 2013), by including *in vivo* IMF (uIMF) measurements in Duroc pigs (Maignel *et al.*, 2010), and, for example, in Angus cattle (Ravagnolo *et al.*, 2010), Nelore cattle (Bonin *et al.*, 2010) or Angus–Brahman crossbred cattle (Elzo *et al.*, 2010).

As '...heritability is a simple dimension less measure of the importance of genetic factors in explaining the differences between individuals, and it allows an immediate comparison of the same trait across populations and of different traits within a population' (Visscher *et al.*, 2008), we use that (additive) genetic variance indicator in the following comparisons of different traits derived from non-invasive techniques (Tables 3 and and4).4). Heritability estimates for uIMF were rather low with h^2=0.12 for Angus in Urugay (Ravagnolo *et al.*, 2010) in contrast with very high (most likely overestimated) h^2 of 0.78 for a very variable Angus–Brahman population in Florida (Elzo *et al.*, 2010). Suther (2009) summarizes a number of earlier studies and shows a similar range of heritability estimates for marbling (IMF) in beef cattle (Table 3).

Table 3: Examples of heritability estimates (h^2, s.e.) for intramuscular fat determined by US *in vivo*

Non-invasive Trait	Technique	Species	Breed	h^2 (s.e.)	Reference
Intramuscular fat (%)	US	Cattle	Angus	0.12 (0.03)	Ravagnolo *et al.* (2010)
			Angus–Brahman	0.78 (0.09)	Elzo *et al.* (2010)
			Diverse	0.12–0.88	Suther (2009)[1]
		Pig	Duroc	0.54 (0.11)	Jiao *et al.* (2014)

US = ultrasound imaging.
[1]References before 2010 in Supplementary Material S1.

Table 4: Examples of heritability estimates (h^2, s.e.) for body composition traits determined by DXA, CT or US [1]

Non-invasive trait	Technique	Species	Breed	h^2 (s.e.)	Reference
Lean meat (%)	CT	Pig	Duroc	0.57 (0.05)	Gjerlaug-Enger *et al.*, 2012
			Landrace	0.50 (0.05)	
Fat (%)	DXA	Pig	F2 with Göttinger Minipig	0.57 (0.14)	Kogelman *et al.*, 2013
Lean Meat (g)	CT	Sheep	Charolais, Suffolk, Texel	0.47, 0.45, 0.46 (0.09)	Jones *et al.* (2004)[2]
			Norwegian White	0.57 (0.16)	Kvame and Vangen (2007)
			Scottish Blackface	0.48 (0.17)	Karamichou *et al.* (2006)
	DXA	Pig	F2 with Göttinger Minipig	0.71 (0.14)	Kogelman *et al.* (2013)
Fat (g)	CT	Sheep	Charolais Suffolk, Texel	0.38, 0.41, 0.40 (0.09)	Jones *et al.* (2004)
			Norwegian White	0.29 (0.13)	Kvame and Vangen (2007)
			Scottish Blackface	0.6 (0.28)	Karamichou *et al.* (2006)
	DXA	Pig	F2 with Göttinger Minipig	0.43 (0.13)	Kogelman *et al.* (2013)
Bone mineral (g)	DXA	Pig	F2 with Göttinger Minipig	0.76 (0.15)	Kogelman *et al.* (2013)
Bone mass (g)	CT	Sheep	Norwegian White	0.51 (0.15)	Kvame and Vangen (2007)
			Scottish Blackface	0.14 (0.11)	Karamichou *et al.* (2006)
Bone mineral density (g/cm²)	DXA	Pig	F2 with Göttinger Minipig	0.92 (0.16)	Kogelman *et al.* (2013)
Loin eye area (cm²)	CT	Sheep	Scottish Blackface	0.33 (0.12)	Karamichou *et al.* (2006)
			Five diverse Austrian	0.24 (0.03)	Maximini *et al.* (2012)
	US	Cattle	Nellore	0.31 to 0.34 (0.03)	Caetano *et al.* (2013)
			Hanwoo	0.09 to 0.24 (0.09 to 0.16)	Lee *et al.* (2014)
		Pig	Duroc, Landrace, Yorkshire	0.21 to 0.22 (<0.01)	Choi *et al.* (2013)
Fat area (cm²)	CT	Sheep	Scottish Blackface	0.5 to 0.76 (0.08 to 0.22)	Karamichou *et al.* (2006)
			five diverse Austrian	0.36 to 0.4 (0.03)	Maximini *et al.*(2012)
Muscle depth (mm, cm)	US	Sheep	Charolais Suffolk, Texel	0.30, 0.32, 0.29 (0.01 to 0.02)	Jones *et al.*, 2004
			Norwegian White	0.28 (0.05)	Kvame & Vangen, 2007
			five diverse Austrian	0.28 (0.05)	Maximini *et al.* (2012)
			Broiler commercial line	0.28 to 0.51 (0.02)	de Genova Gaya (2013)
		Pig	Duroc	0.39 (0.09)	Jiao *et al.* (2014)
Fat depth (mm, cm)	US	Sheep	Charolais, Suffolk, Texel	0.34, 0.35, 0.38 (0.01 to 0.02)	Jones *et al.* (2004)
			Norwegian White	0.44 (0.06)	Kvame and Vangen (2007)
			five diverse Austrian	0.29 (0.05)	Maximini *et al.* (2012)
		Cattle	Angus	0.26 to 0.46 (0.03 to 0.08)	MacNeil and Northcutt (2008)
			Nellore	0.23 (0.02)	Caetano *et al.* (2013)
			Hanwoo	0.05 to 0.47 (0.06 to 0.15)	Lee *et al.* (2014)
		Pig	Duroc	0.58 (0.09)	Jiao *et al.* (2014)
			Duroc, Landrace, Yorkshire	0.32, 0.41, 0.38 (<0.01)	Choi *et al.* (2013)

CT = computed tomography; DXA = dual-energy X-ray absorptiometry; US = ultrasound.
[1]Many more studies with 'US' heritability estimates exist.
[2]References before 2010 in Supplementary Material S1.

Expectedly, real-time US data of muscle depth in sheep or breast muscle thickness in broilers showed medium-to-high direct heritability estimates between 0.2 and 0.51 (Jones *et al.*, 2004, Wolf and Jones, 2007; Grosso *et al.*, 2010, Maximini *et al.*, 2012; Table 4), whereas heritability estimates – depending on age – varied in a similar range between 0.31 and 0.42 for loin muscle (or rib eye) area (between 12th and 13th ribs) in *Bos indicus* (Bonin *et al.*, 2010; Pinheiro *et al.*, 2011), Angus–Brahman (Elzo *et al.*, 2010) and multi-breed beef cattle (Jeyaruban and Johnston, 2014). Heritability estimates for back fat thickness (between 12thand 13th ribs) showed slightly lower values ranging from 0.06 to 0.32 (Bonin *et al.*, 2010; Elzo *et al.*, 2010; Pinheiro *et al.*, 2011) as well as for rump fat thickness with values from 0.26 to 0.29 (alone Pinheiro *et al.*,2011). The advantage of US scanning can be concluded from the above references. US is the only method among the reviewed ones that can be applied in (beef) cattle (Drennan *et al.*, 2009) without size restrictions as there exist for CT, DXA and MRI.

Besides US, CT alone is being used in practical farm animal breeding programs, especially for the selection of body composition traits in pigs and sheep. In this context, CT is, meanwhile, declared as 'part of the routine genetic selection programs in modern times' (Ley, 2013). This, however, is true for only a very few breeding organizations or CT service (research) units in the world so far – such as, for example, for sheep selection in Australia, New Zealand and United Kingdom (e.g. Lambe *et al.*, 2008; Arthur *et al.*, 2011, Bünger *et al.*, 2011) or pig (and sheep) selection in Norway (e.g. Kvame *et al.*, 2006; Kongsro *et al.*, 2008; Gjerlaug-Enger *et al.*, 2012). Gjerlaug-Enger *et al.* 2012 estimated heritabilities for CT lean meat percentage (LMP) between 0.5 and 0.57 (Table 4).

Therefore, the available high additive genetic variance for lean meat percentage in both Norwegian pig breeds based on *in vivo* whole-body CT measurement makes that technique very efficient for selection decisions without having to sacrifice potential breeding animals. At present, the capacity of modern CT machines allows the acquisition of more than 1100 slices per farm animal (e.g. male or female breeding pigs) in an actual whole-body scanning time of less than a minute *in vivo* (Gjerlaug-Enger *et al.*, 2012). Handling, scanning and image analysis for one potential breeding boar or gilt under performance testing takes, in the meantime, only about 15 min. Image analysis is fully automated using MatLab® (The MathWorks Inc., Natick, MA, USA) procedures, especially adapted to CT volume information. A total of 24 boars tested per day is a routine application at Topigs-Norsvin facilities. Information from 1100 slices per potential breeding boar are processed for the body composition phenotypes like lean meat (kg, %), fat (kg, %), bone (kg, %), primal cuts (kg), live and 'carcass' weight (kg), as well as carcass yield (%) (Gjerlaug-Enger *et al.*, 2012).

Slightly modified approaches serve for sheep selection programs at the SRUC (Scotland, UK) in Edinburgh. Routine application at the SRUC covers three important body regions such as thorax (transversal slice at thoracic vertebra 8), loin (lumbar vertebra 5) and gigot (ischium – back of the pelvis) of breeding sheep (Bünger *et al.*, 2011).

Maximini *et al.* (2012) compared genetic (across breeds) parameters for phenotypes derived from either CT or US in five Austrian sheep breeds. They found moderate h^2 estimates for US scan traits for eye muscle depth (0.28) and for fat depth (0.29), whereas CT traits showed higher (across breed) h^2 estimates for fat (0.36 and 0.40), but not for the eye muscle area (0.24) (Table 4). Among other unknown reasons, these slightly problematic across-breed heritability estimates (Visscher *et al.*, 2008) led Austrian sheep breeders to abandon CT in favor for US (Fürst-Waltl and Grill, 2013).

In contrast with Maximini *et al.* (2012), Karamichou *et al.* (2006) found, for all CT tissue areas, moderate-to-high heritability estimates between 0.23 and 0.76. The heritability estimates for CT fat areas started at 0.5, whereas the estimates for loin eye muscle area showed an average value of 0.33 for the univariate model (Table 4). Meanwhile, sheep breeders from New Zealand advocated the combination of US and CT in a stage breeding design because 'selection of meat sheep on CT measurements will increase genetic progress, compared with selection on US measurements alone' (Bünger *et al.*, 2011). In this context, Moore *et al.*(2011) demonstrated that, for sheep, combining CT with US scanning would increase the estimated breeding value (EBV) accuracy by 6% to 20% in comparison with US scanning alone, thus supporting the great benefit of CT. These calculations are derived from US prediction accuracies (R^2) in the order of 0.65 and 0.50 for fat (kg) and muscle (kg), respectively, with heritabilities for US-measured muscle and fat depth of 0.24 to 0.32 and 0.19 to 0.38, respectively (Jones *et al.*, 2004; Bünger *et al.*, 2011; Mortimer *et al.*, 2014), whereas accuracies of CT-based predictions in meat sheep for fat and muscle weight are significantly higher with R^2=0.99 (r.s.d.=434 g) and 0.97 (r.s.d.=611 g), respectively, combined with expected corresponding heritabilities between 0.4 and 0.5 (Young *et al.*, 2001) or between 0.35 and 0.45 (Jones *et al.*, 2004).

Neither DXA nor MRI are being actively used in commercial breeding programs so far, although Kogelman*et al.* (2013) estimated 'heritability' estimates in a F2 pig population originating from crosses of Duroc or Yorkshire and Göttinger Minipig. The heritability estimate for DXA lean mass, of 0.71, is higher than that for DXA fat mass, of 0.43. This observation corresponds with the order of the sheep CT heritability estimates for lean and fat mass, with the exception of h^2 estimates for the Scottish Blackface (Table 4).

HARMONIZATION AND COMPARISON OF TECHNIQUES

Reference (volume) phantoms could help making different CT machines comparable (Christensen and Angel, 2013). The same is true for DXA, because different machines within or among different manufacturers use various settings in order to measure the X-ray attenuation coefficient (*R* value) based on the specific X-ray attenuation of body tissues for high- and low-energy levels (Wood, 2004; Plank, 2005; Lösel *et al.*, 2010). It is even more difficult with MRI, because there are no standardized signal intensities describing one or the other tissue (Baulain and Henning, 2001; Mitchell *et al.*, 2001; Kremer *et al.*, 2012 and2013; Collewet *et al.*, 2013; Addeman *et al.*, 2015; Pérez-Palacios *et al.*, 2014). In addition, not only the velocity of sound depends on the surrounding temperature of body/carcass tissues but also the attenuation of X-rays (Szabó and Babinszky, 2008), and the electromagnetic patterns of protons lead to different results for*in vivo* and *postmortem* (carcass) measurements within the same animal.

The objective of all imaging techniques is to achieve an optimum signal-to-noise ratio combined with small voxel sizes for discrete image segmentation into the body tissues (structures) of interest (Hanna and Cuschieri, 2001). Chemical shift or partial volume effects must be considered when interpreting the accuracy of MRI-derived body/carcass composition estimates (Monziols *et al.*, 2005). Monziols *et al.* (2006) found an increase in estimation accuracy (higher R^2, lower residual standard deviation) for muscle or fat weight and percentage with an increase in body regions (slices) analyzed. If the most relevant body regions are accounted for, or even the whole body, according to the Cavaleri method, MRI, like CT (and DXA), is a very useful tool for growth- or obesity-related studies, because there is no need for serial slaughter anymore (Cavaleri method: Gong *et al.*, 2000; Glasbey and Robinson, 2002; Baulain *et al.*, 2003; Mandarim-De-Lacerda, 2003; Vogt *et al.*, 2007; Szabó and Babinsky, 2009; Arthur *et al.*, 2011).

Only a very few studies exist where DXA measurements are being compared with MRI (Vogt *et al.*, 2007; Brandberg, 2009; Bernau *et al.*, 2015). Vogt *et al.* (2007) found a relationship of R^2=0.95 between whole-body fat measurements performed by DXA (pencil-beam scanner) and by MRI (1.5 T) in human probands, whereas DXA (GE Lunar DPX-L) underestimated the total fat weights compared with both CT −5.23 kg (1.71 kg) and MRI −4.67 kg (2.38 kg) in another study on human probands, summarized in the thesis by Brandberg (2009). Bernau *et al.* (2015) showed for 20 intact boars that both, MRI and DXA, can be used with high accuracy (R^2=0.88 or 0.91, r.m.s.e.=0.9% or 0.82%, respectively) to predict lean meat percentage from dissection. The combination of both techniques resulted in an R^2 of 0.95 (with r.m.s.e.=0.61%). Mitchell and Scholz (2009) could additionally show that the relationship between DXA measurements and the corresponding reference carcass traits was higher than the relationship between US measurements and the same corresponding reference carcass traits with a correlation of r=−0.85 (−0.87) for DXA lean % *v.* carcass fat % in comparison with a corresponding relationship for US fat-free mass % *v.* carcass fat %, with r=−0.69 (−0.74) for pigs with a BW of 110 or 100 kg, respectively. These findings were supported by Suster *et al.*(2004), who reported that DXA measurement values are more closely related with chemically determined carcass values than are carcass P2 back fat measurements performed using a ruler. DXA (or whole-body/carcass MRI/CT) covers the total amount or percentage of body fat or lean meat, whereas US- or ruler-based back fat measurements can only account for subcutaneous fat layers on a limited number of body regions, and therefore cannot account for all the differences in the distribution of fat layers when comparing different farm animal breeds or genotypes, for example.

Table 5 summarizes studies with the 'closest' published relationships between lean meat percentage (LMP) from dissection and the four techniques (DXA, CT, MRI and US) reviewed in this study. All data originate from pigs (carcasses/*in vivo*) for two reasons. First, pigs still have a relatively high variability in body or carcass composition, especially in the subcutaneous fat layer, and, second, reasonable data for the four techniques in the review exist mainly for pigs, followed by sheep and to some extent by poultry or fish.

Table 5: Comparison of non-invasive techniques (reference: lean meat % from dissection)

		Accuracy (alone for pigs)					
		R^2	r.m.s.e.	R^2	r.m.s.e.		
Method	Reference tissue	Carcass		*In vivo*		Scan time whole body	X-radition exposure (mrem)
CT	Lean meat (%)	<0.99	>0.54	<0.94	>1.00	5 to 30 s	9 to 15
MRI	Lean meat (%)	<0.98	>0.62	<0.87	>1.20	15 to 30 min	None
DXA	Lean meat (%)	<0.91	>0.82	<0.72	>1.75	7 to 13 min	0.03 to 0.06
US	Lean meat (%)	<0.77	>0.70	<0.53	>1.95	-	None

CT = computed tomography; MRI = magnetic resonance imaging; DXA = dual-energy X-ray absorptiometry; US = ultrasound.
Carcass: CT data from Judas *et al.* (2005), Romvari *et al.* (2006), Vester-Christensen *et al.* (2009), Monziols *et al.* (2013); MRI data from Baulain and Henning (2001), Mitchell *et al.* (2001), Baulain *et al.* (2003), Collewet *et al.* (2005); Monziols *et al.* (2006); DXA data from Bernau *et al.* (2015); Dunshea *et al.* (2007) (ewes and wheters: % chemical lean: R^2 = 0.94); and US data from Branscheid *et al.* (2011).
In vivo: CT data from Romvari *et al.* (2005) (no error terms provided); MRI data from Baulain and Henning (2001) (R^2 = 0.91, r.m.s.e. = 1.90% in lambs), Mitchell *et al.* (2001), Scholz (2002); DXA data from Scholz *and Förster* (2006), Mitchell *et al.* (2002) (pigs: R^2 = 0.84 for chemical lean %); and US data from Youssao *et al.* (2002), Doeschl-Wilson *et al.* (2005).
References before 2010 in Supplementary Material S1.

CONCLUSION FOR IMAGING/NON-INVASIVE METHODS

If sufficient automatic procedures are available, the 'Cavalieri' method, or even better a whole-body scan, is the preferred CT or MRI imaging procedure, because whole-body information does not require breed, species or age-/weight-specific prediction equations.

For performance testing, a combination of the various methods listed below might be optimal based on cost and accuracy:

- If radiation and the high investment price are not an issue, then use a 'New Generation' spiral, multi-slice CT for the measurement of body/carcass composition.
- If 3D information (e.g. carcass cuts, muscle or fat volumes) is not required, use DXA.
- If radiation is an issue, use MRI. Anesthesia is required in most cases (1–3)!
- If a 'quick' and 'easy' answer is the objective, use A-mode US and for little more B-mode. In all cases (1–4), a scale is very useful!

According to Kallweit (1992), one could still conclude that 'There are advantages and disadvantages of individual systems in their present state...' as summarized in Table 6. '...The rapid progress in technical development may lead to further improvements in the future.' Actually, nothing has changed in the past 20 years.

Table 6: Advantages and disadvantages of non-invasive techniques for the determination of body or carcass composition

	Advantages	Disadvantages
Dual-energy X-ray absorptiometry (DXA)	Easy handling Low radiation Medium price Quick data analysis Regional data analysis	Alone 2D information (so far) No direct data for lean meat (*in vivo*)
Computed tomography (CT)	Very high anatomical resolution High speed Whole-body 3D data Automatic data analysis	X-radiation exposure Expensive
Magnetic resonance imaging (MRI)	Excellent soft tissue differentiation Whole-body 3D data Functional imaging No radiation	Expensive (if high field strength magnet) Rather slow (whole body) Availability (farm animal sector)
Ultrasound imaging (US)	Portable, extensive database for some species Reasonably priced No radiation Real time, online No size limit, no sedation/anesthesia	Less accurate anatomical resolution Image analysis not easily automated No whole-body information

Present and Potential Future Applications – Non-Invasive Measurements of New Exactly Measured 'Phenotypes' to be Associated With New 'Genotypes' and/or Fairer Payment Types

The future of non-invasive techniques or imaging will certainly consider 'new' phenotypes, which are of interest for animal breeders, and need attention for next generations of farm animals, such as lean meat and fat deposition efficiency (Martinsen *et al.*, 2014). Most often these will be traits, which could not be recorded routinely without the application of non-invasive techniques like, for example, traits related to leg health. New selectable leg health traits could be bone mineral content or bone mineral density derived by DXA and/or CT (Charuta *et al.*, 2012; Laenoi *et al.*, 2012; Rangkasenee *et al.*, 2013, Rothammer *et al.*, 2014) or osteochondrosis scores as suggested by Aasmundstad *et al.* (2013), with a promising heritability estimate of 0.31 (±0.09). Several groups are already trying to implement meat and partially fat quality (water proportion in fat) measurements *in vivo*. Beef cattle breeders have been using US imaging in order to measure muscle marbling for several years now, whereas CT scanning is being studied in order to measure the IMF content in sheep and pigs during performance testing *in vivo*, without sacrificing the potential high EBV sire or dam breeding animals. More futuristic, but not less relevant, traits could be the volume of (internal) organs as an indicator of the metabolic capacity of breeding animals or a number of morphological traits under indirect selection pressure by present or future breeding objectives (Kongsro personal communication 2012 to 2014, Carabús *et al.*, 2013). Bünger (personal communication 2013 and 2014) suggests to including more 3D information about muscularity of the gigot, as well as the *longissimus dorsi* muscles or other body parts, for UK sheep breeding programs. Other traits could be, for example, the number of vertebrae counted using CT (Donaldson *et al.*, 2013), the gut or rumen size as an indicator of greenhouse gas output (Goopy *et al.*, 2014) and, for example, pelvic dimensions as indictors for ease of lambing.

In addition, the combination of exact phenotypic data derived from non-invasive techniques in combination with (whole) genome data will provide more knowledge and deeper insight into the control of growth and body/carcass composition of farm animals (Cavanagh *et al.*, 2010; Rothammer *et al.*, 2014). In particular, phenotypic traits, which are difficult and expensive to measure as the ones derived from the non-invasive techniques discussed in this review, will provide extra value for genomic selection (Hayes *et al.*, 2013).

In general, it seems that non-invasive (imaging) methods have become common practice in the growing scientific community and partly in breeding organizations and abattoirs, like the development of an on-line CT for carcass classification in Denmark.

ACKNOWLEDGMENTS

This invited review is based on an invited presentation at the 64rd Annual Meeting of the European Association for Animal Production held in Nantes, France, August 2013, and has been supported by the EU COST action FA1102 (FAIM).

REFERENCES

1. Aasmundstad T, Kongsro J, Wetten M, Dolvik NI and Vangen O 2013. Osteochondrosis in pigs diagnosed with computed tomography: heritabilities and genetic correlations to weight gain in specific age intervals. Animal 7, 1576–1582.
2. Addeman BT, Kutty S, Perkins TG, Soliman AS, Wiens CN, McCurdy CM, Beaton MD, Hegele RA and McKenzie CA 2015. Validation of volumetric and single-slice MRI adipose analysis using a novel fully automated segmentation method. Journal of Magnetic Resonance Imaging 41, 233–241.
3. Anton I, Zsolnai A, Holló I, Repa I and Holló G 2013. Effect of TG gene polymorphism on the intramuscular fat content in cattle examined by CT and Soxhlet methods. Archiv Tierzucht 56, 59–66.
4. Arthur PF, Barchia IM, Giles LR and Eamens GJ 2011. Chemical composition of growing pigs and its relationship with body tissue composition assessed by X-ray-computed tomography. Journal of Animal Science 89, 3935–3944.
5. Ayuso D, González A, Hernández F, Corral JM, Izquierdo M 2013. Prediction of carcass composition, ham and foreleg weights, and lean meat yields of Iberian pigs using ultrasound measurements in live animals. Journal of Animal Science 91, 1884–1892.
6. Barbero A, Astiz S, Lopez-Bote CJ, Perez-Solana ML, Ayuso M, Garcia-Real I and Gonzalez-Bulnes A 2013. Maternal malnutrition and offspring sex determine juvenile obesity and metabolic disorders in a swine model of leptin resistance. Public Library of Science ONE 8, e78424. [PMC free article]
7. Baulain U 2013. Body composition of farm animals by MRI In Farm animal imaging Dublin 2012 (ed. C Maltin, C Craigie and L Bünger), pp. 16–19. Quality Meat Scotland, Ingliston, UK.
8. Baulain U, Schön A, Brandt H and Brade W 2011. Leistungsstand der

niedersächsischen Schafzucht dargestellt anhand von Ergebnissen der stationären Nachkommenprüfung von 1993 bis 2007. Züchtungskunde 83, 439–450.

9. Bernau M, Kremer PV and Scholz AM 2013. Vergleich der Körperzusammensetzung von Schweinen bei 80 kg und 100 kg Lebendmasse mittels Magnetresonanztomographie. Züchtungskunde 85, 430–439.

10. Bernau M, Kremer PV, Lauterbach E, Tholen E, Pappenberger E and Scholz AM 2015. Evaluation of carcass composition of intact boars using linear measurements from performance testing, dissection, dual energy X-ray absorptiometry (DXA) and magnetic resonance imaging (MRI). Meat Science . doi:10.1016/j. meatsci.2015.01.011; in press.

11. Bonin MN, Ferraz JBS, Silva SL, Gomes RC, Cucco DC, Silva Neto PZ, Santana ML Jr, Eler JP, Oliveira ECM, Pedrosa VB, Oliveira PS and Groeneveld E 2010. Genetic parameters for body measurements and ultrasound carcass traits in Nellore cattle. Proceedings 9th World Congress on Genetics Applied to Livestock Production, Leipzig, DE.

12. Branscheid W, Judas M and Höreth R 2011. Zur Klassifizierung von Schweinehälften: Neue Schätzformeln und neue Geräte. Mitteilungsblatt Fleischforschung Kulmbach 50, 9–28.

13. Bünger L, Macfarlane JM, Lambe NR, Conington J, McLean KA, Moore K, Glasbey CA and Simm G 2011. Use of X-ray computed tomography (CT) in UK sheep production and breeding CT scanning – techniques and applications (ed. S Karuppasamy), pp. 329–348. INTECH Europe, Rijeka, Croatia.

14. Caetano SL, Savegnago RP, Boligon AA, Ramos SB, Chud TCS, Lôbo RB and Munari DP 2013. Estimates of genetic parameters for carcass growth and reproductive traits in Nellore cattle. Livestock Science 155, 1–7.

15. Carabús A, Gispert M, Muñoz I, Čandek-Potokar M and Font-i-Furnols M 2013. Testes characteristics of entire and immunocastrated male pigs from 30 to 120 kg live weight measured in vivo with computed tomography In Farm animal imaging Kaposvar 2013 (ed. C Maltin, C Craigie and L Bünger), pp. 38–41. Quality Meat Scotland, Ingliston, UK.

16. Case LA, Wood BJ and Miller SP 2012. The investigation of ultrasound technology to measure breast muscle depth as a correlated trait to breast meat yield in turkey (Meleagris gallopavo). Journal of Animal Science 90, 3410–3417.

17. Cavanagh CR, Jonas E, Hobbs M, Thomson PC, Tammen I and Raadsma HW 2010. Mapping quantitative trait loci (QTL) in sheep. III. QTL for carcass composition traits derived from CT scans and aligned with a meta-assembly for sheep and cattle carcass QTL. Genetics Selection Evolution 42, 36. [PMC free article]

18. Chang J, Jung J, Lee H, Chang D, Yoon J and Choi M 2011. Computed tomographic evaluation of abdominal fat in minipigs. Journal of Veterinary Science 12, 91–94. [PMC free article]

19. Charuta A, Cooper RG, Pierzchała M and Horbańczuk JO 2012. Computed tomographic analysis of tibiotarsal bone mineral density and content in turkeys as

influenced by age and sex. Czech Journal of Animal Science 57, 573–580.

20. Choi JG, Cho CI, Choi IS, Lee SS, Choi TJ, Cho KH, Park BH and Choy YH 2013. Genetic parameter estimation in seedstock swine population for growth performances. Asian-Australasian Journal of Animal Sciences 26, 470–475. [PMC free article]
21. Christensen LB and Angel JAB 2013. Inter-laboratory comparison of medical computed tomography (CT) scanners for industrial applications in the slaughterhouses In Farm animal imaging Kaposvar 2013 (ed. C Maltin, C Craigie and L Bünger), pp. 27–32. Quality Meat Scotland, Ingliston, UK.
22. Clelland N, Bünger L, McLean KA, Knott S and Lambe NR 2013. Prediction of intramuscular fat in Texel lamb loins using different x-ray computed tomography (CT) scanning techniques In Farm animal imaging Kaposvar 2013 (ed. C Maltin, C Craigie and L Bünger), pp. 53–56. Quality Meat Scotland, Ingliston, UK.
23. Collewet G, Bugeon J, Idier J, Quellec S, Quittet B, Cambert M and Haffray P 2013. Rapid quantification of muscle fat content and subcutaneous adipose tissue in fish using MRI. Food Chemistry 138, 2008–2015.
24. Culjat MO, Goldenberg D, Tewari P and Singh RS 2012. A review of tissue substitutes for ultrasound imaging. Ultrasound in Medicine and Biology 36, 861–873.
25. Daumas G, Donko T and Monziols M 2013. Identification of possible and relevant post mortem reference methods for carcass composition In Farm animal imaging Kaposvar 2013 (ed. C Maltin, C Craigie and L Bünger), pp. 14–17. Quality Meat Scotland, Ingliston, UK.
26. de Genova Gaya L 2013. Genetic variability in ultrasound records of breast muscle in a broiler breeding program. Natural Science 5, 1–4.
27. Donaldson CL, Lambe NR, Maltin CA, Knott S and Bunger L 2013. Between and within breed variations of spine characteristics in sheep. Journal of Animal Science 91, 995–1004.
28. Elzo MA, Johnson DD, Lamb GC, Maddock TD, Myer RO, Riley DG, Hansen GR, Wasdin JG and Driver JD 2010. Heritabilities and genetic correlations between postweaning feed intake, growth, and ultrasound traits in a multibreed Angus-Brahman cattle population in the subtropics. Proceedings of the 9th World Congress on Genetics Applied to Livestock Production, Leipzig, DE.
29. Emenheiser JC, Greiner SP, Lewis RM and Notter DR 2010. Longitudinal changes in ultrasonic measurements of body composition during growth in Suffolk ram lambs and evaluation of alternative adjustment strategies for ultrasonic scan data. Journal of Animal Science 88, 1341–1348.
30. England JA, Salas C, Ekmay RD and Coon CN 2012. Dual energy X-ray absorptiometry analysis of broiler breeder eggs for prediction of egg components and evaluation of egg shell quality. International Journal of Poultry Science 11, 316–325.
31. Font-i-Furnols M, Brun A, Tous N and Gispert M 2013. Use of linear regression

and partial least square regression to predict intramuscular fat of pig loin computed tomography images. Chemometrics and Intelligent Laboratory Systems 122, 58–64.

32. Frisullo P, Marino R, Laverse J, Albenzio M and Del Nobile MA 2010. Assessment of intramuscular fat level and distribution in beef muscles using X-ray microcomputed tomography. Meat Science 85, 250–255.

33. Fulladosa E, Santos-Garcés E, Picouet P and Gou P 2010. Prediction of salt and water content in dry-cured hams by computed tomography. Journal of Food Engineering 96, 80–85.

34. Fürst-Waltl B and Grill L 2013. Prerequisites for the implementation of a breeding value prediction for sheep. Universität für Bodenkultur, Wien, Austria: Retrieved April 24, 2014, from https://www.dafne.at/dafne_plus_homepage/download.php?t=ProjectReportAttachment&k=3248.

35. Gaunt AP, Morris RH and Newton MI 2013. Magnetic resonance imaging: a tool for pork pie development. Foods 2, 393–400.

36. Gjerlaug-Enger E, Nordbø Ø and Grindflek E 2014. Genomic selection in pig breeding for improved meat quality. Proceedings of the 10th World Congress of Genetics Applied to Livestock Production, Vancouver, CA.

37. Gjerlaug-Enger E, Kongsro J, Ødegard J, Aass L and Vangen O 2012. Genetic parameters between slaughter pig efficiency and growth rate of different body tissues estimated by computed tomography in live boars of Landrace and Duroc. Animal 6, 9–18.

38. Goopy JP, Donaldson A, Hegarty R, Vercoe PE, Haynes F, Barnett M and Hutton Oddy V 2014. Low-methane yield sheep have smaller rumens and shorter rumen retention time. British Journal of Nutrition 111, 578–585.

39. Grosso JLBM, Eler JP, Balieiro JCC, Ferraz JBS, Mattos EC and Michelan Filho T 2010. Genetic correlation for real-time ultrasound and fleshing of breast muscle with carcass traits in broilers. Proceedings of the 9th World Congress on Genetics Applied to Livestock Production, Leipzig, DE.

40. Gyovai P, Nagy I, Gerencsér Z, Matics Z, Radnai I, Donkó T, Bokor Á, Farkas J and Szendrő Z 2012. Genetic parameters for litter weight, average daily gain and thigh muscle volume measured by in vivo computer tomography technique in Pannon White rabbits. Livestock Science 144, 119–123.

41. Halliwell M 2010. A tutorial on ultrasonic physics and imaging techniques. Journal of Engineering in Medicine 224, 127–142.

42. Harangi S 2013. Application of ultrasound measurement technique to evaluate the meat producing ability and carcass value of beef cattle. PhD Thesis, University of Debrecen, Hungary.

43. Hausman GJ, Basu U, Du M, Fernyhough-Culver M and Dodson MV 2014. Intermuscular and intramuscular adipose tissues – bad v. good adipose tissues. Adipocyte 3 (4), 1–14. [PMC free article]

44. Hayes BJ, Lewin HA and Goddard ME 2013. The future of livestock breeding:

genomic selection for efficiency, reduced emissions intensity, and adaptation. Trends in Genetics 29, 206–214.

45. Haynes FEM, Greenwood PL, Siddell JP, McDonagh MB and Oddy VH 2010. Computer tomography software program 'Osirix' aids prediction of sheep body composition. Proceedings of the Australian Society of Animal Production 28, 49.
46. Ho H, Hunter P and Pearson G 2013. Parametric representation of lamb carcasses based on computed tomography (CT) images In Farm animal imaging Kaposvar 2013 (ed. C Maltin, C Craigie and L Bünger), pp. 76–79. Quality Meat Scotland, Ingliston, UK.
47. Holló G, Barna B and Nuernberg K 2014. Characterisation of carcass composition and meat quality of male suckling buffalo calves kept on natural grassland. Archiv Tierzucht 56, 107.
48. Hunter TE, Suster D, Dunshea FR, Cummins LJ, Egan AR and Leury BJ 2011. Dual energy X-ray absorptiometry (DXA) can be used to predict live animal and whole carcass composition of sheep. Small Ruminant Research 100, 143–152.
49. Jay NP, van de Ven RJ and Hopkins DL 2013. Relationship in lambs between lean weight estimated using CT scanning and VIASCAN®. Proceedings of the 59th International Congress of Meat Science and Technology, August 18–23, Izmir, Turkey.
50. Jeyaruban MG and Johnston DJ 2014. Estimated additive and non-additive breed effects and genetic parameters for ultrasound scanned traits of a multi-breed beef Population in tropical Australia. Proceedings of the 10th World Congress of Genetics Applied to Livestock Production, Vancouver, CA.
51. Jiao S, Maltecca C, Gray KA and Cassady JP 2014. Feed intake, average daily gain, feed efficiency, and real-time ultrasound traits in Duroc pigs: I. Genetic parameter estimation and accuracy of genomic prediction. Journal of Animal Science 92, 2377–2386.
52. Judas M and Petzet A 2013. Statistical dissection – for which tissues does PLS regression work? In Workshop on Farm Animal and Food Quality Imaging, DTU Compute Technical Report-2013-12, Espoo, Finland, June 17, 11–16.
53. Kogelman LJA, Kadarmideen HN, Mark T, Karlskov-Mortensen P, Bruun CS, Cirera S, Jacobsen MJ, Jørgensen CB and Fredholm M 2013. An F2 pig resource population as a model for genetic studies of obesity and obesity-related diseases in humans: design and genetic parameters. Frontiers in Genetics | Livestock Genomics 4, 1–14. [PMC free article]
54. Kongsro J and Gjerlaug-Enger E 2013. In vivo prediction of intramuscular fat in pigs using computed tomography. Open Journal of Animal Sciences 3, 321–325.
55. Kremer PV 2013. In vivo Phänotypisierung von Schlachtkörpermerkmalen bei Schwein und Schaf mittels Magnetresonanztomographie. Habilitation thesis, University of Munich, DE.
56. Kremer PV, Förster M and Scholz AM 2013. Use of magnetic resonance imaging (MRI) to predict the body composition of pigs in vivo . Animal 7, 879–884.

57. Kremer PV, Fernández Fígares I, Förster M and Scholz AM 2012. In vivo body composition in autochthonous and conventional pig breeding groups by dual energy X-ray absorptiometry and magnetic resonance imaging under special consideration of Cerdo Ibérico. Animal 6, 2041–2047.

58. Laenoi W, Rangkasenee N, Uddin MJ, Cinar MU, Phatsara C, Tesfaye D, Scholz AM, Tholen E, Looft C, Mielenz M, Sauerwein H, Wimmers K and Schellander K 2012. Association and expression study of MMP3, TGFβ1 and COL10A1 as candidate genes for leg weakness-related traits in pigs. Molecular Biology Reports 39, 3893–3901.

59. Laursen LF, Bærentzen JA, Igarashi T, Petersen MK, Clemmensen LKH, Ersbøll BK and Christensen LB 2013. PorkCAD: case study of the design of a pork product prototype. In Proceedings of the 5th International Congress of International Association of Societies of Design Research, Tokyo, Japan, August 26–30. Retrieved February 19, 2015, from http://design-cu.jp/iasdr2013/papers/1134-1b.pdf.

60. Lee JH, Lee YM, Oh S-H, Son HJ, Jeong DJ, Whitley N and Kim JJ 2014. Estimation of genetic parameters for real-time ultrasound measurements for Hanwoo cows at different ages and pregnancy status. Asian Australasian Journal of Animal Science 27, 155–160. [PMC free article]

61. Ley C 2013. Pig genetics is all about sustainable profitability. Pig International 43, 4–7.

62. Liao JJZ and Capen R 2011. An improved Bland-Altman method for concordance assessment. The International Journal of Biostatistics 7, Art. 9, 1–17.

63. Lösel D, Kremer P, Albrecht E and Scholz AM 2010. Comparison of a GE Lunar DPX-IQ and a Norland XR-26 dual energy X-ray absorptiometry scanner for body composition measurements in pigs – in vivo. Archiv Tierzucht 53, 162–175.

64. Magnusson M, Malusek A, Muhammad A and Alm Carlsson G 2011. Determination of quantitative tissue composition by iterative reconstruction on 3D DECT volumes. Proceedings of the 11th International Meeting on Fully Three-Dimensional Image Reconstruction in Radiology and Nuclear Medicine, Potsdam, DE, 120–123.

65. Maignel L, Daigle JP, Gariépy C, Wilson D and Sullivan B 2010. Prediction of intramuscular fat in live pigs using ultrasound technology and potential use in selection. Proceedings of the 9th World Congress on Genetics Applied to Livestock Production, Leipzig, DE.

66. Martinsen KH, Ødegård J, Olsen D and Meuwissen THE 2014. Genetic variation in efficiency to deposit lean meat and fat in Norwegian Landrace pigs. Proceedings of the 10th World Congress of Genetics Applied to Livestock Production, Vancouver, Canada.

67. Maximini L, Brown DJ, Baumung R and Fuerst-Waltl B 2012. Genetic parameters of ultrasound and computer tomography scan traits in Austrian meat sheep. Livestock Science 146, 168–174.

68. Milisits G, Donkó T, Dalle Zotte A, Sartori A, Szentirmai E, Emri M, Opposits G,

Orbán A, Pőcze O, Repa I and Sütő Z 2013. Application of computed tomography to assess the effect of egg yolk ratio on body composition in chickens of different genotype and gender at hatch and during the rearing period. British Poultry Science 54, 611–619.

69. Mitchell AD, Ramsay TG and Scholz AM 2012. Measurement of changes in body composition of piglets from birth to 4 kg using quantitative magnetic resonance (QMR). Archiv Tierzucht 55, 64–67.

70. Monziols M, Rossi A and Daumas G 2013. Impact of pig population (light or heavy) on computed tomography (CT) and dissection relationship for lean meat percentage measurement In Farm animal imaging Kaposvar 2013 (ed. C Maltin, C Craigie and L Bünger), pp. 22–26. Quality Meat Scotland, Ingliston, UK.

71. Moore K, McLean KA and Bunger L 2011. The benefits of computed tomography (CT) scanning in UK sheep flocks for improving carcase composition. Proceedings of the British Society of Animal Science and the Association of Veterinary Teaching and Research Work, April 4–5, Nottingham. Advances in Animal Biosciences 2, 10 .

72. Mortimer SI, Swan AA, Brown DJ and van der Werf JHJ 2014. Genetic parameters revisited for ultrasound scanning traits in Australian sheep. Proceedings of the 10th World Congress of Genetics Applied to Livestock Production, Vancouver, CA.

73. Nagy I, Gyovai P, Radnai I, Matics Z, Gerencsér Z, Donkó T and Szendrő Z 2010. Genetic parameters of growth in vivo CT based and slaughter traits in Pannon white rabbits. Proceedings of the 9th World Congress on Genetics Applied to Livestock Production, Leipzig, DE.

74. Navajas EA, Richardson RI, Fisher AV, Hyslop JJ, Ross DW and Prieto N 2010. Predicting beef carcass composition using tissue weights of a primal cut assessed by computed tomography. Animal 4, 1810–1817.

75. OECD/Food and Agriculture Organization of the United Nations 2014. OECD-FAO agricultural outlook 2014. OECD Publishing, 2, rue André-Pascal, 75775 PARIS CEDEX 16: Retrieved February 19, 2015, from http://dx.doi.org/10.1787/agr_outlook-2014-en.

76. Olsen EV and Christensen LB 2013. Comparison of accuracy of reference methods based on CT and manual dissection In Farm animal imaging Kaposvar 2013 (ed. C Maltin, C Craigie and L Bünger), pp. 18–21. Quality Meat Scotland, Ingliston, UK.

77. Pathak V, Singh VP and Sanjay Y 2011. Ultrasound as a modern tool for carcass evaluation and meat processing: a review. International Journal of Meat Science 1, 83–92.

78. Pérez-Palacios T, Caballero D, Caro A, Rodríguez PG and Antequera T 2014. Applying data mining and computer vision techniques to MRI to estimate quality traits in Iberian hams. Journal of Food Engineering 131, 82–88.

79. Picouet PA, Teran F, Gispert M, Font-i-Furnols M 2010. Lean content prediction in pig carcasses, loin and ham by computed tomography (CT) using a density

model. Meat Science 86, 616–622.

80. Pillen S and van Alfen N 2011. Skeletal muscle ultrasound. Neurological Research 33, 1016–1024.

81. Pinheiro TR, Mercadante MEZ, Albuquerque LG, Cyrillo JNSG and Branco RH 2011. Phenotypic and genetic parameters compared during repeated measures of longissimus muscle area and subcutaneous fat thickness in Nelore cattle. Genetics and Molecular Research 10, 2944–2952.

82. Rangkasenee N, Murani E, Brunner R, Schellander K, Scholz AM, Luther H, Hofer A, Ponsuksili S and Wimmers K 2013. KRT8, FAF1 and PTH1R gene polymorphisms are associated with leg weakness traits in pigs. Molecular Biology Reports 40, 2859–2866.

83. Ravagnolo O, Brito G, Aguilar I, Ciappesoni G and Lema M 2010. Genetic parameters for ultrasound live traits in pasture fed Angus cattle. Proceedings of the 9th World Congress on Genetics Applied to Livestock Production, Leipzig, DE.

84. Ribeiro FRB, Tedeschi LO, Rhoades RD, Smith SB, Martin SE and Crouse SF 2011. Evaluating the application of dual X-ray energy absorptiometry to assess dissectible and chemical fat and muscle from the 9th-to-11th rib section of beef cattle. The Professional Animal Scientist 27, 472–476.

85. Rothammer S, Kremer PV, Bernau M, Fernandez-Figares I, Pfister-Schär J, Medugorac I and Scholz AM 2014. Genome-wide QTL-mapping of nine body composition and bone mineral density traits in pigs. Genetics Selection Evolution 46, 68. [PMC free article]

86. Salas C, Ekmay RD, England J, Cerrate S and Coon CN 2012. Determination of chicken body composition measured by dual energy X-ray absorptiometry. International Journal of Poultry Science 11, 462–468.

87. Scholz AM and Mitchell AD 2010. Body composition: indirect measurement In Encyclopedia of animal science, 2nd edition (ed. DE Ullrey, CK Baer and WG Pond), pp. 152–156. CRC Press, Boca Raton, FL, USA.

88. Scholz AM, Kremer PV, Wenczel R, Pappenberger E and Bernau M 2013. Body composition in farm animals by dual energy X-ray absorptiometry In Farm animal imaging Dublin 2012 (ed. C Maltin, C Craigie and L Bünger), pp. 9–14. Quality Meat Scotland, Ingliston, UK.

89. Simeonova ML, Todorov NA and Schinckel AP 2012. Review of in vivo-methods for quantitative measurement of protein deposition rate in animals with emphasize on swine. Bulgarian Journal of Agricultural Science 18, 455–481.

90. Szendrő Z, Metzger S, Nagy I, Szabó A, Petrási Z, Donkó T and Horn P 2012. Effect of divergent selection for the computer tomography measured thigh muscle volume on productive and carcass traits of growing rabbits. Livestock Science 149, 167–172.

91. Szentirmai E, Milisits G, Donkó T, Budai Z, Ujvári J, Fülöp T, Repa I and Sütő Z 2013. Comparison of changes in production and egg composition in relation to

in vivo estimates of body weight and composition of brown and white egg layers during the first egg-laying period. British Poultry Science 54, 587–593.

92. Van de Sompel D, Sasportas LS, Dragulescu-Andrasi A, Bohndiek S and Gambhir SS 2012. Improving image quality by accounting for changes in water temperature during a photoacoustic tomography scan. Public Library of Science ONE 7, e45337. [PMC free article]

93. Wang ZM, Heymsfield SB, Chen Z, Zhu S and Pierson RN 2010. Estimation of percentage body fat by dual-energy x-ray absorptiometry: evaluation by in vivo human elemental composition. Physics in Medicine and Biology 55, 2619–2635. [PMC free article]

94. Zurl B, Tiefling R, Winkler P, Kindl P and Kapp KS 2014. Hounsfield units variations: impact on CT-density based conversion tables and their effects on dose distribution. Strahlentherapie und Onkologie 190, 88–93.

2
CHAPTER

OVERVIEW OF NEW TRAITS AND PHENOTYPING STRATEGIES IN DAIRY CATTLE WITH A FOCUS ON FUNCTIONAL TRAITS

C. Egger-Danne, J. B. Cole, J. E. Pryce, N. Gengler, B. Heringstad, A. Bradley and K. F. Stock

Keywords: Phenotypes, novel traits, dairy cows, functional traits, genomics

IMPLICATIONS

The advances that have been made in breeding goals have been based on the needs of breeders; however, they are also influenced by consumers and societal needs related to food safety and efficient and environmentally sound production systems. New functional traits are growing in importance because of recent declines in animal health and fitness. There is also growing competition for high-quality, plant-based sources of energy and protein; thus, it is important to use those resources very efficiently in animal production. Successful programmes for animal improvement will require a balance between the effort needed to record data and the resulting benefits to farmers.

INTRODUCTION

As a consequence of negative genetic correlations between milk yield and fitness traits, a decline in many functional traits has been observed (e.g. Jones *et al.*, 1994; Lucy, 2001). In recent years, there has been a stabilization or even increase in genetic

Citation: Egger-Danner C, Cole JB, Pryce JE, et al. overview of new traits and phenotyping strategies in dairy cattle with a focus on functional traits. Animal. 2015;9(2):191-207. doi:10.1017/S1751731114002614.

trends of functional traits (e.g. Miglior *et al.*, 2012; ZuchtData, 2014). An example for the increase in genetic trend for longevity is shown for Holstein Friesian in Figure 1 (Fuerst C., 2014, personal communication). Herd management has been challenged to compensate for antagonistic effects and to balance selection for production while maintaining fertility, udder health and resistance to metabolic diseases in order to maximize profit without compromising welfare. As a result, there is increasing demand around the world for novel traits that can be used for herd management and genetic improvement, as well as for monitoring parameters of public interest.

Figure 1: Average estimated breeding value for longevity by birth year and country for Holstein Friesian (Fuerst, 2014).

The work needed to record additional data must provide immediate benefits to the producer in order to motivate the extra effort required. Individual producers place particular value on tools that can help them better manage their herds. If other stakeholders, such as veterinarians, are to be encouraged to record, for example, treatments and diagnoses, they also need to accrue benefits from the effort that they put in. As it is more and more challenging to balance high levels of production with reproduction and health, there is a growing need for data that can be used to detect problems at an early stage so that early interventions can be made.

The inclusion of functional traits in genetic improvement programmes is important for the long-term development of dairy populations. Genetic gains are cumulative, and small improvements that build up over time will provide ongoing savings to all farmers, particularly those using selection indices to combine information from many different traits. In response to these needs, there has been a worldwide shift towards more emphasis on functional traits in selection programmes over the past 15 years. Complex breeding goals that include up to 43 functional traits/country are encompassed by total merit indices (TMI) according to a survey carried out by ICAR in 2012 (Stock *et al.*, 2012), and the responses from 23 countries showed that genetic evaluations for calving ease, fertility, longevity, feet and legs and indirect health traits are very common. However, a wider range of health traits are actually being recorded, and there are plans in a number of countries to further expand their recording programmes to include additional reproduction, metabolic and efficiency traits. Public interest in the use of parameters for monitoring food safety, tracking disease outbreaks and documenting animal welfare is increasing. National monitoring programmes may therefore be developed to meet the demands of producers, consumers and regulators. The objectives of this paper are to give an overview of expected developments and challenges related to the availability of novel traits, with emphasis on new health and fitness traits in the near future, and to suggest strategies for phenotyping of new traits.

BACKGROUND

The world population is estimated to reach 9 billion people in 2050 according to the latest demographic reports from the United Nations (2014), which will require a big increase in food production. Demand for limited resources will increase and prices for energy and concentrates will rise. Disruptions in global environments due to climate change (see, e.g. Walthall *et al.*, 2012) may also drive a greater emphasis on novel traits, especially those that optimize resource use efficiency. Cattle producers will be challenged to find ways to reduce greenhouse gas emissions throughout the production cycle, either by increasing production efficiency and reducing wastage or by using new technologies to decrease pollution. In this context, there is need for novel traits that can be used in herd management and breeding. Functional traits, such as direct information of cow health, have also increased in importance because of growing concerns about animal welfare and consumer demands for healthy and natural products. Consumers are also concerned about the impact of drugs used in veterinary medicine on the spread of antibiotic-resistant strains of bacteria that can negatively impact human health. These requirements are also a part of legal regulations in the European Union (Pavon, 2013).

According to a scientific report on the effects of farming systems on dairy cow welfare and disease (EFSA-Q-2006-113, 2009), the major diseases associated with poor welfare in dairy cows were leg disorders, mastitis and reproductive disorders. Pavon (2013) stated that genetic selection of dairy cattle over many years is considered by the European Commission to be a major factor leading to poor welfare in dairy cows, which has driven a need for welfare indicators. However, only healthy cows can perform at high levels of production for a long time, and it is important to ensure that consumers and policy makers understand this.

A key requirement for the recording of data is the motivation of the stakeholders involved. Several countries have conducted surveys with farmers to understand their needs (Steininger *et al.*, 2012; Roessler *et al.*, 2013; Schwarzenbacher, 2013). A consistent observation in these surveys has been that farmers place priority on the genetic progress of functional traits, aiming at robust cows that are easy to handle. Increasing milk yield is no longer ranked among the most important traits to select for, but it is quite possible that survey respondents are indicating a desire for healthy, long-lived cows that are also highly productive. Growing emphasis on functional traits indicates that they are perceived as being more important than in the past – for example, because herd sizes and labour costs are increasing – but that does not necessarily mean that production is unimportant.

According to Bo (2009), a breeding goal should include the following aspects: increased income (higher production of milk/beef); reduced costs (better fertility, fewer diseases, reduced culling rates); ease of management (temperament, milking speed); and advantages regarding the sale of products (animal welfare, ethics, consumer concerns). Complex breeding goals also require information on a wide range of relevant traits that can be measured economically. In dairy cattle, these traits include efficiency, health, fertility and functional conformation. Many countries now use a diverse group of economically important traits in the TMI to rank cattle for genetic selection (Figure 2). However, recording has to have benefits beyond genetic improvement programmes, and the additional effort required for documentation must result in added value. Electronic systems that make data capture easy are a key to long-term success.

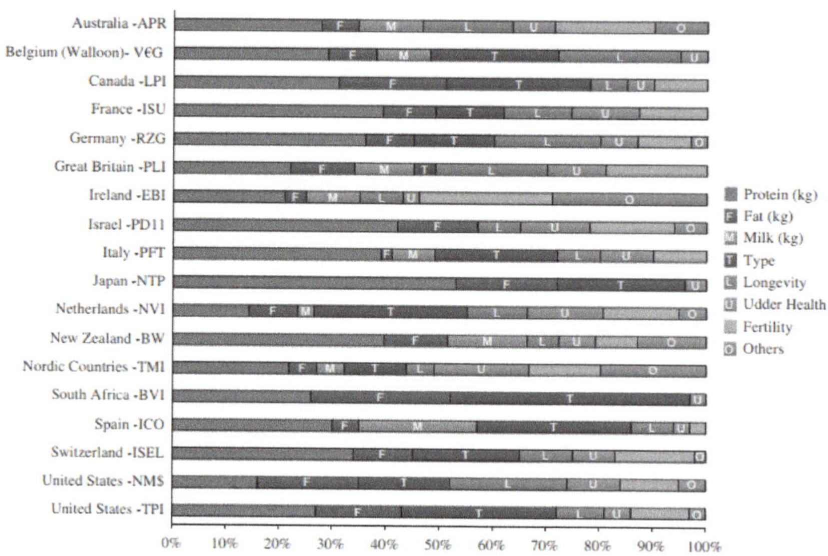

Figure 2: Traits included in the total merit indices of 17 countries. The data used to construct this figure were provided by F. Miglior of the Canadian Dairy Network.

Advances in technology will have an impact on the future definition and availability of phenotypes. The growing use of SNP-based genotyping may lead to routine genomic evaluations of new traits with relatively few daughter records, particularly if producers are willing to use breeding values with lower reliabilities than they are accustomed to. Phenotypes that are expensive to measure, such as direct feed intake, may be handled by establishing small reference populations with very thorough data recording. Those data could be used to develop predictors for the broader population based on traits that are recorded in most herds (e.g. Cole *et al.*,2014). Recent technological developments will provide more knowledge about the biological background and genetic architecture of traits (RNAseq, genome sequencing, genomics, proteomics, metabolomics, etc.).

Sequencing technologies can be used to identify causal mutations, rather than SNP in linkage disequilibrium with causal mutations, which should lead to increased accuracies of genomic prediction (Meuwissen *et al.*,2013). Metabolomics (the study of cellular metabolites present within an organism) and proteomics (the study of protein structure, function and patterns of expression within an organism) should lead to a better description of phenotypes, which could enhance selection opportunities.

NOVEL FUNCTIONAL TRAITS

Direct Health Traits

Several studies show that the use of direct health traits (e.g. mastitis diagnoses) is more effective than indicator traits alone in breeding programmes (Heringstad *et al.*, 2007; Egger-Danner *et al.*, 2012a; Parker Gaddis *et al.*, 2014). The Nordic countries have a well-established history of health recording in dairy cows. In Norway, each case of veterinary treatment has been registered on an individual cow basis since 1975 (Heringstad and Østerås, 2013; ICAR, 2013). Similar recording systems were established in Finland, Sweden and Denmark during the 1980s. Routine genetic evaluations for direct health traits have been in place in Austria and Germany since 2010, in France since 2012 and in Canada since December 2013. Systems for recording of diagnoses are presently being established in other countries as well and will be more widely available in the near future. Some countries will use veterinary diagnoses, whereas others will focus on producer-recorded data. Developments towards integrated use of health data require standardization of diagnoses to ensure that phenotypes are comparable. Gernand *et al.* (2012), Govignon-Gion *et al.* (2012), Parker Gaddis *et al.* (2012) and Pryce *et al.* (1998) have shown that, for use in genetic evaluations, common health disorders recorded by farmers are of a similar quality as those documented by veterinarians. Combined use of health data from farmers and diagnosis documented by veterinarians may be an option to improve coverage of direct health data. In 2012, ICAR approved guidelines for Recording, Evaluation and Genetic Improvement of Health Traits (ICAR, 2012). A hierarchical system with a very comprehensive key of diagnoses (>900 entries), a reduced key of diagnoses (60 to 100 entries) and a simple key with about 10 diagnoses was described, enabling multidisciplinary use ranging from detailed information for veterinarians to simple recording of health-related observations by farmers (Table 1). International co-operation for comparability of results across countries is important.

Table 1: Example of use of the hierarchical health key published by ICAR (2012)

	Comprehensive key of diagnoses	Reduced key of diagnoses	Simple key of diagnoses
Number of diagnosis	>900	60–100	10–15
Source	Veterinarian	Veterinarian	Producer
Recording	Electronic submission (veterinarian)	Veterinarian, performance record, producer	Producer
Example	Central key for health data recording: mastitis catarrhalis acuta, mastitis catarrhalis chronica, mastitis apostematosa	e.g. AUT: acute mastitis chronic mastitis	Mastitis

Udder Health

Mastitis is the most common trait related to udder health. In most countries, somatic cell count (SCC) is obtained from samples processed in milk-recording programmes and used as an indirect measure of mastitis; the genetic correlation between mastitis and SCC is much lower than 1 (e.g. Heringstad *et al.*, 2006; Negussie *et al.*, 2008). Heritability and genetic correlation estimates are presented in Table 2. Recent research has focussed on alternative definitions of SCC, such as prolonged elevated SCC (e.g. de Haas *et al.*, 2008; Koeck *et al.*, 2010a; Urioste *et al.*, 2010). These novel definitions showed higher correlations with mastitis and demonstrated that temporal changes in SCC provide additional information about mastitis resistance. However, such analyses may require more frequent collection of samples than is common in many milk-recording programmes. Genetic evaluations for clinical mastitis (CM) have been computed in the Nordic countries for >35 years, followed by Austria, and are now common in other countries as well (see the 'Direct health traits' section). Udder health indices combining SCC and CM (and udder conformation traits) are published in some countries.

Table 2: Heritabilities for novel traits

Traits	Breed	Heritability	Sources	Remarks
Udder health				
Clinical mastitis (CM)	HF	0.07–0.08	Urioste *et al.* (2012)	Genetic correlations between CM and SCC 0.62–0.74
	FL	0.02–0.06	Koeck *et al.* (2010a and 2010b)	
	NR	0.05–0.09	Heringstad *et al.* (2004)	
	HF	0.05–0.09	Stock *et al.* (2013)	
Improved SCC – definitions (e.g. prolonged elevated SCC)	HF	0.12–0.17	Urioste *et al.* (2012)	Genetic correlations between CM and improved SCC 0.67–0.82
	FL	0.09–0.13	Koeck *et al.* (2010a)	Genetic correlations between CM and improved SCC 0.64–0.77
	HF	0.01–0.13	De Haas *et al.* (2008)	Genetic correlations between CM and improved SCC 0.55–0.93
Electrical conductivity (EC)	BS	0.23	Povinelli *et al.* (2005)	
	HF	0.12–0.36	Norberg (2005)	Genetic correlations between EC and CM 0.65–0.89
Pathogen information	NR	0.04–0.14	Haugaard *et al.* (2012);	
	HF	0.04–0.09	Sorensen *et al.* (2009)	
			De Haas *et al.* (2002)	
Near IR spectroscopy, PCR and IR thermography for detection of mastitis		0.02–0.10	Polat *et al.* (2010); Haussermann and Hartung (2012)	Information on specificity and sensitivity of methods
Lactoferrin		0.22	Arnould *et al.* (2009); Soyeurt *et al.* (2012a)	Lactoferrin predicted from Mid IR (MIR) spectroscopy
Minerals		0.50 (Ca)	Soyeurt *et al.* (2009)	Minerals predicted from MIR spectroscopy
		0.34 (Na)	Soyeurt *et al.* (2012c)	
		0.52 (Mg)		
		0.48 (K) 0.55 (P)		
Reproduction				
Fertility-related diseases (cystic ovaries, retained placenta, metritis, silent heat, etc.)		0.01–0.07	Heringstad (2010)	Genetic correlation of early reproductive disorders to NR56 0.396
		0.01–0.14	Koeck *et al.* (2010b)	
Multiple ovulation, ovarian cysts, retained placenta, metritits, silent heat		0.006–0.26	Berry *et al.* (2014)	Review
Interval from calving to commencement of luteal activity	HF	0.13–0.21	Berry *et al.* (2014)	
Retained placenta, ovary cycle disturbances	HF	0.04	Stock *et al.* (2013)	
Metabolism				
Ketosis, milk fever	FL	0.01, 0.03	Fuerst-Waltl *et al.* (2014)	
Ketosis, displaced abomasum	HF	0.02, 0.04	Koeck *et al.* (2013)	
Milk fever, ketosis	NR	0.09–0.13, 0.14–0.16	Heringstad *et al.* (2005)	
Ratio fat and protein content (first 2 test days)	FL	0.16	Fuerst-Waltl *et al.* (2014)	Genetic correlation to ketosis 0.38
Feet and legs				
Lameness	HF	0.04	Berry *et al.* (2010)	Low genetic correlation to claw health
	HF	0.02	Koeck *et al.* (2014)	
Disorders based on veterinarian diagnoses	FV	0.02	Fuerst-Waltl *et al.* (2012)	Lower frequency (only severe cases)
Disorders based on data from hoof trimming data	HF	0.02–0.13	Häggmann and Juga (2013)	
	HF	0.01–0.09	Chapinal *et al.* (2013)	
	NR	0.04–0.23	Ødegård *et al.* (2013 and 2014)	
Other novel traits				
Temperament (general temperament, milking temperament)	BS	0.12–0.20	Kramer *et al.* (2013)	
General temperament, aggressiveness, milking temperament	HF	0.38, 0.12, 0.04	Gautam and Nakao (2009)	
Suckling behaviour (allowing suckling)	FL	0.04	Fuerst-Waltl *et al.* (2010)	
Milkability from AMS (average flow rate)	HF and SR	0.38–0.42	Carlström *et al.* (2013a and 2013b)	Genetic correlation of 0.93 and 1 between milkability between conventional systems and AMS
	NR	0.11–0.30	Heringstad and Bugten (2014)	
Behaviour traits from AMS	HF	0.06–0.31	Rinell (2013)	
Activity data	HF	0.03–0.27	Schöpke und and Weigel (2014)	
Fatty acids		0.18–0.44	Bastin *et al.* (2011a)	
Feed efficiency and methane				
Residual Feed Intake	HF	0–0.32	Pryce *et al.* (2014a)	Cows
	HF	0.40	De Haas *et al.* (2011)	
Residual Feed Intake (RFI)	HF	0.22–0.38	Pryce *et al.* (2014a)	Heifers
Methane predicted from RFI	HF	0.35	De Haas *et al.* (2011)	
Methane predicted from MIR	HF	0.09–0.12	Kandel *et al.* (2013)	Daily heritability unit: g/day
Methane intensity predicted from MIR	HF	0.12–018	Kandel *et al.* (2013)	Daily heritability unit: g/kg of FPCM

SSS = somatic cell count; AMS = automated milking systems; FPCM = fat and protein corrected milk.

Several new methods for the detection of mastitis have been proposed. Electrical conductivity (EC) is available from some automated milking systems (AMS) and can be used as an indicator of mastitis. Collecting and implementing EC information in a breeding programme may be a challenge (Norberg, 2005), and recent literature on that topic is limited. Haeusermann and Hartung (2012) analysed data from near IR spectroscopy, real-time PCR and IR thermography to determine whether they are useful indicators of mastitis. They concluded that the development and testing of new detection methods for mastitis will require further work. On-farm or in-line utilization of the data is not applicable for all of the techniques examined, and there is the challenge of defining a 'gold standard' for the prevention of mastitis. Recent advances in the capacity of large-scale prediction of lactoferrin (Soyeurt *et al.*, 2012a) and mineral content indicator traits (Soyeurt *et al.*, 2009), both known to be associated with mastitis, could also offer new opportunities to select for improved mastitis resistance (Soyeurt *et al.*, 2012a and 2012b). Research has also been carried out on the use of pathogen-specific information in the genetic evaluation of mastitis (de Haas *et al.*, 2002; Sorenson *et al.*, 2009; Haugaard *et al.*, 2012). Different bacteria may cause different immune responses, and the severity of infection and response to therapeutics can differ between pathogens. It has been shown that prevention programmes for heifer mastitis can be more efficient when information about pathogens is known (De Vliegher *et al.*, 2012). The most significant challenge is in obtaining microbial cultures for a large number of animals because the data are expensive to collect relative to SCC or producer-reported diagnoses of CM.

Reproduction

Reproduction consists of several different traits. To establish a recording scheme for female fertility, the following data are desirable: (1) calving dates; (2) all artificial insemination dates as well as natural mating dates wherever possible; (3) information on fertility disorders; (4) pregnancy test results; (5) body condition score (BCS); and (6) hormone assays (ICAR, 2014). Routinely used fertility traits are mainly based on calving and insemination data. However, the use of fertility-related diagnoses is increasing, which envisions improved selection for reproductive health. Physical activity increases during oestrus. In addition, there are other behavioural changes, such as standing heat and mounting behaviour. These signs can be used to detect oestrus and can be used to calculate traits such as interval between calving and resumption of oestrus. It is probably unrealistic to routinely collect such data without automating heat detection. For example, pedometers and more sophisticated activity monitors are now being used routinely on many farms as part of a herd management package. The pedometer information needs to be compared with a baseline for the same cow, and algorithms have been developed to interpret the data collected. The efficiency of the oestrus detection rate has been reported to range between 50% and 100% depending on the criteria of success (At-Taras and Spahr, 2001). The 'gold standard' for oestrus detection is serum progesterone, and comparisons of pedometer – with progesterone-determined oestrus – has shown that activity monitors cannot detect silent behavioural oestrus (Lovendahl and Chagunda, 2010). However, an advantage of both progesterone- and activity-determined oestrus is that they do not require direct observations by farm workers. Novel predictors of fertility, such as pedometers and BCS, are growing in popularity (Fogh *et al.*, 2013). Several researchers have shown that BCS has a favourable relationship with fertility (e.g. Pryce *et al.*, 2001; Berry *et al.*, 2003). Research on the use of mid-infrared (MIR) predicted fatty acids as indicator traits for fertility (Bastin *et al.*,2012) and of MIR for pregnancy status testing is also underway (Gengler *et al.*, 2013).

Labial position is associated with urovagina, a condition in which urine accumulates in the cranial portion of the vagina, which has harmful effects on fertility (Gautam and Nakao, 2009).

Genomics offers new possibilities for the discovery of harmful recessive effects on fertility based on the absence of homozygous haplotypes (VanRaden *et al.*, 2011). A number of causative mutations have been identified using SNP genotypes and whole-genome sequence data (Adams *et al.*, 2012; Fritz *et al.*, 2013; Sonstegard *et al.*, 2013; McClure *et al.*, 2014), and a deletion with positive effects on milk yield and negative effects on fertility has recently been identified in Nordic Red cattle (Kadri *et al.*, 2014). In Fleckvieh (Simmental) and Brown Swiss, deleterious mutations compromising reproductive and rearing success in cattle have been discovered (Pausch *et al.*, 2014a and 2014 b; Schwarzenbacher *et al.*, 2012).

These methods are amenable to automation, and in the United States and several other countries the process of searching for new recessives is now a part of the genomic evaluation system.

Feet and Legs

Foot and leg conditions rank among the three most frequent reasons for culling after reproduction and udder health. The conformation of feet and legs is recorded routinely by linear-type classification systems that are often part of the services offered by breed societies. Feet and leg conformation traits may be useful indicator traits for claw health. They cannot replace direct measures of claw health because their genetic correlations with claw disorders are low, but they can be used to increase the reliability of EBVs (Häggmann and Juga,2013; Chapinal *et al.*, 2013). Koenig *et al.* (2005 and 2006), Koenig and Swalve (2006) and Linde *et al.* (2010) showed that the efficiency of breeding for improved claw health increases considerably when claw health data are included. Routine electronic recording systems for claw health at the time of trimming exist in the Nordic countries (Johansson *et al.*, 2011; Häggmann *et al.*, 2013; Ødegård *et al.*, 2013), and have been introduced in other countries as well (Kofler, 2013). Ødegård *et al.* (2013) showed, based on Norwegian data, that breeding for claw health is possible, and genetic evaluations based on data from claw trimming will be implemented in the breeding programme for Norwegian Red cattle in 2014.

The most frequently available information is locomotion scoring suggest that lameness might be a useful indicator for claw and leg health. Automated lameness detection based on activity sensors that measure lying time, number of lying bouts combined with milking and feeding data based on a cow-specific model could be a useful tool for dairy management, according to De Mol *et al.* (2013). Giuliana *et al.* (2014) showed that lameness results in behavioural changes in dairy cows in AMS. Lame cows had less feeding time and visited the robot less frequently compared with sound cows. In the Nordic countries, Austria and Southern Germany, veterinary diagnosis of feet and leg traits is recorded routinely. Several studies found that genetic evaluations based on veterinary diagnoses are valuable (e.g. Fuerst-Waltl *et al.*,2012). As veterinarians are typically consulted only in the most severe cases, incidence rates based on veterinary diagnoses are much lower than those based on hoof trimming information. For effective improvement of the feet and legs complex, it is important to establish systems that allow the collection of data from hoof trimmers.

Metabolic Status

To date, direct traits for metabolic disorders are only rarely used in genetic evaluation. Germany and Austria publish EBVs for milk fever for Fleckvieh (Simmental) and Brown Swiss. The Nordic countries include this information in the EBV for general health. In herd management, milk content traits such as fat and protein percentage, fat : protein ratio, and milk urea nitrogen are used for early detection of problems associated with metabolism. Recent research has focussed on using this information for genetic improvement as well (Koeck*et al.*, 2013; Negussie *et al.*, 2013). One of the limitations with this approach is that if the intervals between milk recording are too long, the predictive ability is limited. Automated milk-recording systems could fill this gap by providing more frequent observations (Nicolazzi *et al.*, 2012). Subclinical metabolic disorders like ketosis are thought to account for substantial economic losses; thus it may be important to detect subclinical signs at an early stage. These could be based on serum parameters like non-esterified fatty acids or *β*-hydroxybutyric acid (BHBA) concentration (Robert *et al.*, 2012; Van der Drift *et al.*, 2012). BHBA is the 'gold standard' of many cowside tests. The availability of such subclinical information will depend on the specificity and sensitivity of the tests as well as on ease and costs of implementation on-farm. Other potential indicator traits include MIR spectroscopy-based traits (Bastin *et al.*, 2011b; McParland *et al.*, 2011) or rumen activity, and BW change (Fogh *et al.*, 2013). Melzer *et al.* (2013) showed that there might be potential in the identification of important metabolites that can help monitor the metabolic profile of a cow. Roche *et al.* (2010) have shown a relationship between BCS and metabolic disorders. The risk of metabolic disorders is higher when BCS is below 2.5 or above 3.5.

Calving Traits

Breeding values related to calving ease and stillbirth are commonly available (Interbull, 2013). Genetic evaluations usually differentiate between direct and maternal and first- and later calvings, and there is substantial variation among countries in the models used for evaluations. Although most countries use single-trait models (summarized in Cole *et al.*, 2007, Table 1), Eaglen *et al.* (2012) showed that multiple-trait models may be better than single-trait models, and suggest that aggregate breeding values may be an effective way of ensuring that producers place emphasis on both direct and maternal traits. Breeds differ with respect to rates of dystocia and stillbirth, and evaluations are sometimes available only for some of the breeds in a country (e.g. Cole *et al.*, 2005; Yao *et al.*, 2014). Health disorders related to calving difficulties are recorded in some countries.

Recent studies showed the potential of decreasing losses during the rearing period. Fuerst-Waltl and Fuerst (2010) and Fuerst-Waltl and Sorensen (2010) investigated the genetic background of postnatal mortality in calves and replacement heifers in different age groups until first calving in Danish Holsteins and Austrian Simmentals. Genetic and phenotypic variation seems to be sufficiently high to genetically improve the trait of calf and heifer mortality. Some research also has been conducted on the health traits of Holstein calves in the United States (Mousa *et al.*, 2013), but heritability estimates were low, ranging from 0.01 to 0.06. In a genetic analysis of respiratory disease in Norwegian Red calves, Heringstad *et al.* (2008) found that reasonably precise genetic evaluations of sires for calf diseases could be calculated, despite the low frequency of those diseases.

In this context, information about raising stock is important. Heifers that cycle and do not get pregnant or heifers that are inseminated but are culled before calving are of interest. Reasons also may include conformational or behavioural issues (Bethard, 2008). The trait longevity is included in the total merit index in most of the countries. However, losses during rearing are generally not considered, or are included in a very simple manner. Mortality in dairy cattle is relevant not only with regard to economic losses but also with regard to animal health and welfare. McCorquodale *et al.* (2013) showed the potential of using calfhood diseases and survival in future dairy cattle breeding programmes.

Feed Efficiency

An expected increase in prices for concentrates and energy will increase the focus on feed efficiency. Efficiency can be defined as output per input unit at the farm level, as well as for individual animals. Residual feed intake (RFI), also known as net feed efficiency, is growing in popularity as a measure of feed efficiency. RFI is generally calculated as the difference between actual and predicted dry matter intake (DMI). Efficient animals have lower DMI at the same level of milk production and the same weight. The greatest obstacle in the widespread use of feed intake and feed efficiency measures in breeding objectives is the availability of large quantities of feed intake data on individual animals (Berry and Crowley, 2013). There are two promising options for enabling selection for RFI: (1) predictor traits (Fogh *et al.*, 2013) and (2) genomic prediction (Pryce and Berry, 2014; Pryce *et al.*, 2014a and 2014b).

Rumen activity might be a future indicator trait for feed efficiency. Other measures in feed, faeces and urine samples are being developed – for example, metagenomic predictions using the rumen microbiome (symbiotic microorganisms) of enteric methane in cattle and body mass index in humans are showing promise (Ross *et al.*, 2013) and could possibly also be used for predicting RFI (De Marchi *et al.*, 2014). In a recent review by Pryce *et al.* (2014a), which included seven studies of beef and dairy cattle, the accuracy of genomic predictions of RFI, energy balance and DMI ranged between 0.20 and 0.43. In theory, a genomic reference population can be assembled using a sub-population of animals with genotypes and phenotypes used to generate a genomic prediction equation (Berry *et al.*, 2014; Pryce *et al.*, 2014a and 2014b). The genomic prediction equation can then be applied to the rest of the population. However, farmers are unlikely to accept accuracies in the range currently estimated (<16%). This means that much larger populations need to be assembled. An international collaboration is endeavouring to build a data set of >6000 individuals with genotypes and phenotypes (from nine contributing partners) (Berry *et al.*, 2014; Pryce and Berry, 2014; Pryce *et al.*, 2014a and 2014b). Selection for RFI, or RFI, has been reported to lead to reductions in methane emissions of ~13.45 g CH_4/kg RFI (Hegarty *et al.*, 2007) and 18.2 g CH4/ kg RFI (Nkrumah *et al.*, 2006). A similar potential for abatement has been observed in Holstein dairy cattle, 17.5 g CH_4/kg RFI (Pryce *et al.*, unpublished data). Reducing RFI by 1 kg could lead to between 4% and 8% reduction in methane emissions. If RFI was to be included in the Australian national breeding objective, the Australian Profit Ranking, then it is predicted that RFI would reduce by 1.76 kg/cow per year (Gonzalez-Recio *et al.*, 2014a). Direct measurement of methane is expensive and is unlikely to be available in sufficient numbers for breeding in the near future, unless novel ways of measuring methane on large numbers of animals that are currently being tested prove to be successful. Chagunda *et al.* (2009) reported that a laser methane detector can be used to accurately estimate enteric methane output in dairy cows without affecting their normal activity. Dehareng *et al.* (2012) and McParland *et al.* (2014) reported the potential capacity to provide MIR spectra-based methane indicators. Such systems may be an attractive way to collect many phenotypes if they are cost-effective and durable.

OTHER NOVEL TRAITS

Behavioural Traits

Other traits, such as temperament, are of growing interest because farmers want cows that are easy to handle. From the aspect of animal welfare, these traits are gaining interest as well. So far, little information about heritabilities of behavioural traits is available, but commercial dairymen cull animals because of poor temperament (Berry *et al.*, 2005). Temperament or behaviour is often subjectively scored. According to Kramer *et al.* (2013), heritabilities based on phenotypes scored by independent people were lower than when scored by farmers (0.12 *v.* 0.20). Kramer *et al.* (2013) analysed traits including general temperament, milking temperament, aggressiveness, rank order in herd, milking speed and position of labia for Brown Swiss. General temperament was scored by farmers using a 5-point scale (1=very nervous and 5=very calm) as described by Juga (1996). The heritability for general temperament was 0.38, that for aggressiveness was 0.12 and that for milking temperament was 0.04.

Traits Out of Data from AMS

AMS offer the possibility to record traits related to workability, which are described by Rinell (2013). Temperament or behaviour, ease of handling and milkability are examples of traits that could be derived from measures routinely recorded in AMS. A measure of temperament from the AMS system could be teat cup attachment failures. Rinell (2013) estimated genetic correlations of r=−0.38 and r=−0.50 between two teat cup attachment failure traits and temperament. Different approaches concerning milking speed are used. Data from AMS offer new possibilities to record such traits automatically (Byskov *et al.*, 2012; Carlström *et al.*,2013a and 2013b). Carlström *et al.* (2013b) compared milkability from conventional milking parlours with AMS systems. They showed genetic correlations of 0.93 and 1.00 between the two systems and concluded that the inclusion of only one trait in breeding programmes is sufficient. Lovendahl *et al.* (2012) also discuss milking efficiency, in addition to milking speed. Milking efficiency describes the yield of ECM per minute in the milking box and is of interest for automated milking. This includes time for entry, cleaning, attachment, milking, exit and unit cleaning.

Adaptation to Climate Change

Exposure of animals to extreme weather conditions can be used as proxy for future climate change. Such exposure will trigger autonomous adaptations by the animals, but these adaptations are not perfect and therefore residual impacts or vulnerabilities will remain (Smit *et al.*, 1999). By quantifying these non-adaptations, novel traits can be defined to assess the degree of resilience of an animal to climate change. Resilience can be defined in many ways, the simplest of which may be the individual response of an animal to a stressor variable linked to the weather. A commonly used stressor variable is the temperature heat index, initially developed by Thom (1959) as a heat index for human comfort, and is commonly used as indicator of heat stress. Reported research in this field addressed individual responses in both production and functional traits and reported clear trait differences, fertility and intake being often considered the most strongly affected (e.g. Kadzere *et al.*, 2002; De Rensis and Scaramuzzi, 2003; West, 2003). Recent research also reported responses in some novel traits having a direct biological meaning. Gengler (2014) reported, in this context, the reaction of MIR-predicted fatty acid C18:1-*cis*9 content in milk to heat stress. Several authors have shown that this trait reflects very well the equilibrium between mobilization and intake – heat-stressed animals eating less and mobilizing more (e.g. De Rensis and Scaramuzzi, 2003). This element is of large importance when considering both adaptation to and mitigation of climate change in climate conscious breeding objectives. More detailed physiologically based adaptation traits such as heat shock proteins have been proposed (Lewis *et al.*, 1999), but large-scale phenotyping is obviously more difficult to organize and these traits are therefore less useful for practical breeding.

Currently, despite promising research results, in cattle, no large-scale breeding programme directly includes adaptation traits. However, current breeding objectives are indirectly affecting adaptation given the unfavourable correlations between resilience and production traits (e.g. West, 2003). Although profound scientific knowledge about the detailed relationships of resilience traits with other traits might still be missing, it can be hypothesized that more resilient animals are also more robust animals (i.e. more fertile) in general.

Sucking Traits

Another trait of interest is suckling behaviour. Cross-suckling and inter-suckling are considered abnormal behaviours in cattle and constitute a common problem in dairy farming (Fuerst-Waltl *et al.*, 2010). De Passillé (2001) noted that deprivation of the opportunity to suck may interfere with digestive processes and used an artificial feeding system to examine the effects of milk flow rate on calf behaviour. In a review of dairy cow behaviour, Von Keyserlingk *et al.* (2009) concluded that calf feeding systems that allow the expression of normal behaviours, such as sucking on a teat, are best for the animals. Studies of oral behaviours in calves are typically based on direct observations, and the general consistency of independent studies suggests that the methodology used to record those behaviours is adequately standardized (e.g. Margerison *et al.*, 2003).

Milk Fatty Acids

Fine milk composition traits, and in particular fatty acid profiles, can be considered another group of traits of interest (Gengler and Soyeurt, 2010). The importance of this type of traits is double: first, directly as indicators of milk quality (Bauman *et al.*, 2006) and, second, also as indicators of the physiological and health status of the animals (e.g. Bastin *et al.*, 2011b). Some studies established a link between milk composition and the environmental footprint of cows for nitrogen (Jonker *et al.*, 1998) and for methane (Mohammed *et al.*, 2011), these emissions being linked to fatty acid profiles. Unfortunately, measuring fine milk composition, and especially fatty acid composition, is not feasible on a large scale in an economical fashion as reference chemical analyses are very expensive and time-consuming. Recently, through the use of MIR spectroscopy, the prediction of most major fatty acids has become feasible on a large scale (Soyeurt *et al.*, 2011), creating new opportunities for the recording and use of novel fine milk composition traits, in particular fatty acids.

Milk Coagulation Properties

Several physical characteristics of milk that are of importance in cheese manufacture have been described in the literature, including rennet coagulation time, curd firmness 30 min after rennet addition, and curd-firming time (Bittante *et al.*, 2012). These traits are largely genetically independent of milk yield (Ikonen *et al.*,2004), and there is interest in improving those traits in many breeds (Ikonen *et al.*, 1999; Cecchinato *et al.*,2011). One limiting factor in providing such evaluations is the time and expense involved in collecting the phenotypes, but recent research suggests that MIR spectroscopy may provide reasonable predictors that can be measured on many samples affordably (Cecchinato *et al.*, 2009).

Beef Traits

Carcass traits are routinely recorded at slaughter houses and are available for genetic evaluation. This includes traits like dressing percentage, trading score and information about net daily gain. According to a survey carried out by Gene2Farm, tenderness is another trait that is more widely recorded for beef breeds. The availability of beef quality traits is still limited and under research (e.g. Gene2Farm). As the collection of beef quality traits is very expensive, genomics offers the chance to set up registration of these traits for a limited number of genotyped animals. A detailed review of this topic is outside the scope of this study.

RECORDING SYSTEMS AND SOURCES OF DATA FOR NOVEL TRAITS

For accurate breeding values there are two major possibilities: either cost-effective phenotypes are required on a large scale or investment must be made in the genotyping and collection of high-cost phenotypes for a limited number of females. Owing to advances in technologies, new traits based on lab data or automation are offering new possibilities. Many precision dairy farming technologies, including daily milk yield recording, milk component monitoring (e.g. fat, protein and SCC), pedometers, automatic temperature recording devices, milk conductivity indicators, automatic oestrus detection monitors and daily BW measurements, are already being utilized by dairy producers (Bewley, 2010). To invest in this equipment is expensive and will only pay for large farms. According to the review by Rutten *et al.* (2013), sensor systems for mastitis and reproduction are brought to a high level of development, but still need to improve detection performance.

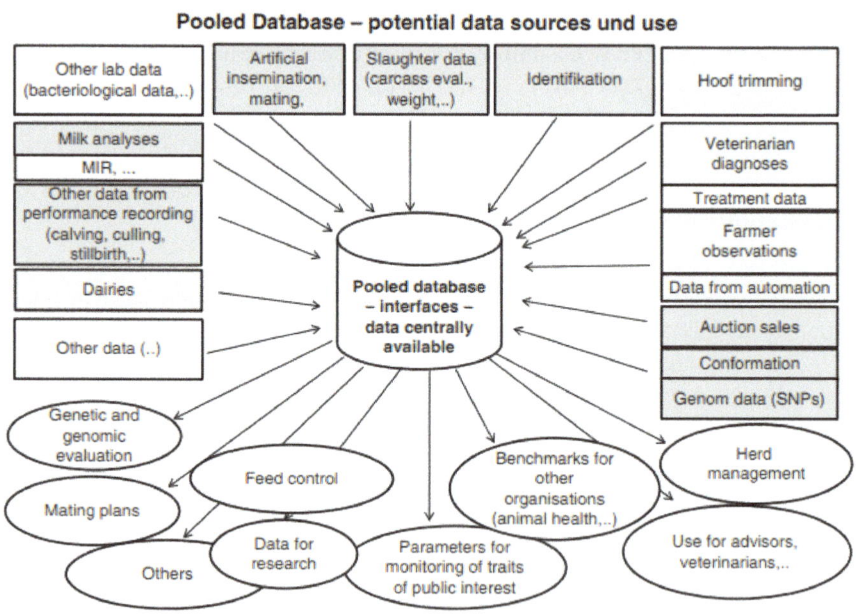

Figure 3: Pooled database with potential data sources and examples of use of data. grey=commonly used data; light grey=partly used data; white=data sources of interest.

Laboratory data include indirect health data with a spectrum of traits according to sampling protocols and testing requests (e.g. microbiological testing, metabolite analyses, hormone tests, virus/bacteria DNA, IR-based measurements) (Soyeurt *et al.*, 2009). The advantage of lab data is that, in general, they are automated or semi-automated recording systems that produce objective measurements, and many data can be obtained at low cost. The disadvantage might be that, for example, for bacteriological findings or hormonal assays – data might be only from preselected animals.

To use synergies and enlarge the range of possible phenotypes for genetic collaboration with other organizations and institutions is an option (Figure 3). Data security issues have to be considered, and standardized protocols for ensuring security are necessary. Slaughter houses, dairy plants, health organizations and other entities are routinely recording data for their own purposes. The monitoring of welfare is gaining increasing importance. The central role of dairy cattle health in the context of animal welfare and consumer protection implies that farmers and veterinarians are obligated to maintain high-quality records, emphasizing the particular sensitivity of health data. The legal requirement to monitor the development of different indicator traits offers the possibility of synergies for recording of traits. One example is direct health data, which are partly recorded based on legal documentation requirements (Stege *et al.*,2003; Østerås *et al.*, 2007; Egger-Danner *et al.*, 2012b).

Genotyping Females in Research Herds

As several novel traits currently under investigation are expensive to record, using research herds for phenotyping in combination with genomics are an attractive source of data, as are herds that engage in intensive data collection. These farms are of particular interest for feed efficiency, methane emission and high-cost health traits. As several thousand phenotypes are required to accurately determine the characteristics of traits in terms of heritability and correlations to other traits of interest in the breeding goal (Calus *et al.*, 2013a), there is a clear need to invest in phenotyping of many animals for the new traits, as well as possible indicator traits (Gonzalez-Recio *et al.*, 2014b).

In breeding programmes worldwide, more and more young bulls with genomic expected breeding values (GEBVs) are used. The reliability of the GEBVs depends to a very high extent on the size of the reference population. To achieve genetic response for the novel trait, the reliability of the trait is crucial. The limitation for novel traits where phenotyping has only recently started is that it will be very difficult or impossible to establish a large reference population based on bulls, as the number of bulls used within breeding programmes has been reduced because of genomic selection. Methods to increase the accuracy of genomic predictions are as follows: enlarging the reference population, using chips of higher density, imputing from lower densities to higher densities and the inclusion of ungenotyped cows in the reference population (Pszcola *et al.*, 2013).

Enlarging the reference population by adding females is also a possibility. Simulation studies (e.g. Buch,2011; Pszcola *et al.*, 2013; Gonzalez-Recio *et al.*, 2014b) show that for novel traits and limited resources it is more effective to genotype females than only males. The impact of the cow reference population also depends on the sampling strategy. Including only elite dams will only result in minor increases of reliabilities and may lead to biased estimates. Exploring the variation of the trait by sampling extreme cows (top tails and bottom tails of distribution) will increase the benefit (Jimenez-Montero *et al.*, 2012). However, it is desirable to either randomly select cows or select those that will contribute the most data, which is the strategy being applied in the Australian genomic information nucleus known as Ginfo (Australia's genomic information nucleus). Traits with medium to high heritability gain more from genotyping cows than direct health traits with low heritability (Egger-Danner *et al.*, 2014). Bolignon *et al.* (2012) showed that reference population animals with extreme yield deviations are the most informative for genomic selection. De Haas *et al.* (2012) have shown that it is possible to increase the number of genotypes and phenotypes by merging data from various research herds worldwide. Validation of equations in completely independent populations showed promising results (Pryce *et al.*, 2014b). Different studies stress the optimum design of female reference populations to use resources efficiently (Buch, 2011; Pszczola *et al.*, 2012; Thomasen *et al.*, 2014). It is important that the variance of the trait can be explored and that the relationship within the reference population is low but closely related to the evaluated population. Therefore, it is important to update the reference population continuously.

Experiences from the United States showed that there is some gain from the inclusion of cows in the predictor population (Wiggans *et al.*, 2011), but adjustments to their records may be needed to avoid potential biases (Dassonneville *et al.*, 2012; Wiggans *et al.*, 2012). The use of cows in predictor populations for novel traits that lack a large group of bulls with high-reliability evaluations will be necessary. Table 3 gives an overview of gEBV reliabilities for novel traits. So far, experience is limited with real data. However, it is unclear whether or when low-reliability cow data should be removed from the predictor population as the reliability of sire EBV increases.

Table 3: Reliabilities of genomic EBVs for novel traits

Traits	Novel traits	Size of calibration group	Reliability/ accuracy*	Sources	Remarks
Udder health	CM	2563 bulls	0.26	Heringstad *et al.* (2011)	
		7800 bulls + 10 000 cows	0.17/0.23	NAV Routine Evaluation (2014)	Increase related to pedigree index (RDC) (bulls/ bulls + cows in reference)
Reproduction	Fertility-related disorders	3363 bulls	0.17–0.65*	Haugaard *et al.* (2014)	Correlation GEBV and EBV
Feet and legs	Claw health	967 bulls	0.29–0.35*	Ødegård *et al.* (2014)	Correlation GEBV and DYD
		7800 bulls + 10 000 cows	0.24/0.33	NAV Routine Evaluation (2014)	Increase related to pedigree index (RDC) (bulls/ bulls + cows in reference)
Feed efficiency	RFI	Various	0.40–0.43*	Pryce *et al.* (2014a)	
	Energy balance		0.29*	Pryce *et al.* (2014a)	
	Dry matter intake		0.20–0.35*	Pryce *et al.* (2014a)	
Other diseases	Other diseases	7800 bulls +10 000 cows	0.17/0.17	NAV Routine Evaluation (2014)	Increase related to pedigree index (RDC) (bulls/ bulls + cows in reference)

GEVB = genomic estimated breeding value; RDC = red dairy cattle; DYD = daugher yield deviation; RFI = residual feed intake; * = accuracy.

Cost efficiency is an important criterion for the success and sustainability of breeding programmes. Therefore, it might be too expensive to continuously record data extensively on research herds. Indictor traits with known genetic correlation to the direct trait of interest may be used instead. According to Pszcola *et al.* (2013), easily recorded predictor traits could be used to increase the reliability of scarcely recorded traits – for example, fat- and protein-corrected milk yield and live weight for DMI – and the bias of genomic breeding values of scarcely recorded traits could be reduced. As described under the various trait complexes, this approach could be applicable also for metabolism, feet and legs and other traits (Axelsson *et al.*, 2013). Genomic selection programmes for indicator traits also benefit from the inclusion of cows in the reference population (Calus *et al.*, 2013a and 2013b).

CHALLENGES

Predictive Biology

Biomarkers, such as metabolites or milk components, may be rich sources of information about novel phenotypes (Gengler *et al.*, 2013). Standard analyses undertaken by milk-recording laboratories, such as by MIR spectroscopy, generate spectral data that reflect many characteristics of milk. Research is currently underway to investigate the use of this data for prediction of indicator traits (RobustMilk, OptiMIR, PhenoFinlait, GplusE, etc.). The main challenge posed by these data is the lack of reference data for the detection of disease, such as mastitis. A reliable pool of 'healthy' and 'sick' animals is required. For traits with low heritability and limited reliability and repeatability, this is even more difficult. The reference data set has to account for different production circumstances. There is the need for comparability of spectra from different instrument manufacturers (Foss, Bentley, Delta, etc.), models within brands, and stability over time. The repeatability of results across different breeds and production environments, as well as logistical challenges related to data collection and transfer, must be considered. This includes extracting data from spectrometers and storing of MIR spectra, standardizing spectra, complex computation of indicators and finally the implementation into routine milk-recording work flow (Soycurt *et al.*, 2012b and 2012c; Gengler*et al.*, 2013).

Predictive biology is also a topic for other sensor techniques. According to a review by Rutten *et al.* (2013), research is still needed. In the future, genomic data could be integrated to improve the prediction. Different environmental effects, such as feeding and management, can influence the prediction equations. Therefore, there is also a need for calibration across breeds and countries (De Marchi *et al.*, 2014). Prediction equations are derived using reliable phenotypes from a large sample of animals, with the ultimate goal of predicting phenotypes directly from the MIR spectra. In Figure 4, the system of prediction is described. The prediction equation is derived based on reliable phenotypes for novel traits (e.g. mastitis) from a larger sample of animals. The ultimate goal is to predict phenotypes directly from, for example, MIR spectra. Advances could be that these equations are improved by taking the genomic information into account.

Egger-Danner, Cole, Pryce, Gengler, Heringstad, Bradley and Stock

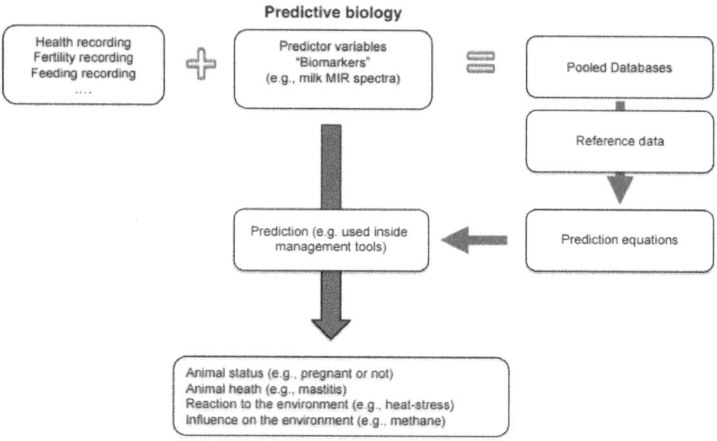

Figure 4: Overview about the system of predictive biology to determine traits based on prediction equations.

Standardization and Integration of Relevant Data Sources

There are many mechanical and electronic systems on dairies that can be used as sources of new phenotypes (Figure 5). Automatic milking and feeding systems, for example, routinely generate many observations. Among them, which data are actually stored is again different. There is a need for harmonization in trait definitions across systems and devices. An additional challenge is the availability of data. Many systems use process computers that do not communicate with external databases. This results in data that are 'stranded' in different systems that cannot easily be integrated into a single database for analysis. When such data are available in a central database, they can also be used for benchmarking, which offers an opportunity to compare performance both between and within farms over time. A single parameter does not provide an appropriate benchmark, and indices based on a basket of parameters may provide a better marker of overall performance (Bradley *et al.*, 2013). Benchmarks also provide farmers with rapid feedback about the effect of management changes on farm performance, and may encourage ongoing participation in data collection programmes.

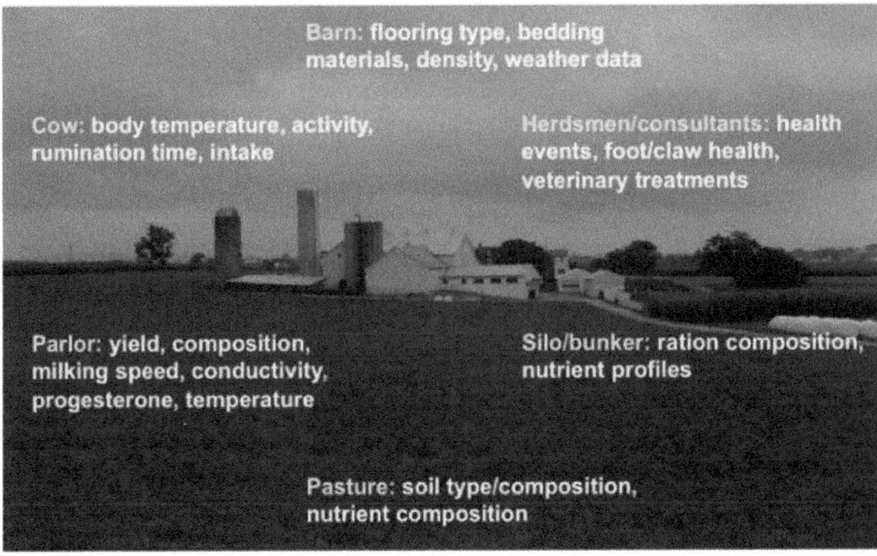

Figure 5: Sources of on-farm information that can be used to collect health and fitness phenotypes

Simplicity and Ease-Of-Use for Farmers

The future availability of phenotypic data will depend heavily on the motivation of the farmers. It is expected that the traits and data available will further increase. The challenge is to present those data in a way that does not overwhelm farmers when providing useful tools. Easy handling and simple access to data is essential because time is a limiting factor on most farms.

CONCLUSIONS

Conditions of production, as well as consumer demands, will have an impact on the future availability of phenotypes. Traits connected with animal health, animal welfare, food safety and efficiency will become more important. Advances in technology will enable more precise trait definitions, with phenotypes closer to the genotype. Technological advances will permit the closer integration of existing data sources. New technologies are expected to provide better indicator traits for fertility, mastitis, metabolism and energy efficiency. There are still challenges to solve before some technologies can be widely applied on a routine basis, including the predictive biology of indicator traits, the harmonization of data from automated systems, the motivation of farmers to record high-quality phenotypes and the genomic gap for novel traits. To overcome the phenomic gap between traditional traits and novel traits, expanded genotyping of cows should be considered. International co-operation will ensure that trait definitions are consistent across countries and will support the research necessary to deliver new management and selection tools to farmers.

ACKNOWLEDGEMENTS

This review is based on an invited presentation at the 64th Annual Meeting of the European Association for Animal Production held in Nantes, France, August 2013. Excellent cooperation within ICAR Working Group Functional Traits. Federal Ministry of Agriculture, Forestry, Environment and Water Management (BMLFUW) in Austria and the Federation of Austrian Cattle Breeders for the support within the projects 'Efficient cow' and 'OptiGene'. Gene2Farm (EU-FP7-KBBE-2011-5-PNr.: 289592) for provision of information. J. B. Cole was supported by appropriated project 1265-31000-096-00, 'Improving Genetic Predictions in Dairy Animals Using Phenotypic and Genomic Information', of the Agricultural Research Service of the United States Department of Agriculture. N. Gengler was supported by the INTERREG IVB NWE Project 190G OptiMIR: new tools for a more sustainable dairy sector and the Ministry of Agriculture of the Walloon Region of Belgium (Service Public de Wallonie, Direction générale opérationnelle 'Agriculture, Ressources naturelles et Environnement' - DGARNE) for the support within the project 'MethaMilk' (D31-1248 and D31-1304). J. E. Pryce acknowledges financial support from the Department of Environment and Primary Industries (Melbourne, Australia).

REFERENCES

1. Adams HA, Sonstegard T, VanRaden PM, Null DJ, Van Tassell CP and Lewin H 2012. Identification of a nonsense mutation in APAF1 that is causal for a decrease in reproductive efficiency in dairy cattle. Proceedings of the Plant & Animal Genome XX Conference, 14–18 January 2014, San Diego, CA, USA, abstract. P0555.
2. Arnould VMR, Soyeurt H, Gengler N, Colinet FG, Georges MV, Bertozzi G, Portelle D and Renaville R 2009. Genetic analysis of lactoferrin content in bovine milk. Journal of Dairy Science 92, 2151–2158.
3. At-Taras EE and Spahr SL 2001. Detection and characterization of estrus in dairy cattle with an electronic heatmount detector and an electronic activity tag. Journal of Dairy Science 84, 792–798.
4. Axelsson H, Fikse WF, Kargo M, Sorensen AC, Johansson K and Rydhmer L 2013. Genomic selection using indicator traits to reduce the environmental impact of milk production. Journal of Dairy Science 96, 5306–5314.
5. Bastin C, Gengler N and Soyeurt H 2011a. Phenotypic and genetic variability

of production traits and milk fatty acid contents across days in milk for Walloon Holstein first-parity cows. Journal of Dairy Science 94, 4152–4163.

6. Bastin C, Soyeurt H, Vanderick S and Gengler N 2011b. Genetic relationship between milk fatty acids and fertility of dairy cows. Interbull Bulletin 44, 195–199.

7. Bastin C, Berry DP, Soyeurt H and Gengler N 2012. Genetic correlations of days open with production traits and contents in milk of major fatty acids predicted by mid-infrared spectrometry. Journal of Dairy Science95, 6113–6121.

8. Bauman DE, Mather IH, Wall RJ and Lock AL 2006. Major advances associated with the biosynthesis of milk. Journal of Dairy Science 89, 1235–1243.

9. Berry DP and Crowley JJ 2013. Cell biology symposium: genetics of feed efficiency in dairy and beef cattle.Journal of Animal Science 91, 1594–1613.

10. Berry DP, Wall E and Pryce JE 2014. Genetics and genomics of reproductive performance in dairy and breed cattle. Animal 8 (suppl. 1), 105–121.

11. Berry DP, Harris BL, Winkelman AM and Montgomerie W 2005. Phenotypic associations between traits other than production and longevity in New Zealand dairy cattle. Journal of Dairy Science 88, 2962–2974.

12. Berry DP, Buckley F, Dillon P, Evans RD, Rath M and Veerkamp RF 2003. Genetic relationships among body condition score, body weight, milk yield, and fertility in dairy cows. Journal of Dairy Science 88, 2193–2204.

13. Bethard G 2008. Quantifying a heifer culling decision. Dairy Calf and Heifer Association Conference, 1–3 April. Rochester, MN, USA. Retrieved 20 March 2014, fromhttp://sroc.cfans.umn.edu/prod/groups/cfans/@pub/@cfans/@sroc/@outreach/documents/article/cfans_article_117790.pdf

14. Bewley J 2010. Precision dairy farming: advanced analysis solutions for future profitability. Retrieved 28 March 2014, from http://www.precisiondairy2010.com/proceedings/s1bewley.pdf

15. Bittante G, Penasa M and Cecchinato A 2012. Invited review: genetics and modeling of milk coagulation properties. Journal of Dairy Science 95, 6843–6870.

16. Bo N 2009. Practical cattle breeding in the future – commercialised or co-operative, across borderlines between countries and organisations. Retrieved 28 March 2014, fromhttps://journal.interbull.org/index.php/ib/article/viewFile/1122/1113

17. Bolignon AA, Long N, Albuquerque LG, Weigel KA, Gianola D and Rosa GJ 2012. Comparison of selective genotyping strategies for prediction of breeding values in a population undergoing selection. Journal of Animal Science 90, 4716–4722.

18. Bradley AJ, Breen JA, Hudson CD and Green MJ 2013. Benchmarking for health from the perspective of consultants. Challenges and benefits of health data recording in the context of food chain quality, management and breeding. Proceedings of the ICAR Health Data Conference, 30–31 May 2013, Aarhus, Denmark, ICAR Technical Series no. 17, 159–166.

19. Buch LH 2011. Genetic improvement of functional traits in dairy cattle breeding

schemes with genomic selection. PhD, Aarhus University, Aarhus, Denmark.

20. Byskov K, Buch LH and Aamand GP 2012. Possibilities of implementing measures from automatic milking systems in routine evaluations of udder conformation and milking speed. Interbull Bulletin 46, 28–32.
21. Calus MPL, de Haas Y, Pszczola M and Veerkamp RF 2013a. Predicted accuracy of and response to genomic selection for new traits in dairy cattle. Animal 7, 183–191.
22. Calus MPL, Bijma P, van Arendonk J and Veerkamp RF 2013b. Economics of using genomic selection at the farm level. Book of Abstracts of the 64th Annual Meeting of the European Association for Animal Science. 26 August 2013, Nantes, France, 184pp.
23. Carlström C, Petterson G, Johansson K, Strandberg E, Stålhammar H and Philipsson J 2013a. Feasibility of using automatic milking system data from commercial herds for genetic analysis of milkability. Journal of Dairy Science 96, 5324–5332.
24. Carlström C, Strandberg E, Johansson K, Pettersson G, Stålhammar H and Philipsson J 2013b. Genetic evaluation of in-line recorded milkability from milking parlors and automatic milking systems. Journal of Dairy Science 97, 497–506.
25. Cecchinato A, De Marchi M, Gallo L, Bittante G and Carnier P 2009. Mid-infrared spectroscopy predictions as indicator traits in breeding programs for enhanced coagulation properties of milk. Journal of Dairy Science92, 5304–5313.
26. Cecchinato A, Penasa M, De Marchi M, Gallo L, Bittante G and Carnier P 2011. Genetic parameters of coagulation properties, milk yield, quality, and acidity estimated using coagulating and noncoagulating milk information in Brown Swiss and Holstein-Friesian cows. Journal of Dairy Science 94, 4205–4213.
27. Chagunda MGG, Ross D and Roberts DJ 2009. On the use of a laser methane detector in dairy cows.Computers and Electronics in Agriculture 68, 157–160.
28. Chapinal N, Koeck A, Sewalem A, Kelton DF, Mason S, Cramer G and Miglior F 2013. Genetic parameters for hoof lesions and their relationship with feet and leg traits in Canadian Holstein cows. Journal of Dairy Science 96, 2596–2604.
29. Cole JB, Wiggans GR and VanRaden PM 2007. Genetic evaluation of stillbirth in United States Holsteins using a sire-maternal grandsire threshold model. Journal of Dairy Science 90, 2480–2488.
30. Cole JB, Goodling RC Jr, Wiggans GR and VanRaden PM 2005. Genetic evaluation of calving ease for Brown Swiss and Jersey bulls from purebred and crossbred calvings. Journal of Dairy Science 88, 1529–1539.
31. Cole JB, Waurich B, Wensch-Dorendorf M, Bickhart DM and Swalve HH 2014. A genome-wide association study of calf birth weight in Holstein cattle using single nucleotide polymorphisms and phenotypes predicted from auxiliary traits. Journal of Dairy Science 97, 3156–3172.

32. Dassonneville R, Baur A, Fritz S, Boichard D and Ducrocq V 2012. Inclusion of cow records in genomic evaluations and impact on bias due to preferential treatment. Genetics Selection Evolution 44, 40.[PMC free article]

33. De Haas Y, Barkema HW and Veerkamp RF 2002. Genetic parameters of pathogen-specific incidence of clinical mastitis in dairy cows. Animal Science 74, 233–242.

34. De Haas Y, Ouweltjes W, ten Napel J, Windig JJ and de Jong G 2008. Alternative somatic cell count traits as mastitis indicators for genetic selection. Journal Dairy Science 91, 2501–2511.

35. De Haas Y, Windig JJ, Calus MPL, Dijkstra J, de Haan M, Bannink A and Veerkamp RF 2011. Genetic parameters for predicted methane production and potential for reducing enteric emissions through genomic selection. Journal of Dairy Science 94, 6122–6134.

36. De Haas Y, Calus MP, Veerkamp RF, Wall E, Coffey MP, Daetwyler HD, Hayes BJ and Pryce JE 2012.Improved accuracy of genomic prediction for dry matter intake of dairy cattle from combined European and Australian data sets. Journal Dairy Science 95, 6103–6112.

37. De Marchi MD, Toffanin V, Cassandro M and Penasa M 2014. Invited review: mid-infrared spectroscopy as phenotyping tool for milk traits. Journal Dairy Science 97, 1–16.

38. De Mol RM, Andre G, Bleumer JB, van der Werf JTN, de Haas Y and Reenen CG 2013. Applicability of day-to-day variation in behavior for the automated detection of lameness in dairy cows. Journal Dairy Science 96, 3703–3712.

39. De Passillé AM 2001. Sucking motivation and related problems in calves. Applied Animal Behaviour Science 72, 175–187.

40. De Rensis F and Scaramuzzi RJ 2003. Heat stress and seasonal effects on reproduction in the dairy cow – a review. Theriogenology 60, 1139–1151.

41. De Vliegher S, Fox LK, Piepers S, McDougall S and Barkema HW 2012. Invited review: mastitis in dairy heifers: nature of the disease, potential impact, prevention, and control. Journal Dairy Science 95, 1025–1040.

42. Dehareng F, Delfosse C, Froidmont E, Soyeurt H, Martin C, Gengler N, Vanlierde A and Dardenne P 2012.Potential use of milk mid-infrared spectra to predict individual methane emission of dairy cows. Animal 6, 1694–1701.

43. Eaglen S, Coffey MP, Wooliams JA and Wall E 2012. Evaluating alternate models to estimate genetic parameters of calving traits in United Kingdom Holstein-Friesian dairy cattle. Genetics Selection Evolution44, 2 Retrieved 28 March 2014, from http://www.gsejournal.org/content/44/1/23 [PMC free article]

44. EFSA-Q-2006-113 2009. Scientific report on the effects of farming systems on dairy cow welfare and disease. European Food Safety Authority. Scientific Report of EFSA Question number:EFSA-Q-2006-113. Retrieved 28 March 2014, from http://www.efsa.europa.eu/en/efsajournal/pub/1143r.htm

45. Egger-Danner C, Schwarzenbacher H and Willam A 2014. Short communication:

genotyping of cows to speed up availability of GEBV for direct health traits in Austrian Fleckvieh (Simmental)cattle – genetic and economic aspects. Journal of Dairy Science 97, 4552–4556.

46. Egger-Danner C, Willam A, Fuerst C, Schwarzenbacher H and Fuerst-Waltl B 2012a. Effect of breeding strategies using genomic information on fitness and health. Journal of Dairy Science 95, 4600–4609.
47. Egger-Danner C, Fuerst-Waltl B, Obritzhauser W, Fuerst C, Schwarzenbacher H, Grassauer B, Mayerhofer M and Koeck A 2012b. Recording of direct health traits in Austria-experience report with emphasis on aspects of availability for breeding purposes. Journal of Dairy Science 95, 2765–2777.
48. Fogh A, Aamand GP, Hjorto L and Lauritsen U 2013. Recording of data and identification issues. ICAR Conference, Aarhus, Denmark. Retrieved 28 March 2014, fromhttp://www.icar.org/documents/aarhus_2013/Manuscripts/Fogh.pdf
49. Fritz S, Capitan A, Djari A, Rodriguez SC, Barbat A, Baur A, Grohs C, Weiss B, Boussaha M, Esquerré D, Klopp C, Rocha D and Boichard D 2013. Detection of haplotypes associated with prenatal death in dairy cattle and identification of deleterious mutations in GART, SHBG and SLC37A2. PLoS One 8, e65550. [PMC free article]
50. Fuerst-Waltl B and Fuerst C 2010. Mortality in Austrian dual purpose Fleckvieh calves and heifers.Livestock Science 132, 80–86.
51. Fuerst-Waltl B and Sorensen MK 2010. Genetic analysis of calf and heifer losses in Danish Holstein. Journal of Dairy Science 93, 5436–5442.
52. Fuerst-Waltl B, Fuerst C and Egger-Danner C 2012. Claw health diagnoses in the routine health monitoring system of Austrian Fleckvieh cattle. 63rd Annual Meeting of the Association of European Animal Production, 27 August, Bratislava, Slovakia. Retrieved 28 March 2014, fromhttp://www.eaap.org/Previous_Annual_Meetings/2012Bratislava/Papers/Published/07_Fuerst-Waltl.pdf
53. Fuerst-Waltl B, Rinnhofer B, Fuerst C and Winckler C 2010. Genetic parameters for abnormal sucking traits in Austria Fleckvieh heifers. Journal of Animal Breeding and Genetics 127, 113–118.
54. Fuerst-Waltl B, Egger-Danner C, Zollitsch W and Fuerst-Waltl B 2014. Metabolic disorders and their relationships to milk production traits in Austrian Fleckvieh. Retrieved 9 August, fromhttp://www.icar.org/Documents/Berlin_2014/PPTs/Presented/Fuerst_Waltl.pdf
55. Gautam G and Nakao T 2009. Prevalence of urovagina and its effects on reproductive performance in Holstein cows. Theriogenology 71, 1451–1461.
56. Gengler N 2014. Mitigation strategies versus adaptation strategies. Invited presentation at the ASGGN Workshop. Retrieved 17 August, from http://orbi.ulg.ac.be/bitstream/2268/171555/1/NGENGLER_VF.pdf
57. Gengler N and Soyeurt H 2010. Interest, recording and possible use of new phenotypes from fine milk composition. Proceedings 9th World Congress of Genetics Applied to Livestock Production, Leipzig, Germany.

58. Gengler N, Berry DP and Bastin C 2013. Use of automated systems for recording of direct and indirect data with special emphasis on the use of MIR milk spectra (OptiMIR project). Challenges and benefits of health data recording in the context of food chain quality, management and breeding. Proceedings of the ICAR Health Data Conference, 30–31 May 2013, Aarhus, Denmark. ICAR Technical Series no. 17, pp. 55–61.
59. Gernand E, Rehbaein P, von Borstel UU and König S 2012. Incidences of and genetic parameters for mastitis, claw disorders, and common health traits recorded in dairy cattle contract herds. Journal of Dairy Science 95, 2144–2156.
60. Giuliana GM-P, Kaler J, Remnant J, Cheyne L, Abbott C, French AP, Pridmore TP and Huxley JN 2014.Behavioural changes in dairy cows with lameness in an automatic milking system. Applied Animal Behavioural Science 150, 1–8.
61. Gonzalez-Recio O, Pryce JE, Haile-Mariam M and Hayes BJ 2014a. Incorporating heifer feed efficiency in the Australian selection index using genomic selection. Journal of Dairy Science 97, 3883–3893.
62. Gonzalez-Recio O, Coffey MP and Pryce JE 2014b. On the value of phenotypes in the genomic era. Journal of Dairy Science (in press).
63. Govignon-Gion A, Dassonneville R, Baloche G and Ducrocq V 2012. Genetic evaluation of mastitis in dairy cattle in France. Interbull Bulletin 46, Retrieved 28 March 2014, fromhttps://journal.interbull.org/index.php/ib/article/view/1276
64. Häggmann J and Juga J 2013. Genetic parameters for hoof disorders and feet and leg conformation traits in Finnish Holstein cows. Journal of Dairy Science 96, 3319–3325.
65. Haugaard K, Heringstad B and Whist AC 2012. Genetic analysis of pathogen-specific clinical mastitis in Norwegian Red cows. Journal of Dairy Science 95, 1545–1551.
66. Haugaard K, Svendsen M and Heringstad B 2014. Genomic predictions of fertility related disorders in Norwegian Red using 30 years of data. Proceedings 10th World Congress of Genetics Applied to Livestock Production, Vancouver, Canada.
67. Haussermann A and Hartung E 2012. Detection of mastitis during milking – current solutions and prospective ideas. 63rd Annual Meeting of the Association of European Animal Production, 27 August, Bratislava, Slovakia. Retrieved 28 March 2014, fromhttp://www.eaap.org/Previous_Annual_Meetings/2012Bratislava/Papers/Published/07_Haeussermann.pdf
68. Hegarty RS, Goopy JP, Herd RM and McCorkell B 2007. Cattle selected for lower residual feed intake have reduced daily methan production. Journal of Animal Science 85, 1479–1486.
69. Heringstad B, 2010. Genetic analysis of fertility related diseases and disorders in Norwegian Red cows.Journal of Dairy Science 93, 2751–2756.
70. Heringstad B and Østerås O 2013. More than 30 years of health recording in Norway. ICAR 2013 Health Data Conference: challenges and benefits of health

data recording in the context of food chain quality, management and breeding. 30–31 May 2013, Århus, Denmark.

71. Heringstad B and Bugten HK 2014. Genetic evaluations of milkability in Norwegian Red based on data from Automatic Milking Systems. Proceedings 10th World Congress of Genetics Applied to Livestock Production, Vancouver, Canada.

72. Heringstad B, Klemetsdal G and Steine T 2007. Selection responses for disease resistance in two selection experiments with Norwegian red cows. Journal of Dairy Science 90, 2419–2426.

73. Heringstad B, Chang YM, Gianola D and Klemetsdal G 2004. Multivariate threshold model analysis of clinical mastitis in multiparous Norwegian dairy cattle. Journal of Dairy Science 87, 3038–3046.

74. Heringstad B, Chang YM, Gianola D and Klemetsdal G 2005. Genetic analysis of clinical mastitis, milk fever, ketosis, and retained placenta in three lactations of norwegian red cows. Journal of Dairy Science 88, 3273–3281.

75. Heringstad B, Chang YM, Gianola D and Østerås O 2008. Short communication:genetic analysis of respiratory disease in Norwegian Red calves. Journal of Dairy Science 91, 367–370.

76. Heringstad B, Gianola D, Chang YM, Ødegård J and Klemetsdal G 2006. Genetic associations between clinical mastitis and somatic cell score in early first-lactation cows. Journal of Dairy Science 89, 2236–2244.

77. Heringstad B, Su G, Solberg TR, Guldbrandtsen B, Svendsen M and Lund MS 2011. Genomic predictions based on a joint reference population for Scandinavian red breeds. 62nd Annual Meeting of the Association of European Animal Production, 29 August. Retrieved 6 September 2014, fromhttp://www.eaap.org/Previous_Annual_Meetings/2011Stavanger/Papers/Published/S04_Heringstad.pdf

78. ICAR2012. ICAR guidelines for recording, evaluation and genetic improvement of health traits. Retrieved 10 December 2013, fromhttp://www.icar.org/Documents/Rules%20and%20regulations/Amendments%202012/Recording,%20Evaluation%20and%20Genetic%20Improvement%20of%20health%20traits.pdf

79. ICAR 2013. Challenges and benefits of health data recording in the context of food chain quality, management and breeding. ICAR Technical Series no. 17. Retrieved 25 March 2014, fromhttp://www.icar.org/Documents/technical_series/tec_series_17_Aarhus.pdf

80. ICAR 2014. ICAR guidelines for recording, evaluation and genetic improvement of female fertility in dairy cattle. Retrieved 24 January 2014, fromhttp://www.icar.org/Documents/Rules%20and%20regulations/Guidelines/Guidelines_Female_Fertility_for_approval.pdf

81. Ikonen T, Ahlfors K, Kempe R, Ojala M and Ruottinen O 1999. Genetic parameters for the milk coagulation properties and prevalence of noncoagulating milk in Finnish dairy cows. Journal of Dairy Science 82, 205–214.

82. Ikonen T, Morri S, Tyrisevä A-M, Ruottinen O and Ojala M 2004. Genetic and phenotypic correlations between milk coagulation properties, milk production traits, somatic cell count, casein content, and pH of milk. Journal of Dairy Science 87, 458–467.

83. Interbull 2013. National genetic evaluation forms provided by country. http://www.interbull.org/ib/geforms

84. Jimenez-Montero JA, Gonzales-Rezio O and Alenda R 2012. Genotyping strategies for genomic selection in small dairy cattle populations. Animal 6, 1216–1224.

85. Johansson K, Eriksson J-Å, Nielsen US, Pösö J and Aamand GP 2011. Genetic evaluation of claw health in Denmark, Finland and Sweden. Interbull Bulletin 44, 224–228.

86. Jones WP, Hansen LB and Chester-Jones H 1994. Response of health care to selection for milk yield of dairy cattle. Journal of Dairy Science 77, 3137–3152.

87. Jonker JS, Kohn RA and Erdman RA 1998. Using milk urea nitrogen to predict nitrogen excretion and utilization efficiency in lactating dairy cows. Journal of Dairy Science 81, 2681–2692.

88. Juga J 1996. Evaluation methods of subjectively scored functional traits in Finland. Interbull Bulletin 14, 155–160.

89. Kadri NK, Sahana G, Charlier C, Iso-Touru T, Guldbrandtsen B, Karim L, Nielsen US, Panitz F, Aamand GP, Schulman N, Georges M, Vilkki J, Lund MS and Druet T 2014. A 660-Kb deletion with antagonistic effects on fertility and milk production segregates at high frequency in Nordic Red cattle: additional evidence for the common occurrence of balancing selection in livestock. PLoS Genetics 10, e1004049.[PMC free article]

90. Kadzere CT, Murphy MR, Silanikove N and Maltz E 2002. Heat stress in lactating dairy cows: a review.Livestock Production Science 77, 59–91.

91. Kandel PB, Vanrobays ML, Vanlierde A, Dehareng F, Froidmont E, Dardenne P, Lewis E, Buckley F, Deighton MH, McParland S, Gengler N and Soyeurt H 2013. Genetic parameters for methane emission predicted from milk mid-infrared spectra in dairy cows. Advances in Animal Biosciences 4, 279.

92. Koeck A, Miglior F, Jamrozik J, Kelton DF and Schenkel FS 2013. Genetic associations of ketosis and displaced abomasum with milk production traits in early first lactation of Canadian Holsteins. Journal of Dairy Science 96, 4688–4696.

93. Koeck A, Heringstad B, Egger-Danner C, Fuerst C, Winter P and Fuerst-Waltl B 2010a. Genetic analysis of clinical mastitis and somatic cell count traits in Austrian Fleckvieh cows. Journal of Dairy Science 93, 5987–5995.

94. Koeck A, Egger-Danner C, Fuerst W, Obritzhauser W and Fuerst-Waltl B 2010b. Genetic analyses of reproductive disorders and their relationship to fertility and milk yield in Austrian Fleckvieh dual-purpose cows. Journal of Dairy Science 93, 2185–2194.

95. Koeck A, Loker S, Miglior F, Kelton DF, Jamrozik J and Schenkel FS 2014. Genetic relationships of clinical mastitis, cystic ovaries, and lamness with milk yield and somatic cell score in first-lactation Canadian Holsteins. Journal of Dairy Science 97, 1–8.

96. Koenig S and Swalve HH 2006. Modelkalkulationen zu zuechterischen moeglichkeiten auf klauengesundheit beim milchrind. Züchtungskunde 78, 345–356.

97. Koenig S, Köhn F, Kuwan K, Simianer H and Gauly M 2006. Use of repeated measures analysis for evaluation of genetic background of dairy cattle behavior in automatic milking systems. Journal of Dairy Science 89, 3636–3644.

98. Koenig S, Scharifi AR, Wentrot H, Landmann D, Eise M and Simianer H 2005. Genetic parameters of claw and foot disorders estimated with logistic models. Journal of Dairy Science 88, 3316–3325.

99. Kofler J 2013. Computerised claw trimming database programs as the basis for monitoring hoof health in dairy herds. Veterinary Journal 198, 358–361.

100. Kramer M, Erbe M, Bapst B, Bieber A and Simianer H 2013. Estimation of genetic parameters for novel functional traits in Brown Swiss cattle. Journal of Dairy Science 96, 1–11.

101. Lewis S, Handy RD, Cordi B, Billinghurst Z and Depledge MH 1999. Stress proteins (HSP's): methods of detection and their use as an environmental biomarker. Ecotoxicology 8, 351–368.

102. Linde C, de Jong G, Koenen DPC, Eding H 2010. Claw health index for Dutch dairy cattle based on claw trimming and conformation data. Journal Dairy Science 93, 4883–4891.

103. Lovendahl P and Chagunda MG 2010. On the use of physical activity monitoring for estrus detection in dairy cows. Journal of Dairy Science 93, 249–259.

104. Lovendahl P, Lassen J and Chagunda MG 2012. Genetic variation in milking efficiency: a novel trait for milkability in automatic milking systems. 63rd Annual Meeting of the Association of European Animal Production, 27 August, Bratislava, Slovakia. Retrieved 28 January 2014, fromhttp://www.eaap.org/Previous_AnnualMeetings/2012Bratislava/Papers/Published/22Lovendahl.pdf

105. Lucy MC 2001. Reproductive loss in high-producing dairy cattle: where will it end? Journal of Dairy Science 84, 1277–1293.

106. Margerison JK, Preston TR, Berry N and Phillips CJC 2003. Cross-sucking and other oral behaviours in calves, and their relation to cow suckling and food provision. Applied Animal Behaviour Science 80, 277–286.

107. McClure MC, Bickhart D, Null D, VanRaden P, Xu L, Wiggans G, Liu G, Schroeder S, Glasscock J, Armstrong J, Cole JB, Van Tassell CP and Sonstegard TS 2014. Bovine exome sequence analysis and targeted SNP genotyping of recessive fertility defects BH1, HH2, and HH3 reveal a putative causative mutation in SMC2 for HH3. PLoS One 9(3), e92769 .10.1371/journal.pone.0092769 [PMC free article] [Cross Ref]

108. McCorquodale CE, Sewalem A, Miglior F, Kelton DF, Robinson A, Koeck A and Leslie KE 2013.Analysis of health and survival in a population of Ontario Holstein heifer calves. Journal of Dairy Science96, 1880–1885.

109. McParland S, Banos G, Wall E, Coffey MP, Soyeurt H, Veerkamp RF and Berry DP 2011. The use of mid-infrared spectrometry to predict body energy status of Holstein cows. Journal of Dairy Science 94, 3651–3661.

110. McParland S, Lewis E, Kennedy E, Moore S, McCarthy B, O'Donovan M, Butler S, Pryce JE and Berry D 2014. Mid-infrared spectrometry of milk as a predictor of energy intake and efficiency in lactating dairy cows. Journal of Dairy Science 97, 5863–5871.

111. Melzer N, Wittenburg D, Harwig S, Jakubwoski S, Kesting U, Willmitzer L, Lisec J, Reinsch N and Repsilber D 2013. Investigating associations between milk metabolite profiles and milk traits of Holstein cows. Journal of Dairy Science 96, 1521–1534.

112. Meuwissen THE, Hayes BJ and Goddard ME 2013. Accelerating improvement of livestock with genomic selection. Annual Review of Animal Biosciences 1, 221–237.

113. Miglior F, Chesnais J and Van Doormaal BJ 2012. Genetic improvement: a major component of increased dairy farm profitability. Retrieved March 28, 2014, fromhttp://www.icar.org/cork_2012/Manuscripts/Published/Miglior%20A1.pdf

114. Mohammed R, McGinn SM and Beauchemin KA 2011. Prediction of enteric methane output from milk fatty acid concentrations and rumen fermentation parameters in dairy cows fed sunflower, flax, or canola seeds.Journal of Dairy Science 94, 6057–6068.

115. Mousa MS, Seykora AJ, Chester-Jones H, Ziegler DM and Cole JB 2013. Heritability estimates of performance and health traits of Holstein calves. Journal of Dairy Science 96 (suppl. 2), 291. (abstract W194).

116. NAV Routine Evaluation 2014. Nordic cattle genetic evaluation. Retrieved 18 August 2014, fromhttp://www.nordicebv.info/News/NewsNAVRoutineEvaluationAugust12th2014.htm?wbc_purpose=Basic&WBCMODE=PresentationUnpublished?Mode=Print&Site=DLINavSkin1

117. Negussie E, Strandén I and Mäntysaari EA 2008. Genetic association of clinical mastitis with test-day somatic cell score and milk yield during first lactation of Finnish Ayrshire cows. Journal of Dairy Science 91, 1189–1197.

118. Negussie E, Stranden I and Mäntysaari EA 2013. Genetic associations of test-day fat:protein ratio with milk yield, fertility, and udder health traits in Nordic Red cattle. Journal of Dairy Science 96, 1237–1250.

119. Nicolazzi E, Biscarini F and Stella A and ProZoo team 2012. Automated milk-recording systems: an experience in Italian dairy cattle farms. 63rd Annual Meeting of the Association of European Animal Production, August, Bratislava, Slovakia. Retrieved 20 December 2014, fromhttp://www.eaap.org/Previous_Annual_Meetings/2012Bratislava/Papers/Published/12_Biscarini.pdf

120. Nkrumah JD, Okine EK, Mathison GW, Schmid K, Li C, Basarab JA, Price MA, Wang Z and Moore SS 2006. Relationships of feedlot feed efficiency, performance, and feeding behavior with metabolic rate, methane production, and energy partitioning in beef cattle. Journal of Animal Science 84, 145–153.

121. Norberg E 2005. Electrical conductivity of milk as a phenotypic and genetic indicator of bovine mastitis: a review. Livestock Production Science 96, 129–139.

122. Ødegård C, Svendsen M and Heringstad B 2013. Genetic correlations between claw health and feet and leg conformation in Norwegian Red cows. Journal of Dairy Science 97, 4522–4529.

123. Ødegård C, Svendsen M and Heringstad B 2014. Accuracy of genomic breeding values for corkscrew claw in Norwegian Red. Proceedings 10th World Congress of Genetics Applied to Livestock Production, Vancouver, Canada.

124. Østerås O, Solbu H, Refsdal AO, Roalkvan T, Filseth O and Minsaas A 2007. Results and evaluation of thirty years of health recordings in the Norwegian dairy cattle population. Journal of Dairy Science 90, 4483–4497.

125. Parker Gaddis KL, Cole JB, Clay JS and Maltecca C 2012. Incidence validation and causal relationship analysis of producer-recorded health event data from on-farm computer systems in the U.S. Journal of Dairy Science 95, 5422–5435.

126. Parker Gaddis KL, Cole JB, Clay JS and Maltecca C 2014. Genomic selection for producer-recorded health event data in US dairy cattle. Journal of Dairy Science 97, 3190–3199.

127. Pausch H, Schwarzenbacher H, Burgstaller J, Flisikowski K, Wurmser C, Jansen S, Schnieke A, Wittek T and Fries R 2014a. Homozygous haplotype deficiency reveals deleterious mutations compromising reproductive and rearing success in cattle. Submitted in September 2014. [PMC free article]

128. Pausch H, Kölle S, Wurmser C, Schwarzenbacher H, Emmerling R, Jansen S, Trottmann M, Fuerst C, Götz K and Fries R 2014b. A nonsense mutation in TMEM95 endcoding a nondscript transmembrane protein causes idiophathic male subfertility in cattle. PLoS Genetics (submitted). [PMC free article]

129. Pavon S 2013. Future challenges concerning animal breeding and consumer protection (regulations of interest for this topics and expected developments). ICAR 2013 Health Data Conference, 30–31 May, Århus, Denmark. Retrieved 20 March 2014, fromhttp://www.icar.org/documents/aarhus_2013/powerpoint/HealthConference/Session1/Pavon.pdf

130. Polat B, Colak A, Gengiz M, Yanmaz LE, Oral H, Bastan A, Kaya S and Hayirli A 2010. Sensitivity and specificity of infrared thermography in detection of sublinical mastitis in dairy cows. Journal of Dairy Science93, 3525–3532.

131. Povinelli M, Gallo L, Carnier P, Marcomin D, Zotto RD and Cassandro M 2005. Genetic aspects of milk electrical conductivity in Italian Brown cattle. Italian Journal of Animal Science 4 (suppl. 3), 169–171.

132. Pryce JE and Berry DP 2014. Genomic breeding values for novel traits such as feed efficiency through female only reference populations. ICAR Conference

Berlin, 19–23 May, Berlin, Germany.

133. Pryce JE, Coffey MP and Simm G 2001. The relationship between body condition score and reproductive performance. Journal of Dairy Science 84, 1508–1515.

134. Pryce JE, Wales WJ, de Haas Y, Veerkamp RF and Hayes BJ 2014a. Genomic selection for feed efficiency in dairy cattle. Animal 8, 1–10.

135. Pryce JE, Gonzalez-Recio O, Thornhill JB, Marett LC, Wales WJ, Coffey MP, de Haas Y, Veerkamp RF, Hayes BJ 2014b. Short communication: validation of genomic breeding value predictions for feed intake and feed efficiency traits. Journal of Dairy Science 97, 537–542.

136. Pryce JE, Esslemont RJ, Thompson R, Veerkamp RF, Kossaibati MA and Simm G 1998. Estimation of genetic parameters using data from a health and fertility management recording system. Animal Science 66, 577–584.

137. Pszczola M, Strabel T, Mulder HA and Calus M 2012. Reliability of direct genomic values for animals with different relationships within and to the reference population. Journal of Dairy Science 95, 389–400.

138. Pszczola M, Veerkamp RF, de Haas Y, Wall E, Strabel T and Calus M 2013. Effect of predictor traits on accuracy of genomic breeding values for feed intake based on a limited cow reference population. Animal 7, 1759–1768.

139. Rinell E 2013. A genetic analysis of traits recorded by automatic milking systems – the possibility for a new method to evaluate temperament of dairy cows. Master's Thesis, 30 HEC. Swedish University of Agricultural Sciences, Department of Animal Breeding and Genetics, Uppsala, 420pp.

140. Roberts T, Chapinal N, LeBlanc SJ, Kelton DF, Dubuc J and Duffield TF 2012. Metabolic parameters in transition cows as indicators for early-lactation culling risk. Journal of Dairy Science 95, 3057–3063.

141. Roche JR, Friggens NC, Kay JK, Fischer MW, Stafford KJ and Berry DP 2010. Invited review: body condition score and its association with dairy cow productivity, health and welfare. Journal of Dairy Science92, 5769–5801.

142. Roessler R, Herold P, Weidele A and Zarate AV 2013. Definition nutzerspezifischer Zuchtziele für Braunvieh und Hinterwälder Rind in Baden-Württemberg. Züchtungskunde 3, 173–187.

143. Ross EM, Moate PJ, Marett LC, Cocks BG and Hayes BJ 2013. Metagenomic predictions: from microbiome to complex health and environmental phenotypes in humans and cattle. PLoS One 9, e92769.[PMC free article]

144. Rutten CJ, Velthuis AGJ, Steeneveld W and Hogeveen H 2013. Invited review: sensors to support health management on dairy farms. Journal of Dairy Science 96, 1928–1952.

145. Schöpke K and Weigel K 2014. Use of accelerometer data for genetic evaluation in dairy cattle. Interbull Open Meeting Berlin. Retrieved on 21 August 2014, fromhttp://www.interbull.org/web/static/presentations/Berlin/Tuesday/12_18_Schoepke.pdf

146. Schwarzenbacher H 2013. Survey within Gene2Farm.Gene2Farm 2014. Next

generation European system for cattle improvement and management (EU-FP7-KBBE-2011-5-PNr.:289592). Project from the 7th Framework Program. Retrieved 28 March 2014, from http://www.gene2farm.eu

147. Schwarzenbacher H, Fuerst C, Fuerst-Waltl B and Dolezal M 2012. A genome-wide search for harmful recessive haplotypes in Brown Swiss and Fleckvieh cattle. 63rd Annual Meeting of the Association of European Animal Production, 27 August, Bratislava, Slovakia. Retrieved 6 September 2014, fromhttp://www.eaap.org/Previous_Annual_Meetings/2012Bratislava/Papers/Published/22_Schwarzenbacher.pdf
148. Smit B, Burton I, Klein RJT and Street R 1999. The science of adaptation: a framework for assessment.Mitigation and Adaptation Strategies for Global Change 4, 199–213.
149. Sonstegard TS, Cole JB, VanRaden PM, Van Tassell CP, Null DJ, Schroeder SG, Bickhart D and McClure MC 2013. Identification of a nonsense mutation in CWC15 associated with decreased reproductive efficiency in Jersey cattle. PLoS One 8, e54872. [PMC free article]
150. Sorenson LP, Madsen P, Mark T and Lund MS 2009. Genetic parameters for pathogen-specific mastitis resistance in Danish Holstein Cattle. Animal 3, 647–656.
151. Soyeurt H, Bruwier D, Romnee JM, Gengler N, Bertozzi C, Veselko D and Dardenne P 2009. Potential estimation of major mineral contents in cow milk using mid-infrared spectrometry. Journal of Dairy Science92, 2444–2454.
152. Soyeurt H, Dehareng F, Gengler N, McParland S, Wall E, Berry DP, Coffey M and Dardenne P 2011. Mid-infrared prediction of bovine milk fatty acids across multiple breeds, production systems, and countries.Journal of Dairy Science 94, 1657–1667.
153. Soyeurt H, Bastin C, Colinet FG, Arnould VMR, Berry DP, Wall E, Dehareng F, Nguyen HN, Dardenne P, Schefers J, Vandenplas J, Weigel K, Coffey M, Théron L, Detilleux J, Reding E, Gengler N and McParland S 2012a. Mid-infrared prediction of lactoferrin content in bovine milk: potential indicator of mastitis. Animal6, 1830–1838.
154. Soyeurt H, Dehareng F, Gengler N and Dardenne P 2012b. Capitalizing on mid-infrared to improve nutritional and environmental quality of milk. World Holstein Conference, Toronto, Canada. Retrieved 20 February 2014, from http://www.whff.info/info/documents/17_NicolasGengler.pdf
155. Soyeurt H, Dehareng F, Romnée JM, Gengler N and Dardenne P 2012c. Genetics of the mineral contents in bovine milk predicted by mid-infrared spectrometry. 63rd Annual Meeting of the Association of European Animal Production, 27 August, Bratislava, Slovakia. Retrieved 28 March 2014, fromhttp://www.eaap.org/Previous_Annual_Meetings/2012Bratislava/Papers/Published/03_Soyeurt.pdf
156. Stege H, Bager F, Jacobsen E and Thougaard A 2003. VETSTAT – the Danish system for surveillance of the veterinary use of drugs for production

animals. Preventive Veterinary Medicine 57, 105–115.

157. Steininger F, Fuerst-Waltl B, Pfeiffer C, Fuerst C, Schwarzenbacher H and Egger-Danner C 2012.Participatory development of breeding goals in Austrian dairy cattle. Acta Agriculturae Slovenica 01/2012 (suppl. 3), 143–147.

158. Stock KF, Agena D, Schafberg R and Reinhardt F 2013. Health monitoring concepts for long-term improvement of dairy health. 64th Annual Meeting of the Association of European Animal Production, 26–30 August, Nantes, France. Abstract 17060.

159. Stock KF, Cole J, Pryce J, Gengler N, Bradley A, Andrews L and Egger-Danner C 2012. Survey on the recording and use of functional traits in dairy management and breeding. Proceedings ICAR Annual Meeting, 30 May, Cork/Ireland.

160. Thom EC 1959. The discomfort index. Weatherwise 12, 57–60.

161. Thomasen JR, Sorensen AC, Lund MS and Guldbrandtsen B 2014. Adding cows to the reference population makes a small dairy population competitive. Journal of Dairy Science 97, 1–11.

162. United Nations 2014. World population projected to reach 9.6 billion by 2050 – UN report. Retrieved 13 March 2014, from http://www.un.org/apps/news/story.asp?NewsID=45165

163. Urioste JI, Franzen J and Strandberg E 2010. Phenotypic and genetic characterization of novel somatic cell count traits from weekly and monthly observations. Journal of Dairy Science 93, 5930–5941.

164. Urioste JI, Franzen J, Windig JJ and Strandberg E 2012. Genetic relationships among mastitis and alternative somatic cell count traits in the first 3 lactations of Swedish Hosteins. Journal of Dairy Science 95, 3428–3434.

165. Van der Drift SGA, Jorritsma R, Schonewille JT, Kmnijn HM and Stegeman JA 2012. Routine detection of hyperketonemia in dairy cows using Fourier transform infrared spectroscopy analysis of ß-hydroxybutyrate and acetone in milk in combination with test-day information. Journal of Dairy Science 95, 4886–4898.

166. VanRaden PM, Olson KM, Null DJ and Hutchison JL 2011. Harmful recessive effects on fertility detected by absence of homozygous haplotypes. Journal of Dairy Science 94, 6153–6161.

167. Von Keyserlingk MAG, Rushen J, de Passillé AM and Weary DM 2009. Invited review: the welfare of dairy cattle – key concepts and the role of science. Journal of Dairy Science 92, 4101–4111.

168. Walthall CL, Hatfield J, Backlund P, Lengnick L, Marshall E, Walsh M, Adkins S, Aillery M, Ainsworth EA, Ammann C, Anderson CJ, Bartomeus I, Baumgard LH, Booker F, Bradley B, Blumenthal DM, Bunce J, Burkey K, Dabney SM, Delgado JA, Dukes J, Funk A, Garrett K, Glenn M, Grantz DA, Goodrich D, Hu S, Izaurralde RC, Jones RAC, Kim S-H, Leaky ADB, Lewers K, Mader TA, McClung A, Morgan J, Muth DJ, Nearing M, Oosterhuis DM, Ort D, Parmesan C, Pettigrew WT, Polley W, Rader R, Rice C, Rivington M, Rosskopf E, Salas WA, Sollenberger LE, Srygley R, Stöckle C, Takle ES, Timlin D, White JW, Winfree R,

Wright-Morton L, Ziska LH 2012. Climate change and agriculture in the United States: effects and adaptation. USDA Technical Bulletin 1935, Washington, DC. Retrieved 28 March 2014, fromhttp://www.usda.gov/oce/climate_change/effects.htm

169. Weber A, Stamer E, Junge W and Thaller G 2013. Genetic parameters for lameness and claw and leg diseases in dairy cows. Journal of Dairy Science 96, 3310–3318.
170. West JW 2003. Effects of heat-stress on production in dairy cattle. Journal of Dairy Science 86, 2131–2144.
171. Wiggans GR, VanRaden PM and Cooper TA 2012. Technical note: adjustment of all cow evaluations for yield traits to be comparable with bull evaluations. Journal of Dairy Science 95, 3444–3447.
172. Wiggans GR, Cooper TA, VanRaden PM and Cole JB 2011. Technical note: adjustment of traditional cow evaluations to improve accuracy of genomic predictions. Journal of Dairy Science 94, 6188–6193.
173. Yao C, Weigel KA and Cole JB 2014. Short communication: genetic evaluation of stillbirth in US Brown Swiss and Jersey cattle. Journal of Dairy Science 97, 2474–2480.
174. ZuchtData 2014. Yearly report 2013, Vienna, Austria. Retrieved 9 August 2014, fromhttp://zar.at/Downloads/Jahresberichte/ZuchtData-Jahresberichte.html

3
CHAPTER

PHYSIOLOGICAL RESPONSE TO A BREED EVALUATION FIELD TEST IN ICELANDIC HORSES

G. J. Stefánsdóttir, S. Ragnarsson, V. Gunnarsson and A. Jansson

Keywords: Exercise physiology, hematological parameters, lactate, heart rate, Icelandic horse

IMPLICATIONS

The Icelandic horse breed is widespread and its popularity is increasing as a leisure and sport horse for gait competitions. However, knowledge is lacking on the physiological response of Icelandic horses to exercise and competitions. Therefore, it is an animal welfare issue to generate such knowledge, so that training programmes and competition content can be adjusted to support the general welfare of the breed.

INTRODUCTION

The Icelandic horse is a purebred gaited riding horse originating from Iceland (International Federation of Icelandic Horse Associations (FEIF), 2002), but located in 33 countries outside Iceland and bred in 18 countries (WorldFengur, 2013). The breed evaluation system and the studbook WorldFengur (2013) are also international (FEIF, 2002). The official breeding goal includes 15 traits (eight conformation and seven riding ability traits) that are assessed individually at breeding shows, hereafter referred to as the breed evaluation field test (BEFT), which is standard for all countries (FEIF, 2002). In recent years (2008 to 2012), 1700 to 2800 horses attended a BEFT

Citation: Stefánsdóttir GJ, Ragnarsson S, Gunnarsson V, Jansson A. Physiological response to a breed evaluation field test in Icelandic horses. Animal. 2014;8(3):431-439. doi:10.1017/S1751731113002309.

every year, of which 1000 to 1800 were in Iceland (FEIF, 2013). For comparison, the number of live foals registered in FEIF countries annually in 2008 to 2012 ranged from 10 559 to 14 866, of which 4861 to 7399 were in Iceland (WorldFengur, 2013).

The BEFT consists of three parts: (1) objective body measurements, (2) judging of conformational traits and (3) riding abilities. Scores for conformation and riding abilities are weighed to give the final total score. Riding abilities (performance in walk, tölt, trot, flying pace (pace) and canter/gallop, as well as spirit and general impression) account for 60% of the total score and conformation for 40% (FEIF, 2002). It is known that stallions obtain higher total scores than mares and that total score increases with age from 4 to 6 years and older (Árnason, 1984). This implies that fitness and also physiological characteristics related to age and sex could be important for performance. It is known from other breeds that performance is higher in stallions than in mares (Buttram et al., 1988; Persson, 1997) and increases with age, at least to a certain limit (Seeherman and Morris, 1991) and that sex, age (Ronéus et al., 1991; Ronéus, 1993) and training (Ronéus et al., 1992) influence, for example, muscle fibre composition and also blood volume and haematocrit (Htc) (Persson et al., 1996).

To the best of our knowledge, no published data are available on the physiological response of Icelandic horses to a BEFT. In fact, the physiological responses of this breed to exercise have not been assessed, except in a few pilot studies (unpublished results). However, speed observations from competitions (WorldFengur, 2013) and heart rate (HR) and blood lactate responses from the pilot studies indicate that Icelandic horses are subjected to high workloads. Therefore, the aim of this study was to describe the response in terms of heart and respiratory rate (RR), Htc, rectal temperature (RT), and some plasma variables in stallions and mares of different ages performing a BEFT. Our starting hypotheses were that the BEFT is a high-intensity exercise and that some physiological responses could be affected by sex and age.

MATERIAL AND METHODS

The study was approved by the National Animal Research Committee of Iceland. The study was conducted at BEFT in Hella in southern Iceland on 30 May to 2 June and 6 to 9 June 2011. Owners and trainers were informed that participation in the study was voluntary.

BEFT and Design of the Study

The riding abilities of a horse are judged in two separate assessments. In the first and main assessment, the horse is expected to show all its gaits, but exactly how this is done is decided by the rider. The horse is ridden alone on a straight field track (250 to 300 m long, 4 to 6 m wide) and 200 m of the track is used for judging and 50 m at each end to turn around. The horse has to be ridden along the track a minimum of six times and a maximum of 10 times, that is, three to five times in each direction (FEIF, 2002). The second assessment is voluntary and is carried out if the rider/trainer/owner would like to try to improve the scores from the first assessment. In the present study, only the results from the first assessment were included.

A horse must be at least 4 years old in the calendar year to be judged for riding abilities in BEFT. Stallions and mares are shown in separate sex and age classes (4, 5, 6 and $\geq$ 7 years old) and geldings are shown in one group irrespective of age (FEIF, 2002), as very few are judged annually (WorldFengur, 2013). Only mares and stallions were included in the present study. According to the judging scale used in a BEFT, each trait is scored within the range 5.0 to 10.0, with 0.5 increments. A score of 5.0 is given if a gait is not shown and 10.0 is given for an excellent trait (FEIF, 2002). Three internationally accepted judges generally work in a committee and give a joint score for each trait (FEIF, 2002). During the present study six judges worked, three at a time. Immediately after a horse has finished the riding assessment, there is also an obligatory shoe and health check (legs and mouth), which lasts for about 1 to 2 min.

Horses

In total, 396 horses were shown at the BEFT and the study included 266 of these, 86 stallions and 180 mares, leaving 130 in a 'non-study' group. The horses were mainly fed roughage 5.7±1.5 kg (range: 0 to 12 kg, n=254) and some (at least 10%) were also grazed (0.5 to 6.0 h/day), whereas one horse was grazed only. Most horses (68%) were given some concentrates (1.1±0.5 kg, range: 0.2 to 4 kg, n=182). The inclusion criteria were that the owners/trainers were willing to take part, and that the horses were judged for both conformation and riding abilities and had completed both parts of the BEFT. One of the horses in the study was later disqualified and did not get a score because of deviation from the rules of legal shoes.

Data Collection

Before the riding assessment, the BW of each horse and rider and the weight of a spot sample of saddles (23 saddles and blankets) were recorded using an electronic livestock scale (Smartscale 300; Gallagher, Hamilton, New Zealand). The ratio of rider BW (riding tack not included) to horse BW was calculated (BWR) and used as a variable in the statistical analyses. The riders answered a questionnaire about the horse regarding name, age, feeding, travel time to the showground and the level of preparation for the BEFT (scale 1 to 10, 1=badly and 10=very well prepared). The horses were given a body condition score (BCS) according to the specific BCS-scale devised for Icelandic horses (scale 1 to 5; Stefánsdóttir and Björnsdóttir,2001).

RT (digital thermometer; Disney, Hartmann, Heidenheim, Germany) was measured and RR was recorded by counting breaths both before the horses were warmed up before the riding assessment and within 5 min after the riding assessment. Blood samples were also taken from the jugular vein by Vacutainer in chilled lithium heparinised tubes (9 ml, Vacuette®; Greine-Bio-One, Kremsmuenster, Austria) before the warm-up and the riding assessment and within 5 min after the horse left the track after performing the riding assessment. The same person recorded RT and RR and collected the blood samples.

During the warm-up and riding assessment, and for a 5-min recovery period after the assessment HR (Polar HR Monitor RS800CX and belt with Polar Equine T56H transmitter W.I.N.D., Kempele, Finland), velocity and distance covered (Polar G3 GPS sensor, Kempele, Finland) were recorded. According to the manufacturer, the accuracy of distance measurements is ±2% and that of velocity measurements is ±2 km/h. The HR monitors were set to record in 1 s mode. All horses were videotaped during the riding assessment, using a digital HD video-camera recorder (Sony HDR CX360VE, Tokyo, Japan). All watches used in the study (on HR monitors, the video camera and watches used to record times) were synchronised and the video recordings were used to check and synchronise the start and end time of the riding assessment. The riding track was labelled with bars every 50 m for 300 m. Scores from the riding assessment and information on wither height (WH) were obtained from the official studbook (WorldFengur, 2013).

The weather at the showground was recorded every 5 min (two to four observations per horse) by an automatic weather station (Art. no. 36-3242, Model WH-1080; Clas Ohlson, Insjön, Sweden). The ambient temperature at the track was 10.7±2.0°C (range: 9.0 to 14.1°C), the relative humidity 38±14% (range: 24% to 60%), the wind speed 5.6±2.3 m/s (range: 3.5 to 9.8 m/s) and average strongest wind speed 7.6±2.9 m/s (range: 5.0 to 12.9 m/s).

Blood Analysis

Blood samples were kept chilled and Htc and haemoglobin (Hb) were analysed within 30 min. Htc was analysed using non-heparinised capillary tubes, centrifuged at 18 840×**g** (Sigma 1–15; Laborzentrifugen GmbH, Osterode am Harz, Germany) for 6 min. Triple measurements were performed and a mean value was used for the statistical analyses. Hb concentration was analysed using a Hemocue POC analyser (Helsingborg, Sweden). Plasma was separated by centrifugation (15 min, 520×**g**, Hettich, Tuttlingen, Germany) and stored at −18°C until analysis of lactate, creatine kinase (CK), aspartate amino transferase (AST) and total plasma protein concentration (TPP). Plasma lactate concentration (Lac) was analysed after ~4 months using an enzymatic (l-lactate dehydrogenase and glutamate-pyruvate transaminase) and spectrophotometric method (Boehringer Mannheim/R-Biopharm, Darmstadt, Germany) with a CV of 2.3% according to the manufacturer. All duplicates showing variation >10% were re-analysed until the variation was <10%. The CV of the plasma lactate analyses was 3.3%. CK and AST were analysed after ~4 months using an enzymatic method (spectrophotometer, Architect c4000, Abbott Park, IL, USA). TPP was analysed after ~12 months using a refractometer (Atago, Tokyo, Japan).

Data Handling

All HR recordings were transported using an infrared USB Adapter (IrDA, Kempele, Finland) to the software Polar Pro Trainer 5 Equine Edition (Kempele, Finland) and then transferred to Microsoft Excel 2010 for further analysis. Only HR recordings where a minimum of 80% of the recording for the whole riding assessment was considered reliable were used, which resulted in 102 observations. On average, 6±6% (34±34 s/horse) of the recordings from the 102 horses were excluded from the analysis.

Statistical Analysis

Means and standard deviations were calculated for descriptive data and Pearson's correlations were performed in SAS (Statistical Analysis Systems package 9.2, Cary NC, USA). ANOVA was performed using Proc GLM (model 1 and 2) and Proc MIXED (model 3 and 4) in SAS. The results are expressed as least square means (LS Means) with their root mean square error (RMSE). For comparison, the Tukey test was used and level of statistical significance was set to P<0.05.

The BCS, WH, BW and level of preparation were analysed using model (1): $Y_{ij}=\mu+\alpha_i+\beta_j+e_{ij}$, where Y_{ij} is the observation/parameter, μ the mean value, α_i the fixed effect of sex, β_j the fixed effect of age group and e_{ij} the residuals; $e_{ij}\sim(0, \delta^2)$.

The before and after samples and whether the two groups of horses (study and non-study) differed in total score for riding abilities were analysed using model (2): $Y_{ijk}=\mu+\alpha_i+\beta_j+\varepsilon_k+e_{ijk}$, where Y_{ijk} is the observation/parameter, μ the mean value, α_i the fixed effect of sex, β_j the fixed effect of age group, ε_k the fixed effect of sample/group of horse and e_{ijk} the residuals; $e_{ijk}\sim(0, \delta^2)$.

The variables distance, speed and duration of warm-up, and the BEFT were analysed using model (3): $Y_{ijkr}=\mu+\alpha_i+\beta_j+\varepsilon_k+a_r+e_{ijkr}$, where Y_{ijkr} is the observation/parameter, μ the mean value, α_i the fixed effect of sex, β_j the fixed effect of age group, ε_k the fixed effect of whether a horse was scored for pace or not, a_r the random effect of the rider and e_{ijkr} the residuals; $e_{ijkr}\sim(0, \delta^2)$. Interactions between fixed factors were tested in the model, but no effect was found for any of the variables tested. The random factor 'rider' accounted for 5% to 46% of the random variation in the models for distance, duration and speed, that is, the variation that was not explained by the fixed factors.

The physiological responses (overall effect) to the BEFT and the total score for riding abilities were analysed using model (4): $Y_{ijklmnr}=\mu+\alpha_i+\beta_j+\varepsilon_k+\gamma_l+\eta_m+\tau_n+a_r+e_{ijklmnr}$, where $Y_{ijklmnr}$ is the observation, μ the mean value, α_i the fixed effect of sex, β_j the fixed effect of age group, ε_k the fixed effect of whether a horse was scored for pace or not, γ_l the average velocity of a horse in the BEFT as a continuous variable, η_m the distance ridden in the BEFT as a continuous variable, τ_n the BWR as a continuous variable, a_r the random effect of the rider and e_{ijklmr} the residuals; $e_{ijklmnr}\sim(0, \delta^2)$. The random factor 'rider' accounted for 0% to 18% of the random variation in the models for physiological responses to the BEFT and for total score for riding abilities, that is, the variation that was not described by the fixed factors.

RESULTS

Description of Horses, Riders and Riding Tack Included in the Study

The mean travel time to the BEFT was 31±31 min (range: 3 to 180 min) but two horses had travelled for 6 h on the previous day.

The scores for riding abilities of the horses participating in the study (n=265) and the non-study horses (n=130) were not different (P>0.05) for the whole group (7.69 v. 7.72,

RMSE=0.32). In addition, there were no differences between the age groups within the studied group and non-studied group (4-year-olds: 7.52 v. 7.41, RMSE=0.29, 5-year-olds: 7.71 v. 7.71, RMSE=0.27, 6-year-olds: 7.79 v. 7.89, RMSE=0.33, $\geq$ 7-year-olds: 7.79 v. 7.85, RMSE=0.36) or the sex groups (stallions: 7.82 v. 7.79, RMSE=0.34; mares: 7.59 v. 7.66, RMSE=0.30). The average age of the horses was 5.9±1.4 year (range: 4 to 11 years, n=266; Table 1). The WH was 141.0±2.7 cm (range: 134.0 to 149.0 cm, n=266), and was greater in stallions than in mares (141.9v. 140.4 cm, RMSE=2.6, $P<0.001$), but did not differ ($P>0.05$) between age groups. The BW was 339±19 kg (range: 289 to 397 kg, n=264) and was not affected by sex (stallions: 340 v. mares: 338 kg, RMSE=19,$P>0.05$). However, 4- and 5-year-old horses were lighter than horses $\geq$ 7 years old (333, 335 and 346 kg, respectively, RMSE=19, $P<0.01$), whereas the BW of 6-year-olds (342 kg, $P>0.05$) did not differ from that of the other age groups. The BCS of stallions was lower than that of mares (2.9 v. 3.1, RMSE=0.3, $P<0.001$) and did not differ ($P>0.05$) between age groups.

Table 1: Number of horses participating in the study, divided into sex and age group

	Age group				
	4 years	5 years	6 years	7 years and older	Total
Mares	19	55	55	51	180
Stallions	18	28	19	21	86
Total	37	83	74	72	266

The study involved 69 riders, who each rode 1 to 29 horses. Five riders rode $\geq$ 10 horses and 13 riders $\geq$ 5 horses, whereas 56 riders rode 1 to 4 horses. Rider BW was 83±11 kg (range: 59 to 112 kg, n=67) and weight of saddle and blanket was 8.7±1.0 kg (range: 6.8–10.4 kg, n=23). According to the riders, there was a wide range in the level of preparation (2% scores 1.0 to 3.9, 36% scores 4.0 to 6.9 and 62% scores 7.0 to 10.0, n=248).

Duration, Distance, and Speed of the Warm-up and BEFT

Duration, distance, and speed of the warm-up were not affected ($P>0.05$) by sex, age group or whether the horse showed pace or not in the BEFT (Table 2). The distance of the BEFT was 2.9±0.4 km (range: 1.8 to 3.8 km, n=248), the duration was 9:37±1:22 min:s (range: 5:07 to 15:32 min:s, n=260) and the average speed during the BEFT was 17.8±1.4 km/h (range: 13.2 to 21.3 km/h, n=248). Most horses (41%) were ridden $\geq$ 3.0 km, whereas 20% were ridden <2.5 km and 39% were ridden 2.5 to <3.0 km.

Table 2: Duration, distance, and speed of stallions and mares in warm-up and the breed evaluation field test (BEFT)[1] and effects of sex, age group (age: 4, 5, 6 and $\geq$ 7 years) and whether flying pace (Pace) was shown or not

	Stallions				Mares					*P*-values		
	n	Mean	Min	Max	*n*	Mean	Min	Max	RMSE	Sex	Age	Pace
Warm-up												
Distance (km)	85	2.2	0.3	4.7	164	2.0	0.2	6.3	0.8	0.070	0.542	0.654
Duration (min:s)	85	19:06	03:05	49:02	174	18:48	02:00	50:28	08:47	0.806	0.786	0.088
Average speed in warm-up (km/h)[2]	85	7.2	2.3	16.8	164	6.9	2.0	13.2	2.3	0.356	0.838	0.175
Average speed, moving in warm-up (km/h)[3]	85	10.8	5.7	19.7	164	10.8	5.8	17.0	2.1	0.886	0.949	0.459
Peak speed in warm-up	85	23.5	14.9	38.5	164	23.0	15.4	38.7	2.9	0.217	0.440	0.934
BEFT												
Distance (km)	85	2.9	1.8	3.6	162	2.8	1.8	3.8	0.3	0.027	0.064	0.013
Duration (min:s)	85	09:40	05:17	15:32	174	09:33	06:10	13:27	01:07	0.472	0.050	0.717
Average speed (km/h)	85	18.0	13.2	21.1	162	17.4	13.4	21.3	1.1	0.001	0.212	0.003
Peak speed (km/h)	85	42.1	32.0	56.8	162	42.0	33.5	52.2	3.3	0.826	0.068	0.718
Total for warm-up and BEFT												
Distance (km)	85	5.1	3.5	7.3	162	4.8	2.9	9.4	0.8	0.014	0.288	0.537
Duration (min:s)	85	28:36	12:04	60:07	174	28:13	11:24	60:40	08:50	0.752	0.700	0.084

[1]Values presented as least square means (mean) and root mean standard error (RMSE) and minimum (min) and maximum (max) values.
[2]The average speed in the complete duration of warm-up.
[3]The average speed in the duration of warm-up after excluding the part when horses and riders were standing still and waiting to start the riding assessment.

Stallions covered a longer distance in the BEFT than mares ($P<0.05$) and horses that showed pace (score $\geq$ 5.5 for pace) covered a longer distance than those that did not (score=5.0 for pace) (2.9 v. 2.8 km, RMSE=0.3, $P<0.05$; Table 2). Stallions were ridden at a faster speed than mares ($P<0.001$) and horses awarded a score for pace were faster than horses that received no score for pace (18.0 v. 17.5 km/h, RMSE=1.1, $P<0.01$; Table 2). Stallions covered a longer distance in total (in warm-up and BEFT) than mares ($P<0.05$; Table 2).

There was a tendency for 4-year-old horses to be ridden a shorter distance in the BEFT than 5-, 6- and $\geq$ 7-year-olds (2.7 v. 2.8, 2.9, 2.9 km, RMSE=0.3, P=0.06), for a shorter duration (9:12 v. 9:35, 9:48 and 9:53 min:s, RMSE=1.06, P=0.05) and at lower peak velocity (40.7 v. 42.5, 42.6, 42.5 km/h, RMSE=3.3,P=0.07).

Physiological Responses to the BEFT

Measured RR, RT, Htc, Hb, CK, AST and TPP increased following the BEFT ($P<0.001$; Table 3). The HR during the BEFT was 184±13 b.p.m. (range: 138 to 210 b.p.m., n=102) and during 36% of the BEFT HR was $\geq$200 b.p.m., for 28% it was 180 to 199 b.p.m., for 15% it was 161 to 179 b.p.m. and for 21% it was <160 b.p.m. Peak HR during the BEFT was 224±9 b.p.m. (range: 195 to 238 b.p.m., n=102). Mean HR as a proportion of peak HR during the BEFT was 82% (range: 64% to 90%). The HR during the BEFT was affected by velocity ($P<0.001$) and increased by 4 b.p.m. per km/h. Wind speed was positively correlated with the time when HR was $\geq$200 b.p.m. in the BEFT (r=0.25, $P<0.05$) The strongest wind speed was positively correlated with mean HR in the BEFT (r=0.23, $P<0.05$) and with the time when HR was $\geq$200 b.p.m. in the BEFT (r=0.27, $P<0.01$).

Table 3: Physiological responses in stallions and mares before warm-up and after the breed evaluation field test (BEFT)[1] and effects of sex, age group (age; 4, 5, 6 and ≥7 years), whether flying pace (Pace) was shown or not, distance (Dist.), velocity ...

	Stallions				Mares					*P*-values					
	n	Mean	Min	Max	*n*	Mean	Min	Max	RMSE	Sex	Age	Pace	Dist.	Vel.	BWR
RR before (breaths/min)	84	29	16	72	158	30	12	76	11	0.983	0.904	0.548	0.236	0.311	0.972
RR after (breaths/min)	84	103	40	160	159	103	12	168	28	0.915	0.015	0.363	0.415	0.001	0.398
RT before (°C)	84	37.8	36.7	38.5	158	37.8	36.9	38.6	0.3	0.770	0.419	0.979	0.527	0.546	0.414
RT after (°C)	84	39.4	38.0	40.9	158	39.6	38.4	41.1	0.5	0.041	0.796	0.566	0.015	0.105	0.410
RT change (°C)	84	1.7	0.4	3.4	158	1.8	0.5	3.7	0.5	0.067	0.510	0.512	0.029	0.060	0.144
Haematocrit before (%)	84	39	27	50	159	34	25	42	3	<0.001	0.189	0.327	0.830	0.452	0.086
Haematocrit after (%)	84	49	41	55	159	43	36	50	2	<0.001	<0.001	0.275	0.934	0.933	0.085
Hb concentrationbefore (g/l)	83	140	110	197	149	122	95	151	13	<0.001	0.243	0.839	0.834	0.181	0.030
Hb concentration after (g/l)	84	176	150	202	159	157	132	194	10	<0.001	0.002	0.941	0.339	0.215	0.002
TPP concentration before (g/l)	84	61	54	70	157	59	49	74	4	<0.001	0.314	0.798	0.692	0.215	0.966
TPP concentration after (g/l)	79	65	58	76	149	63	52	80	4	<0.001	0.088	0.219	0.975	0.243	0.877
CK concentration before (U/l)	84	239	117	1633	159	237	105	1913	155	0.946	0.143	0.990	0.167	0.526	0.437
CK concentration after (U/l)	84	298	135	2008	159	295	103	2162	221	0.941	0.113	0.533	0.005	0.095	0.801
AST concentration before (U/l)	84	349	195	625	159	354	158	894	96	0.686	0.890	0.614	0.188	0.001	0.942
AST concentration after (U/l)	84	383	229	643	159	378	124	928	102	0.739	0.838	0.860	0.154	0.005	0.929
Lac concentration after (mmol/l)	84	13.1	2.1	31.3	159	18.6	5.9	34.4	5.0	<0.001	0.200	<0.001	0.385	<0.001	0.005
HR_{av} during BEFT (b.p.m.)[3]	41	178	138	204	56	189	166	210	11	<0.001	0.241	0.163	0.612	0.001	0.277
HR_{peak} in BEFT (b.p.m.)[4]	41	225	195	238	56	224	203	237	8	0.773	0.073	0.108	0.915	0.185	0.568
HR_{min} in BEFT (b.p.m.)[5]	41	115	95	142	56	125	81	154	13	0.001	0.288	0.791	0.463	0.001	0.451
HR at end of BEFT (b.p.m.)	41	194	123	229	56	203	136	229	21	0.067	0.152	0.608	0.248	0.022	0.178
HR 5 min recovery (b.p.m.)[6]	62	108	69	137	114	115	87	136	9	0.001	0.687	0.003	0.079	0.001	0.312
HR 5th min of recovery (b.p.m.)[7]	73	90	49	124	142	96	74	119	9	<0.001	0.745	0.002	0.060	0.001	0.162

RR = respiratory rate; RT = rectal temperature; Hb = Haemoglobin; TPP = total plasma protein; CK = plasma creatine kinase; AST = plasma aspartate amino transferase; Lac = plasma lactate.

[1]Values presented as least square means (mean) and root mean standard error (RMSE) and minimum (min) and maximum (max) values.

[2]The ratio of rider BW (riding tack not included) to horse BW.

[3]Average HR during BEFT.

[4]Peak HR in BEFT.

[5]Minimum HR in BEFT.

[6]Average HR during the first 5 min of recovery after the BEFT.

[7]Average HR during the 5th minute of recovery after the BEFT.

Mean recovery HR during the 5-min recording period was 113±10 b.p.m. (range: 69 to 137 b.p.m., n=187) and HR during the last minute of the 5-min recovery period was 95±10 b.p.m. (range: 49 to 124 b.p.m.,n=228). HR during the 5-min recovery period and HR during the last minute of that period were both affected by velocity (P<0.001) and both increased by 2 b.p.m. per km/h. Horses with scores ≥5.5 for pace had higher recovery HR during the 5-min recovery period (114±1 v. 109±1 b.p.m., P<0.01) and during the last minute of the recovery period (95±1 v. 91±1, P<0.01).

Lac after the BEFT was 18.0±6.5 mmol/l (range: 2.1 to 34.4 mmol/l, n=266) and 72% of the horses had Lac ≥15.0 mmol/l (range: 15.2 to 34.4 mmol/l), 15% had 10.0 to 14.9 mmol/l, 11% had 4.1 to 9.9 mmol/l and 2% had ≥4 mmol/l. Velocity had an effect on Lac (P<0.001), with faster horses having higher values (1.3 mmol/l increase per km/h). The Lac after BEFT increased (P<0.01) by 0.4 mmol/l for every 1% increase in BWR. Horses awarded a score ≥5.5 for pace had higher Lac (17.7±0.6 v. 14.0±0.6 mmol/l, P<0.001) than horses that received no score (=5.0) for pace.

The RR after the BEFT was affected by velocity (Table 3) and increased by five breaths/min per km/h. RT was affected by distance (Table 3) and was 0.2°C higher per km covered in the BEFT. CK was affected by distance (P<0.01) and was 114 U/l lower per km covered. Hb values after the BEFT increased by 0.8 g/l per 1% increase in BWR.

Effect of Sex

Stallions had lower RT than mares after the BEFT ($P<0.05$) and higher Htc, Hb and TPP both before and after the BEFT ($P<0.001$; Table 3). Stallions had lower mean and minimum HR during the BEFT, lower recovery HR during the 5-min recovery period and lower HR during the last minute of the 5-min recovery period than mares ($P<0.001$; Table 3). Stallions also had lower Lac than mares after the BEFT ($P<0.001$; Table 3).

Effects of Age Group

Four-year-old horses had higher RR than 5-, 6- and $\geq$7-year-old horses (117 v. 102, 97, 96 breaths/min, RMSE 28, $P<0.05$) after the BEFT. The 4-year-old horses had lower Htc (45% v. 46%, 46%, 47%, RMSE 2, $P<0.05$) and Hb values (161 v. 166, 168, 170 g/l, RMSE 10, $P<0.01$) than 5-, 6- and $\geq$7-year-old horses after the BEFT. The 5-year-old horses had lower Htc (46% v. 47%, RMSE 2, $P<0.01$) and Hb values (166 v. 170 g/l, RMSE 10, $P<0.05$) than the $\geq$7-year-old horses after the BEFT.

Judgement Scores

Stallions received higher scores than mares (7.76 v. 7.57, RMSE=0.27, $P<0.001$) and 4-year-old horses received lower scores than the older age groups (7.48 v. 7.68, 7.75, 7.74, RMSE=0.27, $P<0.01$).

DISCUSSION

Intensity of the BEFT

The physiological responses observed in our study confirmed that the BEFT is a high-intensity exercise. The high peak, average, average/peak HR and a high proportion of exercise performed at HR 200 b.p.m., as well as marked increases in RR, RT, Htc, Hb, TPP, plasma CK, AST and Lac, show that the work load was strenuous and that anaerobic energy metabolism was necessary. However, there was considerable individual variation in the physiological responses to the BEFT.

It can be speculated that the peak HR observed was the maximal HR, as it was comparable to the maximal HR observed in racing Thoroughbreds (range: 204 to 241 b.p.m.; Krzywanek et al., 1970) and Standardbreds (range: 210 to 238 b.p.m.; Åsheim et al., 1970). The average HR (184 b.p.m.) during the BEFT was probably close to the lactate threshold (Persson, 1983) and, together with the amount of time spent at HR $\geq$200 b.p.m. (~3½ min), explains why the horses had high Lac. The Lac after the BEFT (18.0±6.5 mmol/l) was similar to that observed in Standardbred horses after a race (15 to 43 mmol/l; Ronéuset al., 1999), in 'high goal' polo ponies after a training match (18.7±5.4 mmol/l; Ferraz et al., 2010) and in eventing horses after an advanced 3-day event (19.1±4.2 mmol/l; White et al., 1995). The variation in Lac in the present study could be explained mostly (36%) by sex, age group, distance, velocity, BWR and pace, whereas rider accounted for 14% of the random variation. However, the variation might also be explained by differences in muscle fibre composition and fitness (Lovell and Rose, 1991), although our study did not assess those parameters. Unfortunately, we had limited information on the training background of the horses, but according to the riders most (62%) horses were quite well prepared (preparation score >7.0). However, the only physiological parameter correlated ($r \geq 0.2$, $P<0.05$) to the riders' opinion was minimum HR ($r=-0.25$, $P<0.05$), which indicates that the riders' opinions were poorly reflected in the physiological responses. It is possible that the trainers did not have physical preparation in mind when answering the question about level of preparation, but rather preparation for showing gaits that score high according the judging scale (FEIF, 2002). However, few significant ($P<0.05$) correlations between rider opinion and scores were found and the significant ones were weak. For example, there was a correlation between rider opinion and score for gallop ($r=0.21$, $P<0.001$) and with form under rider ($r=0.18$, $P<0.01$).

Effect of Sex

There was a clear effect of sex in the response to the BEFT. Although stallions covered longer distances (warm-up plus BEFT) and performed at higher speed, they had lower HR and Lac. This indicates that aerobic fitness was higher in stallions than in mares, which is also supported by the higher Htc and Hb values. A higher Htc in Standardbred trotter stallions than in mares has been shown previously (Persson et al., 1996) and a higher Hb concentration has been found in stallions than mares of other breeds (Persson,1967; Čebulj-Kadunc et al., 2002). A significant sex difference in the proportion of type IIB fibres in the gluteal muscle has been found in untrained 4 to 5-year-old Icelandic horses, with a higher proportion in mares than in geldings, 53.5% v. 44.4% (Henckel et al., 1994). In addition, all fibre types in geldings were surrounded by more capillaries than those found in mares (Henckel et al., 1994). A sex differences in fibre composition is also supported by findings in Thoroughbred and Standardbred horses, with stallions having a higher type IIA/IIB ratio in m. gluteus medius than mares (Ronéus et al., 1991; Ronéus, 1993). Horses with higher oxidative capacity usually have higher type IIA/IIB ratio (Ronéus et al., 1992) and, taken together, this indicates that male Icelandic horses have higher oxidative capacity than mares. Our results confirm findings by Persson and Ullberg (1974) in Standardbred horses, where stallions were suggested to have higher aerobic capacity than mares and geldings and findings on Thoroughbred and Standardbred horses, where males are reported to be faster than females (Árnason, 2001; Mukai et al., 2003) and to have lower HR than females (Mukai et al., 2003). The difference in aerobic fitness observed between stallions and mares in the present study might of course also be owing to differences in training practices, but according to the riders stallions were not better prepared than mares (7.3 v. 7.2, RMSE 1.5 and $P=0.66$).

In human athletes, it is commonly accepted that there are gender differences such as greater muscle mass, muscle strength and aerobic capacity in males that contribute to superior performance (Lewis et al., 1986). Another reason for the superior performance observed in stallions in the present study might be lower body fat content compared with mares, as indicated by the lower BCS. In humans, it is known that body fat content is inversely related to both sprint and endurance running performance (Lewis et al.,1986) and similar observations have been made in horses (Kearns et al., 2002). There is no obvious explanation for the higher TPP observed in stallions, but they could have been more excited than mares and increased sympathetic activity can increase hydrostatic pressure and sweat losses, both of which could elevate TPP. Increased sympathetic activity in the stallions is also supported by their higher Htc before exercise. Training can also increase TPP (Fazio et al., 2011), but as earlier mentioned stallions were, according to the riders, not better prepared than mares.

Effects of Age Group

The 4-year-old horses had lower Htc and Hb values and higher post-exercise RR than the older horses, although they were ridden a shorter distance, for a shorter duration and at lower peak velocity ($P<0.1$). This indicates less aerobic capacity and might be a reason for the riders consciously spared their performance. The lower Htc values are in accordance with previous findings on Standardbred horses in training, where the Htc has been shown to increase up to 4 and 5 years in mares and stallions, respectively (Persson et al., 1996). In the present study, Hb and Htc increased numerically up to $\geq$ 7 years of age, indicating that fitness was improved. However, if an increase in age is associated with a more experienced and better-trained horse, more signs of improved fitness could be expected (e.g. reduced HR). The lack of such signs might indicate that the general level of fitness in this population was not improved much after 5 years of age. However, it is also likely that there were more horses in the older age groups that were less talented and might have spent more energy on showing good gaits.

Effect of Velocity and Distance

As could be expected, velocity during the BEFT had a significant effect on several physiological parameters. It affected post-exercise Lac, exercise and post-exercise HR, and post-exercise RR, all of which increased with increased velocity. The higher the velocity, the more fibre types recruited and recruitment of type II fibres, particularly IIB, results in lactate production (Snow and Valberg, 1994). The effect of velocity on AST both before and after exercise could be related to common training practices and higher fitness in faster horses (Fazio et al., 2011).

The distance covered seemed only to affect post-exercise RT and CK concentration. Similarly, Hodgson et al. (1993) showed a linear increase in RT with time and workload. The plasma activity of CK can reflect cell membrane permeability, which could be fatigue dependent (Anderson, 1975). Interestingly, the concentration of CK was lower as the horses were ridden a longer distance, although duration of exercise has been suggested to be an even more important factor than intensity for the release of CK during exercise (Anderson, 1975). Possible explanations could be that fitter horses were ridden longer and/or that horses with higher CK concentrations were not ridden longer because the riders perceived these horses to be already tired. It has been suggested that the higher response of glycolytic metabolism might be reflected in higher values of CK and AST because of permeability changes in muscle fibre membranes (Fazio et al., 2011). CK and AST levels may also be affected by concentrate intake (Ribeiro et al., 2004) and transportation time (Leadon et al., 1989); however, in the present study, no correlations were found between these enzymes and diet and transport time (data not shown).

The distance covered during a BEFT should be 1.8 to 3.0 km (6 to 10 times of maximum 300 m each). In the present study, only 20% of horses were ridden <2.5 km and 41% were ridden $\geq$3.0 km, which indicates that many riders do not comply with the BEFT instructions.

Effect of Pace

Flying pace is a high-speed gait in the BEFT (velocity can reach around 30 to 40 km/h) and horses showing pace not only covered a longer distance but also at a higher average speed. Pace is also a gait likely to increase energy expenditure compared with the most economical gait at this velocity gallop. It has been found that ponies appear to choose a gait that maximises energy efficiency and that if they are forced to use extended gaits the VO_2 can increase by up to 70% (Hoyt and Taylor, 1981). Altogether this may explain why horses that showed the pace gait had significantly higher Lac and recovery HR than horses not showing pace.

Scores

We believe that the horses participating in this study were representative of the population of Icelandic horses attending a BEFT, as there was no difference in total score for riding abilities between horses participating in the study compared with non-study horses. According to Árnason (1984, Þorvaldur Árnason personal communication, 2013), age is important for scores and an increase in total score (conformation and riding abilities) can be expected yearly from 4 to 7 years old. In the present study, we compared the total score for riding abilities, for which 4-year-old horses had lower scores than older horses.

Weather and Correlations with Physiological Parameters

Because of the low ambient temperature, low humidity and windy conditions, heat loss through conduction, convection and evaporation was facilitated. It has long been known that air resistance increases energy expenditure during running and walking in humans (Pugh, 1971), and interestingly in the present study, HR increased as wind speed increased. Actually a few riders decided not to finish the BEFT (horses excluded from the study) because of the strong wind speed (the day with mean 12.9 m/s).

CONCLUSIONS

This study showed that the riding assessment in a BEFT is a high-intensity exercise and maybe even, for some horses, supramaximal exercise. The study also showed that aerobic fitness was higher in stallions and that age had a limited effect on the physiological response. These results should be used as a guide for the development of training programmes and fitness tests in Icelandic horses that would improve both performance and welfare of the horses.

ACKNOWLEDGEMENTS

The authors sincerely thank the riders, riders' assistants, owners of horses and staff at the BEFT at Hella between 30 May and 9 June 2011 for their collaboration. Special thanks to the research team at Hella; Eggert Gunnarsson at Keldur, Heiða Sigurðardóttir, Johanna Karin Knutsson, Pieter de Gouw, Rósa Kristinsdóttir and Sigríður Ólafsdóttir. Thanks to Anna Greta Haglund for analyses of the plasma lactate and Claudia von Brömssen for statistical advice, both at the Swedish University of Agricultural Sciences. Thanks to the Institute for Experimental Pathology at Keldur University of Iceland, the Farmers Association in Iceland, the Horse Breeders Association in Iceland, the horse breeding advisor Guðlaugur Antonsson and the head of the Horse Breeding Council Kristinn Guðnason for their support and help. The project was funded by Developmental Fund of the Icelandic Horse Breeding, the Pálmi Jónsson Natural Conservation Fund and the Stock Protection Fund of the Icelandic Horse breed.

REFERENCES

1. Anderson MG 1975. The influence of exercise on serum enzyme levels in the horse. Equine Veterinary Journal 7, 160–165 [PubMed]
2. Árnason T 1984. Genetic studies on conformation and performance of Icelandic toelter horses. IV. Best linear unbiased prediction of ten correlated traits by use of an 'Animal Model'. Acta Agriculturae Scandinavica 34, 450–462
3. Árnason T 2001. Trends and asymptotic limits for racing speed in Standardbred trotters. Livestock Production Science 72, 135–145
4. Åsheim A, Knudsen O, Lindholm A, Rülcker C and Saltin B 1970. Heart rates and blood lactate concentrations of Standardbred horses during training and racing. Journal of the American Veterinary Medical Association 157, 304–312 [PubMed]
5. Buttram ST, Willham RL and Wilson DE 1988. Genetics of racing peformance in the American Quarter Horse: II Adjustment factors and contemporary groups. Journal of Animal Science 66, 2800–2807
6. Čebulj-Kadunc N, Božič M, Kosec M and Cestnik V 2002. The influence of age and gender on haematological parameters in Lipizzan horses. Journal of Veterinary Medicine Series A 49, 217–221[PubMed]
7. Fazio F, Assenza A, Tosto F, Casella S, Piccione G and Caola G 2011. Training and haematochemical profile in Thoroughbreds and Standardbreds: a longitudinal study. Livestock Science 141, 221–226
8. Ferraz GC, Soares OAB, Foz NSB, Pereira MC and Queiroz-Neto A 2010. The workload and plasma ion concentration in a training match session of high-goal

(elite) polo ponies. Equine Veterinary Journal 42(suppl. 38), 191–195 [PubMed]

9. Henckel P, Ducro B and Sveinsson I 1994. Muscle histochemical profile of the Icelandic horse. Age and sex differences (abstract). Livestock Production Science 40, 78–79
10. Hodgson DR, McCutcheon LJ, Byrd SK, Brown WS, Bayly WM, Brengelmann GL and Gollnick PD 1993. Dissipation of metabolic heat in the horse during exercise. Journal of Applied Physiology 74, 1161–1170 [PubMed]
11. Hoyt DF and Taylor CR 1981. Gait and the energetics of locomotion in horses. Nature 292, 239–240
12. International Federation of Icelandic Horse Associations (FEIF) 2002. FEIF rules for Icelandic Horse Breeding (FIZO 2012). Retrieved April 16, 2013, from http://www.feiffengur.com/documents/fizo12e.pdf
13. International Federation of Icelandic Horse Associations (FEIF) 2013. News, facts and figures. Retrieved April 16, 2013, from http://www.feif.org/FEIF/Factsandfigures/tabid/497/Default.aspx
14. Kearns CF, McKeever KH, Kumagai K and Abe T 2002. Fat-free mass is related to one-mile race performance in elite Standardbred horses. The Veterinary Journal 163, 260–266 [PubMed]
15. Krzywanek H, Wittke G, Bayer A and Borman P 1970. The heart rates of Thoroughbred horses during a race. Equine Veterinary Journal 2, 115–117
16. Leadon D, Frank C and Backhouse W 1989. A preliminary report on studies on equine transit stress. Journal of Equine Veterinary Science 9, 200–202
17. Lewis DA, Kamon E and Hodgson JL 1986. Physiological differences between genders. Implications for sports conditioning, review. Sports Medicine 3, 357–369 [PubMed]
18. Lovell DK and Rose RJ 1991. Changes in skeletal muscle composition in response to interval and high intensity training In Equine exercise physiology. Proceedings of the 3rd International Conference (ed. SGB Persson, A Lindholm and LB Jeffcott), pp. 215–222 ICEEP Publications, Davis, California, USA
19. Mukai K, Takahashi T, Hada T, Eto D, Kusano K, Yokota S, Hiraga A and Ishida N 2003. Influence of gender and racing performance on heart rates during submaximal exercise in Thoroughbred racehorses.Journal of Equine Science 14, 93–96
20. Persson SGB 1967. On blood volume and working capacity in horses. Acta Veterinaria Scandinavica suppl. 19, 101–121 [PubMed]
21. Persson SGB 1983. Analysis of fitness and state of training: evaluation of exercise tolerance and fitness in the performance horse In Equine exercise physiology. Proceedings of the 1st International Conference (ed. DH Snow, SGB Persson and RJ Rose), pp. 441–457 Granta Publications, Cambridge, UK
22. Persson SGB 1997. Heart rate and blood lactate responses to submaximal treadmill exercise in the normally performing Standardbred Trotter – age and sex variations and predictability from the total red blood cell volume. Journal of

Veterinary Medicine Series A 44, 125–132 [PubMed]

23. Persson SGB and Ullberg LE 1974. Blood volume in relation to exercise tolerance in trotters. Journal of the South African Veterinary Association 45, 293–299
24. Persson SGB, Funkquist P and Nyman G 1996. Total blood volume in the normally performing Standardbred Trotter: age and sex variations. Journal of Veterinary Medicine Series A 43, 57–64 [PubMed]
25. Pugh LG 1971. The influence of wind resistance in running and walking and the mechanical efficiency of work against horizontal or vertical forces. The Journal of Physiology 213, 255–276 [PMC free article][PubMed]
26. Ribeiro WP, Valberg SP, Pagan JD and Essén-Gustavsson B 2004. The effect of varying dietary starch and fat content on serum creatine kinase activity and substrate availability in equine polysaccharide storage myopathy. Journal of Veterinary Internal Medicine 18, 887–894 [PubMed]
27. Ronéus M 1993. Muscle characteristics in Standardbreds of different ages and sexes. Equine Veterinary Journal 25, 143–146 [PubMed]
28. Ronéus M, Lindholm A and Åsheim Å 1991. Muscle characteristics in Thoroughbreds of different ages and sexes. Equine Veterinary Journal 23, 207–210 [PubMed]
29. Ronéus M, Essén-Gustavsson B, Lindholm A and Persson SGB 1992. Skeletal muscle characteristics in young trained and untrained Standardbred trotters. Equine Veterinary Journal 24, 292–294 [PubMed]
30. Ronéus N, Essén-Gustavsson B, Lindholm A and Persson S 1999. Muscle characteristics and plasma lactate and ammonia response after racing in Standardbred trotters: relation to performance. Equine Veterinary Journal 31, 170–173 [PubMed]
31. Seeherman HJ and Morris EA 1991. Comparison of yearling, two-year-old and adult Thoroughbreds using a standardized exercise test. Equine Veterinary Journal 23, 175–184 [PubMed]
32. Snow DH and Valberg SJ 1994. Muscle anatomy, physiology and adaptations to exercise and training InThe athletic horse (ed. DR Hodgson and RJ Rose), pp. 145–179 WB Saunders, USA
33. Stefánsdóttir GJ and Björnsdóttir S 2001. Body condition scoring of horses (Mat á holdafari hrossa), special issue. Eiðfaxi-Ræktun, pp. 60–65
34. White SL, Williamson LH, Maykuth P, Cole S and Andrews F 1995. Heart rate and lactate concentration during two different cross country events. Equine Veterinary Journal 27 (suppl. 18), 463–467
35. WorldFengur 2013. The studbook of origin for the Icelandic horse. Retrieved March 25, 2013, fromwww.worldfengur.com

4
CHAPTER

DIETARY FIBER UTILIZATION AND ITS EFFECTS ON PHYSIOLOGICAL FUNCTIONS AND GUT HEALTH OF SWINE

R. Jha and J. D. Berrocoso

Keywords: Dietary fiber, digestion, fermentation, gut health, pig

IMPLICATIONS

This review provides relevant information on the utilization of dietary fiber (DF) by pigs and its effects on gut physiological functions, microbiota and health. This review delivers in-depth insight on both negative and positive effects of different fibers inclusion in swine diets. The DF lowers nutrient digestibility in swine. But, the fermentation of DF in the gut affects positively by modulating gut environment and potentially favoring 'beneficial bacteria', thereby improving gut health of pigs. These insights will help swine nutritionists and researchers in nutrition programing for better gut health of pigs by utilizing dietary fiber from different sources.

INTRODUCTION

During the last half century, there has been tremendous development in the field of pork production, resulting to more than 109 million tons of pork produced per year (FAO, 2013). This progress has been achieved by an intensification of the swine production systems, coupled with selective breeding programs and a better knowledge of pig nutrition. Feeding strategies based on antibiotics used as growth promoters aimed at improving the pig growth rate by improving its gut health status. However, there is a

Citation: Jha R, Berrocoso JD. Dietary fiber utilization and its effects on physiological functions and gut health of swine. Animal. 2015;9(9):1441-1452. doi:10.1017/S1751731115000919.

growing concern about the resistance of numerous bacteria to antibiotics used in human medicine, and is claimed to be due to the consumption of meat from animals grown with in-feed antibiotics, as meat from antibiotic-fed animals were found to have antibiotic-resistant bacteria (Gebreyes et al., 2006). Dietary fiber (DF) has been found to be an effective alternative to growth promoters (Verstegen and Williams, 2002), to improve gut health (Williams et al., 2001) by modulating gut microbiota, improve growth performance and reduce post-weaning diarrhea of the pigs (Mateos et al., 2006).

In commercial pig production, plant carbohydrates (CHO) represent the main fraction of a pig diet, accounting for more than 2/3rd of the dry matter (DM; Bach Knudsen, 1997) and the single most abundant feed energy in diets for piglets, growing pigs and sows comprising 60.0% to 70.0% of total energy intake (Bach Knudsen et al., 2012). However, part of the CHO is not digested by the digestive enzymes of the small intestine and becomes available as a substrate for bacterial fermentation, mainly in the large intestine. This fraction, that is, DF, reduces nutrient and energy digestibility (Bach Knudsen, 2001; Noblet, 2007; Jhaet al., 2010). Its physico-chemical properties (like solubility, viscosity and water-holding capacity (WHC)) also have a marked effect on nutrient digestibility along the gastro-intestinal tract (Bach Knudsen and Hansen, 1991; Chabeauti et al., 1991; Molist et al., 2014).

Despite its negative impact on nutrient and energy digestibility (Bach Knudsen, 2001; Noblet, 2007; Jha et al., 2010), there is growing interest to include DF in pig diets, due to its possible effects on gut health, welfare and environment. Moreover, during the last decade, there has been dramatic shift in the landscape of the feed industry, in terms of price and availability of feed ingredients for animal feeds. There is increased availability of different alternative feed ingredients and coproducts from distillers and milling industries, which are rich sources of DF as well. However, it is important to know the implication of the use of these relatively new and potential DF sources in pig nutrition. Therefore, nutritionists are attempting to gain a more thorough understanding of inclusion of DF in pig diets.

To address the concerns and have a better understanding of DF and its role, this review has analyzed different aspects of DF in swine nutrition. More specifically, composition of DF in different feed ingredients (grains and coproducts) is first briefly reviewed. The digestibility and fermentability of the DF components in the gastrointestinal tract (GIT) of pigs and their effect on gut physiology, microbial environment and health are highlighted.

DF

DF And Its Components

The term 'dietary fiber' was first used by Hipsley in 1953 (De Vries et al., 1999) for 'the non-digestible constituents that make up the plant cell wall'. However, different definitions have been proposed and used over time. It is now accepted that an accurate definition of DF must include the physiological effects of fiber. Therefore, an important aspect of the definition is that DF consists of CHO that are indigestible by endogenous animal enzymes (AACC, 2001).

Broadly, DF includes cell wall compounds like cellulose, hemicelluloses, mixed linked β-glucan (βG), pectins, gums and mucilages (Davidson and McDonald, 1998). Lignin, a complex phenolic compound, is also included in DF because it is a constituent of the plant cell walls that can greatly affect the digestibility of plant-derived foods (Theander et al., 1989). From a physiological point of view, non-starch polysaccharides (NSP), non-digestible oligosaccharides and resistant starch (RS) have to be included in the soluble DF fraction because they are not hydrolyzed by endogenous enzymes, and consequently, become available as substrates for microbial fermentation in the intestine (Cummings and Stephen, 2007).

Sources of DF and Interest for their use in Swine Diet

The origin and composition of DF could be responsible for large variations in their utilization (Chabeauti et al., 1991). The physico-chemical properties of the DF sources may lead to changes in the gut environment, altering the growth of the gut microflora. The acceptability of the alternative feed ingredients in pig diets depends on several factors, like the DF content, the degree of microbial fermentation in the large intestine and the extent of absorption and utilization of the volatile fatty acid (VFA) produced (Molist et al., 2014). The fiber sources are fermented in the GIT producing VFA, which in turn positively affects gut health (Lindberg,2014). Wellock et al. (2007) noted that gut health might benefit most from diets containing appropriate sources of predominantly soluble NSP rather than insoluble NSP. Soluble DF includes pectins, βG, gum and hemicelluloses, while cellulose and lignin compromise the insoluble fraction (Davidson and McDonald,1998). Thus, it is imperative to know the source and type of fiber being supplied in pig diets.

There is also increasing interest and incentive for the identification and characterization of alternative feed ingredients. These alternative feed ingredients include cereals and legume grains, distillery coproducts, coproducts from oil industry (like canola meal) and wheat flour milling (millrun and bran) and other fibrous feeds.

Cereal grains and their coproducts account for the major part of pig rations as main sources of energy. The DF of cereal grains are mainly composed of NSP (arabinoxylans, **AX**, βG and cellulose) and noncarbohydrate component lignin (Bach Knudsen, 2014). In addition, small amounts of pectin substances found in the stems and leaves of cereals (Choct, 1997). Several workers mentioned that the level of DF in the commonly available feed ingredients vary in relation to type and quality. As a reference to show the variation found in fiber components, the composition of some of the most common cereals and coproducts is presented in Table 1. Rye, wheat, corn and sorghum are all rich in AX, whereas barley and oats contain a high level ofβG. The AX from rye and wheat and βG from barley and oats are to a large extent soluble, whereas the solubility of AX found in corn and sorghum is lower than the other cereals (Bach Knudsen et al., 2014).

Table 1: Types and levels of fiber components in some common cereal grains and coproducts (g/kg DM)

		Barley[2,3,4]		Oats[2,4]								
	Wheat[1]	Hulled	Hulless	Hulled	Hulless	Corn[1]	Sorghum[1]	Rye[2]	Wheat bran[2]	Wheat middlings[1]	DDGS[1]	Sugar beet pulp[5]
Starch	618	587	645	468	557	620	690	613	222	168	86	5
Cellulose	13	39	10	82	14	17	15	15	72	67	58	203
NCP												
Soluble	19	56	50	40	54	25	4	42	29	12	34	290
Insoluble	62	88	64	110	49	38	47	94	273	227	158	207
NSP												
Arabinoxylans	81	12	48	98	36	17	17	89	238	52	61	165
β-glucan	8	43	42	28	41	6	1	20	24	21	63	8
Mannose	2	2	4	3	3	2	1	3	5	19	9	8
Galactose	3	2	3	7	4	8	3	3	9	13	14	38
Uronic acids	4	2	2	10	5	8	4	2	15	16	16	199
Total NSP	95	167	124	232	116	81	66	132	374	250	192	700
Lignin	18	35	9	66	32	8	16	21	75	39	32	37
Dietary fiber	112	202	133	298	148	89	83	153	449	289	322	737

DM = dry matter; NCP = non-cellulosic polysaccharides; NSP = non-starch polysaccharides; DDGS = distillers dried grains with solubles.
[1]According to Jaworski *et al.* (2015).
[2]According to Bach Knudsen (1997).
[3]According to Holtekjolen *et al.* (2006).
[4]According to Jha *et al.* (2011a).
[5]According to Serena and Knudsen (2007).

Corn coproducts are typically rich in DF but variable in starch, amino acid and fat. The concentration and composition of DF of feed ingredients is important, because it may reduce amino acid and energy digestibility (Noblet and Perez, 1993). The insoluble fraction of DF in corn and its coproducts is resistant to fermentation (~40% is fermented in the entire GIT of the pigs), consisting of insoluble NSP such as cellulose, arabinoxylans and lignin (Bach Knudsen, 1997). Increasing the efficiency of starch and oil extraction from the corn grain, resulting in changes in chemical composition of corn coproducts, present a challenge to estimate their nutritional value (Gutierrez et al., 2014). In fact, Fairbairn et al. (1999) reported that NDF or ADF alone accounted for 68.0% and 85.0% of the total variation in digestible energy content of barley, respectively.

Wheat (Triticum aestivum) is primarily used in pig diets as source of energy. However, it contains significant amount of DF, including AX, βG and cellulose. There are several wheat varieties, which vary widely in the amount and type of DF content. For example, in 12 wheat cultivars studied, ADF and NDF content ranged from 1.8% to 3.2% and 7.2% to 9.1%, respectively (Jha et al., 2011c). Similarly, Kim et al. (2005) reported wide variation in the NSP and lignin contents of wheat while reviewing 426 samples dataset. The NSP content varied from 3.5% to 10.6% AX, 0.3% to 1.2% βG and 7.5% to 16.6% total NSP and the lignin content ranged from 0.9% to 1.1%. This variation in amount and type of DF content of wheat can be attributed to different varieties (genetic makeup), growing locations and climate.

Barley (Hordeum vulgare L.) grains are relatively rich in DF such as βG, cellulose and AX. In particular, hulless barleys contain high levels of soluble βG (Izydorczyk et al., 2000), which are associated with health benefits for the gut (Brennan and Cleary, 2005). Even within hulless barleys, the variability in total DF, soluble and insoluble NSP is very wide (Holtekjolen et al., 2006; Jha et al., 2011b). There is also wide variation in βG content within and between cereal types, ranging from ~1.9% to 11.0% in barley, 1.7% to 7.0% in oats and 0.2% to 0.7% in wheat (Lee et al., 1997; Brennan and Cleary, 2005; Jha et al., 2011b). The rate of solubility of βG also varies, for example, Lambo et al. (2005) found that most of the βG in barley fiber were insoluble (about 85.0%) whereas in oat, the opposite is observed, where about 68.0% of βG were found in soluble fiber fractions.

Oat (Avena sativa L.) contains both soluble and insoluble DF, with high levels of βG, AX and cellulose (Johansen et al., 1997; Jha et al., 2011b). Among the DF components of oat, βG play an important role because of their functional properties in the GIT (Pieper et al., 2008; Jha et al., 2010).

Grain legumes are used in pig diets primarily as sources of protein and contain significant amounts of NSP as well. Cellulose and xylans are found in the hulls, whereas pectic polysaccharides are found in the cotyledons (Choct, 1997; Jha and Leterme, 2012).

The wheat flour milling industry generates a number of coproducts (wheat bran (WB), millrun, middlings, shorts, etc.), which are classified based on their DF content (Jha et al., 2012). These coproducts contain much higher DF than wheat as are enriched during processing of wheat to produce flour. In the six types of wheat flour milling coproducts studied, ADF and NDF content varied from 8.0% to 15.5% and 22.9% to 49.2%, respectively (Jha et al., 2012).

Distillers dried grains with solubles (DDGS) is a coproduct from dry mill ethanol plants resulting from the fermentation of starch of cereal grains (corn, wheat, etc.) to produce fuel ethanol and carbon dioxide. Cereal grains are good source of starch, and as most of the starch is converted to ethanol during fermentation, which results in increased concentrations of the protein, oil and fiber that are two to three times higher than in the parent grain. Because of the depletion of most of the starch, the concentrations of CHO is lower than in the parent grain (Widmer et al., 2007) with most of the CHO present as fiber. The concentration of the different fiber fractions (NDF and ADF) and total DF is approximately three times greater in DDGS than in parent grains. However, the nutrient composition of DDGS varies depending on the ethanol plant where they are processed and the DDGS originating from different parent grains (Pedersen et al., 2014; Jha et al., 2015). The digestibility of fiber in DDGS is <20.0% in the small intestine and <50.0% over the entire GIT, and the DF, therefore, contributes relatively little to the energy value of these products. Moreover, low digestibility of DF in distillers coproducts results in increased quantities of manure being excreted from pigs fed these products, and the overall DM digestibility of diets containing distillers coproducts is lower than in corn-based diets (Pedersen et al., 2007). This lower digestibility of DDGS can be attributed to the complex fiber–starch–protein matrix (Jha et al., 2015), which limits the accessibility and action of endogenous enzymes for degradation. However, supplementation of exogenous multi-carbohydrase and protease enzymes may enhance the degradation of DDGS, thereby improving the nutrient release (Jha et al., 2015; Pedersen et al.,2015).

Canola meal is a coproduct derived after oil extraction processing, and can be used as ingredients in animal diets. CHO in canola seed may be categorized into soluble sugars and oligosaccharides, insoluble CHO and fiber. The concentration of soluble CHO in mature seeds is ~10.0% of the oil-free weight, with sucrose ranging from 3.9% to 9.8%, raffinose from 0.3% to 2.6%, stachyose from 0.8% to 1.6%, fructose from 0.1% to 0.5% and glucose from 0.1% to 0.4% (Barthet and Daun, 2011). The concentration of hemicellulose is ~3.0%, cellulose ranges from 4.0% to 5.0%, and starch is ~1.0% (Salunkhe et al., 1992). The concentration of crude fiber, NDF and ADF in canola meal ranges from 10.0% to 12.0%, 22.0% to 30.0% and 15.0% to 20.0%, respectively. Canola meal has relatively high concentration of DF because hulls in canola seeds stay with the meal (Barthet and Daun, 2011). However, canola breeding programs have developed canola varieties with greater oil and protein content than traditional varieties. The new high-protein varieties of canola also contain less DF, and the resulting canola meal, therefore, has a reduced DF concentration compared with conventional canola.

DF AND DIGESTION

Effect of DF on Digestion and Physiological Functions in Gut

Effect of DF on Digestibility

Digestibility of DF is more variable (40.0% to 60.0%) and lower than that of other nutrients like starch, sugars, fat and CP (above 80.0% in general). It is negatively affected by the amount and source of DF content in the diet (Noblet, 2007; Jha et al., 2010). Consequently, the digestible energy content of diets is negatively and linearly affected by DF (Noblet, 2007). DF is better digested in adult sows than in growing pigs. The difference reaches 0.6 MJ/kg DM, on average (Noblet, 2007). This is ascribed to differences in the physiological stage of pigs as there is a higher rate of degradation of DF in the hindgut of sows, compared with growing pigs, due to longer retention time consecutive to their higher GIT volume, combined with a lower feed intake per live weight (Le Goff et al., 2002). However, it is not known if the DF in all cereal grains has similar effects on the digestibility of energy and nutrients in the diet. Just et al. (1983) found that every 1.0% of additional crude fiber in the diet decreases the gross energy digestibility by 1.3% and metabolizable energy by 0.9%. The NSP, both in purified form and embedded within the matrix, also reduce CP digestion in pigs (Bedford et al., 1992; Jha et al., 2010). There is a linear decrease in apparent ileal digestibility of DM and CP with increased levels of purified NDF in the diet (Schulze et al.,1994) and lower organic matter, CP and starch digestibility in diets containing hulled barley, as compared with hulless barley-based diets in pigs. Also, Jha et al. (2010) reported that the lower organic matter and starch digestibility of the hulled barleys and oats was likely due to greater insoluble DF content, which negatively affects accessibility and the action of endogenous enzymes required for insoluble DF digestion in the upper gut and microbial fermentation in the lower gut.

On the other hand, the inclusion of the rice in the diet improves the digestibility of nutrients as compared with other cereal grains (Mateos et al., 2006; Cervantes-Pahn et al., 2014). These improvements in nutrient digestibility, especially gross energy, organic matter, starch and total CHO in rice diet, indicate that in young pigs, energy and nutrients from rice are better digested and absorbed than energy and nutrients from corn (Solá-Oriol et al., 2010).

The digestibility of nutrients in sorghum and wheat is relatively similar to that of corn, but in terms of grain structure and nutrient composition, sorghum is more similar to corn than to wheat (Taylor and Emmambux,2010). But, CP digestibility of sorghum is less than corn. The reduced digestibility of CP in sorghum could be attributed to the binding of tannins to the protein in sorghum, which makes the protein resistant to proteolysis (Duodu et al., 2003). However, tannins content of the sorghum varies depending on the varieties and parts of the sorghum grain. Moreover, Wilfart et al. (2007b) reported decreasing apparent total tract digestibility of DM, organic matter, CP and gross energy when added 0%, 20% and 40% WB (16.5%, 20.9% and 27.0% total DF, respectively) to a wheat–barley–soybean meal diet. These authors suggested that digestion of the WB in the hindgut is affected by the time that the digesta is exposed to fermentation, and a rapid passage of digesta may reduce the efficacy of the digestion process. In fact, WB is one of the most effective fiber sources for increasing the rate of passage in the digestive tract.

Effect of DF on Physiological Functions

The presence of DF in the diet does not only affect digestibility but also other physiological functions in the gut. The latter are affected by the level and type of fiber (Schulze et al., 1995) and their physico-chemical properties, like WHC, solubility and viscosity (Leterme et al., 1996; Molist et al., 2014).

The presence of soluble DF in the diet increases digesta viscosity (Gallaher et al., 1999) and increased viscosity in the digesta can limit the interaction between nutrients and enzymes facilitating the formation of an unstirred water layer in the intestinal surface, thereby creating a physical barrier and consequently, reducing nutrient digestion and absorption. Moreover, insoluble DF sources such as WB are relatively resistant to microbial degradation (Jorgensen et al., 1996) and its inclusion in the diet produces an increase in fecal DM and bulkiness (Wilfart et al., 2007b). On the other hand, DF inclusion in the diet increases the endogenous nitrogen losses depending on DF level, type (Schulze et al., 1995) and physico-chemical properties such as WHC (Leterme et al., 1996). For example, soluble DF increases digesta viscosity and endogenous nitrogen losses (Mariscal-Landin et al., 1995). The high viscosity of the gut chyme stimulates the epithelial cell proliferation and may contribute to some loss of epithelial cells (Gee et al., 1996). In weanling pigs, Schiavonet al. (2004) reported that an increase in intestinal viscosity increased cell exfoliation in the apical part of the intestinal villus, causing atrophy and deeper crypt depth. Consequently, an increase in intestinal viscosity might reduce the digestion and absorption of nutrients in the diet (Molist et al., 2014). In this respect, Lizardoet al. (1997) reported decreasing apparent ileal digestibility in 25-day-old piglets fed a control diet or a same diet supplemented with 12.0% sugar beet pulp (SBP). In presence of DF, there is also an increase in the pancreatic secretions and the number of goblet cells (Schneeman et al., 1982). Moreover, there is increase in mucus secretion in the small intestine (Mariscal-Landin et al., 1995), which might be due to the mechanical effect of DF on the gut wall that affects the integrity of the mucus layer, resulting in superficial cell lesions (Schmidt-Willig et al., 1996). However, Leterme et al. (1998) did not observe any influence of insoluble NSP with high WHC on protein digestion and absorption, as opposed to soluble NSP with high WHC. In the study, the contrasted result of the effect of NSP on endogenous nitrogen losses was due to different sources of fibers used, which supports the findings that the source of DF with different physico-chemical properties behave differently and have different effects on nutrient digestibility and physiological properties in the gut.

It can be concluded that DF content negatively affects nutrient and energy digestibility, which varies according to the amount and source of DF and their physico-chemical properties. Moreover, different types of DF also exert their effect on different physiological functions in the gut.

Degradability of DF in the upper and Lower Gut

Degradability of DF in the Upper Gut

DF escapes enzymatic digestion in the small intestine and becomes available for fermentation by bacteria in the colon. However, substantial degradation of DF may also occur in the small intestine (Jensen and Jorgensen, 1994; Jha et al., 2010; Jha and Leterme, 2012). Fiber-degrading bacteria are present in the stomach and in the proximal small intestine. They can partially disrupt the cell wall components of fiber, which leads to partial digestion (Varel and Yen, 1997). Bach Knudsen et al. (2008) summarized the ileal digestibility of NSP and their components from different studies. The results clearly indicate a wide variation in the digestion of NSP components, within and between different cereal sources. The wide variation in the fiber degradability can be ascribed to the physico-chemical properties of DF, the complexity of digestion/fermentation process, differences in experimental design, sample collection and analytical techniques. Gdala et al. (1997) reported lower digestibility of xylose, arabinose and uronic acids in the small intestine of piglets compared with glucose when fed diets based on cereals and soybean meal. This might be due to the high digestibility of mixed linked βG, which is highly degradable in the upper gut, due to its soluble nature (Bach Knudsen and Hansen, 1991; Jha et al., 2010, 2011b). Also, Jha and Leterme (2012) suggested that the higher insoluble DF content of the different fiber sources such as WB or DDGS affects negatively to the accessibility and action of the endogenous enzymes in the upper gut and microbial fermentation in the lower gut, resulting in lower degradability of DF.

Degradability of DF in the Lower Gut

There is wide variation in the degradation of fiber in the large intestine, ranging from 48% to 95.0% (Bach Knudsen et al., 1993a; Jha et al., 2010; Jha and Leterme, 2012). Similarly, the total tract apparent digestibility of cellulose varies widely (2.0% to 84.0%). Soluble pectin and hemicelluloses are digested to a greater extent than cellulose, while soluble βG from barley are almost completely digested by the end of the gut; the prececal digestibility has been found to range as high as 70.0% to 97.0% (Bach Knudsen et al.,1993a; Jha et al., 2010). On the other hand, insoluble branched-chain AX from wheat, rye and oat are less digestible in the pig gut (Bach Knudsen and Hansen, 1991; Glitsø et al., 1998). There are also noted effects of the DF source on variation in NSP digestibility. Chabeauti et al. (1991) found that the NSP digestibility in growing pigs varies from 16.3% for wheat straw, 43.5% for WB, and 69.5% for SBP to 79.1% for soybean hulls. The poor digestibility of WB is ascribed to their high lignin content that makes the NSP less fermentable compared with highly digestible pectin substances of SBP and soybean hulls (Karr-Lilienthal et al., 2005).

Gdala et al. (1997) analyzed the digestibility of different NSP residues in piglets and found that the rate and overall degradation of the polymers in the large intestine was largely influenced by the chemical nature of the DF, especially its solubility and degree of lignification. Similar results were obtained by Johansen et al.(1997) while studying the degradation of βG and AX from oat bran in the pig gut. However, the total loss of NSP from the anterior to terminal ileum was lower than reported by other workers in old pigs (Bach Knudsen and Hansen, 1991; Bach Knudsen et al., 1993a), possibly due to lower microbial activity in young piglets. Among the NSP components, soluble βG, AX and pectins are rapidly degraded in the cecum and proximal colon while insoluble components of NSP like cellulose and insoluble AX are degraded slowly and at the distal part of the colon (Bach Knudsen et al., 1993a; Glitsø et al., 1998; Canibe and Bach Knudsen,2002). In this respect, Gidenne (2015) mentioned that the amounts of DF entering the cecum of rabbit is not a limiting factor for the fermentation processes, because the digesta retention time in the cecum of rabbit is very short as compared with pigs, and consequently, soluble fiber fractions such as pectins are degraded easily. Moreover, soluble and non-cellulosic mannose and galactose are highly digestible and fermentable compared with the insoluble cellulosic components of NSP (Serena and Knudsen, 2007).

FIBER FERMENTATION IN THE GIT

Fiber Fermentation

The susceptibility of DF to microbial fermentation varies depending on the accessibility of DF to the microbial population in the hindgut (Oakenfull, 2001). In monogastric animals, the large intestine is the most important site of fermentation (Williams et al., 2001). Fermentation of soluble DF is mainly at the proximal colon, whereas fermentation of insoluble DF is sustained until the distal colon (Choct, 1997). However, substantial fermentation of soluble DF has been observed in the pig's small intestine (Jensen and Jorgensen,1994; Jha et al., 2010; Jha and Leterme, 2012). Fiber fermentation is an extremely complex process, affected by many factors in the GIT, including the host, its microflora and their interaction, which takes place between them (Williams et al., 2005).

Fiber Fermentation and Production of Metabolites

Fermentation of DF is more variable than digestion of the macronutrients starch, fat and CP (generally above 80.0%; Bach Knudsen et al., 2008). The variation in fermentability is mainly due to changes in physico-chemical properties of DF such as bulk, viscosity, solubility, WHC and fermentability. DF fermentation results in the production of VFA like acetate, propionate and butyrate, along with some gases like hydrogen, carbon dioxide and methane (Macfarlane and Macfarlane, 1993; Williams et al., 2001). Acetate is the most abundant VFA, comprising about 60.0% of the total VFA produced in the hindgut, whereas propionate and butyrate are produced in smaller quantities (Lunn and Buttriss, 2007). However, the extent of fermentation and the profile in VFA depend on the substrate (Salvador et al., 1993; Jha et al., 2012) while the rate of fermentation of DF in the pig's intestines depends on its composition and physico-chemical properties, degree of lignification and particle size (Le Goff et al., 2003) and transit time in the digestive tract (Wilfart et al., 2007a). Soluble DF has, in general, a higher WHC than insoluble DF that give raise to a larger surface area and thereby large areas for bacterial enzyme attack. Thus, these characteristics are directly dependent on the botanical origin and (or) processing of the DF source (Johansen et al., 1997).

In addition to VFA, other metabolites such as lactate, ethanol and succinate are also produced from bacterial fermentation of DF (Drochner et al., 2004). The majority of these metabolites (possibly except ethanol) are further converted into VFA by cross-feeding mechanisms (Macfarlane and Gibson, 1995).

Fate of Fermentation Metabolites in the Pig Large Intestine

The main site of VFA absorption in pigs is the large intestine (Imoto and Namioka, 1978) where the majority (about 90.0%) of the VFA are absorbed and metabolized (Jorgensen et al., 1997).

The VFA are absorbed in the gut through passive diffusion. The exact mechanism for absorption is still unclear. However, several mechanisms have been proposed, stating their dependence on luminal pH, CO_2, as well as the fluxes of water, protons and inorganic ions (Bugaut, 1987). The fermentation metabolites are taken up by the cells in the intestines and used for bacterial growth (Bach Knudsen, 2001). Although VFA are primarily taken up and metabolized by colonocytes, these are also used as a source of energy by other tissues (Wong et al., 2006). The absorbed VFA are basically metabolized in three major sites: in the ceco-colonic epithelial cells that use butyrate for their energy production pathway; liver cells that metabolize residual butyrate and propionate for gluconeogenesis, as well as 50.0% to 70.0% of acetate; and muscle cells, mainly from skeletal and cardiac muscles that oxidize the residual acetate (Roberfroid, 2007). The energy produced from VFA may contribute up to 15.0% of the maintenance energy requirements of growing pigs (Dierick et al., 1989) and even up to 30.0% in gestating sows (Varel and Yen, 1997).

Absorption of VFA also facilitates absorption of other nutrients from the diet. Water and sodium are absorbed along with VFA (Yen, 2001). Plant lignans, diphenolic compounds similar to endogenous steroid hormones, are also co-transported by VFA (Bach Knudsen et al., 2006).

Effect of Source of Fiber on Metabolite Production

Among the non-digestible oligosaccharides, fructo-oligosaccharides (FOS) are the most extensively studied. The FOS contains 2–70 fructose residues and Bifidobacteria can digest them as they produce the enzyme fructofuranosidase (Gibson et al., 2004). FOS are fermented by bacteria, yielding energy for bacterial growth. Houdijk et al. (2002) evaluated the effect of FOS and transgalacto-oligosaccharides (TOS) in comparison with non-fiber control diets in weaned pigs and found that both FOS and TOS increased VFA production in the gut. However, there were differences in the concentration of VFA in different sections of the gut between FOS and TOS fed pigs, which supports the view that FOS and TOS have different fermentation characteristics in the GIT of the pigs.

Among the DF fractions, βG is gaining more attention as it is a source of easily fermentable energy for intestinal microbiota. It yields higher levels of VFA (Brennan and Cleary, 2005) due to a relatively high concentration, soluble state and high molecular weight and results in several beneficial physiologic effects to the host (Dongowski et al., 2002). Oat bran, a rich source of soluble DF in the form of βG, produces almost twice as much VFA per gram DF as WB in the pig intestines (Bach Knudsen and Hansen, 1991; Bach Knudsen et al., 1993a). However, AX and not βG in the cell walls of oat bran are responsible for the enhanced butyric acid production of oat bran (Bach Knudsen et al., 1993b). Bach Knudsen and Canibe (2000) found higher concentrations and flows of lactic acid in the ileum of cannulated pigs after feeding a diet supplemented with soluble DF from oat bran, which supports the view that βG stimulates the production of lactic acid in the small intestine, which is found to promote the development of Lactobacilli, a family of health-promoting bacteria.

There are noted effects of DF on VFA production and profile, depending on type and source of DF. For example, Freire et al. (2000) compared the effects of inclusion in the diet of 20.0% of WB, SBP, soybean hulls or alfalfa meal on concentration of the total VFA in the cecum and reported that the inclusion of soybean hulls in the weaning diet increased the concentration of total VFA by 11.2%, 30.5% and 27.2% as compared with WB, SBP and alfalfa diets, respectively. The values suggests that soybean hulls is highly degraded in the cecum, and is in agreement with the high digestibility values of the NDF and ADF fractions reported in this research. Also, the lower values of VFA measured in the SBP diet might also be associated with a higher absorption rate of its metabolites in the cecum. Carneiro et al. (2008) compared the effect of two fiber sources, WB and maize cobs in weaned pigs and found no difference in the amounts of VFA in the small intestine. However, there was higher acetic acid and lower butyric acid production in the cecum when WB was replaced with maize cobs. Findings of these studies clearly indicate that not only the amount and type of substrate, but also the source of fiber fraction is important to determine the amount and type of VFA production.

Effect of Fiber Fermentation on Gut Microbiota

The influence of diet on microbial communities in the pig intestines has been of interest for long time. However, the interaction of diet and microbiota in the intestines of the pig are still not well understood.

The GIT microbiota of pig is composed primarily of bacteria. The microbial population increases from 10^3 to 10^5/g of digesta in the stomach to 10^9 to 10^{10} in the distal small intestine, and further to 10^{10} to 10^{11} in the large intestine of pigs, belonging to more than 50 genera and over 500 species of bacteria (Jensen and Jorgensen, 1994; Gaskins, 2001). The majority (about 90.0%) of the cultivable bacteria are Gram-positive, strict anaerobes belonging to the Streptococcus, Lactobacillus, Eubacterium, Clostridium andPeptostreptococcus genus while the remaining 10.0% of total flora belongs to Gram-negative of Bacteroidesand Prevotella groups (Gaskins, 2001). Each bacterial species occupies a particular niche with numerous interrelationships between them (Flint et al., 2008).

Effect on Microbial Composition

The population and activity of bacteria in the gut is influenced by several factors, the main one being diet (Bach Knudsen et al., 2012). More specifically, the structure and composition (Konstantinov et al., 2004; Bindelle et al., 2010), solubility (Hogberg and Lindberg, 2004) and amount and type of substrate available (Macfarlane and Macfarlane, 1993) affects the gut microbial ecology. Among the different constituents of diets, DF is found to affect the gut environment (Awati et al., 2005; Jhaet al., 2010; Jha and Leterme, 2012). The source of DF affects the digestion site and gut environment, thereby affecting the conditions for the proliferation of microbiota in the gut (Hogberg and Lindberg, 2004). Moreover, DF serves as an energy source for microbes and supports their proliferation. Gidenne (2015) reported that the energy provided by the cecal VFA could reach up to 50% of the maintenance energy in growing rabbit. In pigs, Bach Knudsen et al. (1991) reported 5.5 times (as measured by ATP concentration) increased microbial activity in the GIT of pigs when fed with a high-fiber diet. In addition, there was increased (five to nine times) carbon dioxide and methane production, suggesting increased microbial fermentation that takes place in the GIT of pigs fed a high-fiber diet. Similar increased microbial activity was observed in the intestines of pigs fed pea fiber and pectin, as indicated by higher bacterial counts, ATP concentration, adenylate energy charge and low pH (Jensen and Jorgensen, 1994). However, Varel et al.(1982) noted that there was initially a decrease in the bacterial population of the pig intestines when the animals were fed with high-fiber diet (50.0% alfalfa meal) in lean genotype pigs. The microbial population, however, increased after continuous fiber-feeding for 17 weeks. It suggests that there is some kind of adaptation of the microbiota in the pig intestines when fed with high DF diets.

DF affects fermentation in the GIT by stimulating the growth or metabolism of special bacterial species (Williams et al., 2001). These increased numbers of cellulolytic bacteria enhance the hindgut fermentation and production of VFA, which decreases the pH of the gut content. A decrease in pH promotes growth of beneficial bacteria (e.g. Bifidobacteria spp., Lactobacilli spp.), at the expense of pathogenic ones likeClostridium or Salmonella, which contribute to enhance the health of host species (Bouhnik et al., 2004). This phenomenon is termed as 'prebiotic effect' (Gibson and Roberfroid, 1995).

The potential 'prebiotic effect' of DF has been studied in several monogastric species, including swine. The results are quite variable from one study to another in terms of their effect on microbial population, diversity and gut health, which can be ascribed to the type of substrate available for fermentation and the gut environment of the host. At the increased level of DF in the large intestine, there is an increase in activity of the entire microbial community. However, some types of DF may have selective effects and stimulate particular niches of microorganisms (Louis et al., 2007).

Estrada et al. (2001) found increased numbers of Bifidobacteria spp. and decreased numbers of total anaerobes and Clostridia in feces of pigs fed with diets containing 0.5% FOS in conjunction withBifidobacterium longum. Drew et al. (2002) compared the effect of CHO sources (corn, wheat and barley) in weaned pigs and found that the bacterial population was significantly related with ADF and NDF contents of the diets. There were increased Lactobacilli spp. and decreased Enterobacteria spp. populations in barley-fed pigs, as compared with corn-fed pigs. Moreover, barley-based diets increased Lactobacilli spp. andBifidobacterium spp. in the cecum, compared with corn-based diets. One possible explanation might be the higher amount of βG in barley-based diets, compared with corn-based diets, which is found to enhance the growth of beneficial bacteria at the cost of harmful bacteria. In fact, Pieper et al. (2008) reported that dietaryβG in barley-based diets increases colonic lactobacilli spp. and bifidobacteria spp. promoting butyrate-producing bacteria. Interestingly, wheat-based diets had higher numbers of Bifidobacterium spp. and lower numbers of total aerobes and Clostridium spp., compared with barley-based diets. Similarly, Nielson et al. (2014) reported that pigs fed with diet rich in AX resulted in higher levels of Bifidobacterium spp. andLactobacillus spp. in the feces. These can be ascribed to the complexity of the various bacterial species and fermentable substrates present, especially the monomers of the fiber, which are selective for certain microbes. Konstantinov et al. (2004) found Ruminococcus-like species in the feces of pigs fed diets containing fibers (SBP and FOS), but not in pigs fed a control diet, suggesting that these bacteria may play a role in the utilization of DF. Moreover, there was a specific response of a novel and abundant Lactobacillus amylovorus-like phylotype to dietary oligosaccharides in the gut of weaning pigs (Konstantinov et al., 2004). Similarly, Owusu-Asiedu et al. (2006) reported increased Bifidobacteria spp. and Enterobacteria spp. populations in the ileal digesta of growing pigs fed diets supplemented with guar gum or cellulose to a standard diet.

The effects of NSP compounds such as βG and AX have mainly been studied in isolated form, whereas in swine diets, these compounds are present as part of the grain matrix. The fermentation rate of these CHO in the intestinal tract will thus depend on their composition, form and physical properties (Le Goff et al., 2003). As a consequence, cereal NSP in isolated form or within a matrix may act differently in the GIT. In this respect, Pieper et al. (2008) conducted an experiment with weaned pigs and found that hulless barley varieties with high soluble NSP content favored xylan- and βG-degrading bacteria, whereas βG-supplemented hulled barleys favored Lactobacilli spp. Moreover, there was a decrease in the number ofLactobacilli spp. in the ileum of pigs fed hulless/high βG barley-based diets. This suggests that both type and form of βG affect the bacterial population in the pig intestines. Processed fibers are also found to exert a positive response on health-promoting characteristics. As an example, oat fiber and βG isolates, fermented ropy oat-based products containing both native and microbial βG, stimulated Bifidobacteria spp. in the GIT apart from other health benefits like reduced blood cholesterol level to host (Martensson et al., 2005). Like the NSP, RS is also found to affect the bacterial population in the pig intestines. Brown et al. (1997) found higher Bifidobacteria spp. counts in feces of pigs fed with a high amylose cornstarch diet, than in feces of pigs fed with a low amylose cornstarch. Part of the high amylose corn-starch becomes RS, which is fermented in the large intestine and exerts prebiotic effects.

Effect on Gut Health

The maintenance of gut health is complex phenomenon and relies on a delicate balance between the diet, the commensal microflora and the mucosa, including the digestive epithelium and the mucus overlying the epithelium (Montagne et al., 2003). DF plays an important role in the function of the pig GIT. It is evidenced by several studies reporting the positive role of DF in controlling bacterial infections, particularly reducing post-weaning diarrhea (Williams et al., 2001; Mateos et al., 2006; Molist et al., 2014), which is a major problem for the pig industry in many parts of the world. In this respect, Lizardo et al. (1997) reported improved digestive function in 39-day-old pigs when 12.0% SBP was included in the diet. Probably, soluble fiber sources such as SBP, are easily fermentable by the microflora in the large intestine, which could help to create a stable environment within the GIT reducing the incidence of post-weaning diarrhea. Thomsen et al. (2007) observed that the inclusion of DF, like fructan-rich chicory roots and sweet lupins completely protected against the development of swine dysentery. Similarly, the inclusion of DF in piglet diets enhanced intestinal populations of Lactobacilli spp. and reduced the incidence and severity of diarrhea (Edwards, 1996). These studies support the view that diets supplemented with fiber can protect pigs against swine dysentery. However, there is continuous debate on whether fiber exerts beneficial or detrimental effects on the development of post-weaning enteric dysentery. Pluske et al. (2003) noted increased incidence of clinical swine dysentery in growing pigs and diarrhea in weanling pigs fed with diets high in fermentable NSP and RS. Similar negative effect on gut health was also noticed with supplementation of isolated soluble fiber. Nursery pig diets supplemented with 0.025% βG increased growth performance but also increased the susceptibility to Streptococcus suis infection (Dritz et al., 1995). These authors suggest that a complex interaction exists between growth performance and disease susceptibility in pigs fed βG.

There is some information available on fractions of DF of other grains like peas, chickpeas, faba beans and lupins and their effect on microbial population and gut health. Queiroz-Monici et al. (2005), while working with rats, found that peas and chickpeas have a bifidogenic effect due to the DF and RS present in these legumes. RS is considered a good source of butyrate production, which is utilized by the colonic cells as fuel, thus strengthening the first line of defense in the gut (Topping and Clifton, 2001). Thus, RS sources with the potential to stimulate butyrate production have a potential to improve gut health.

It can be summarized that the presence of fiber in the gut significantly affects the gut microbial environment, creates more favorable lumen conditions for gut health by stimulating the growth of 'beneficial bacteria' at the cost of 'harmful bacteria', with the possibility of some negative impact on gut health, which depends on the type of fiber substrate available for fermentation. However, there is no straightforward answer of the benefits of DF on gut health and direct evidence for enhanced resistance to unfavorable conditions is still lacking.

CONCLUSION

There is wide variation found in the composition of DF in different feeds and feed ingredients, which has a major impact on their physico-chemical properties and digestion and fermentation characteristics in the GIT. In general, DF affects nutrients and energy digestibility in pigs negatively but with differences between the different types: (1) soluble DF fractions are fermented faster than insoluble fractions, produce higher amounts of VFAs and lower ammonia concentrations; (2) some DF constituents, for example RS, may stimulate butyrate production to a higher degree than others thereby contributing to an improved gut health; and (3) some DF constituents may exert 'prebiotic effects', enhancing 'beneficial bacteria' at the cost of 'harmful bacteria' in the pig gut. Therefore, strategic selection of DF in diets can be used as a nutritional strategy to modulate the intestinal health of swine.

ACKNOWLEDGEMENT

This work was partially supported by USDA National Institute for Food and Agriculture, Hatch Project HAW02030-H administered by the College of Tropical Agriculture and Human Resources, University of Hawaii at Manoa.

REFERENCES

1. AACC 2001. The definition of dietary fiber. AACC report. American Association of Cereal Chemists 46, 112–126.
2. Awati A, Konstantinov SR, Williams BA, Akkermans ADL, Bosch MW, Smidt H and Verstegen MWA 2005. Effect of substrate adaptation on the microbial fermentation and microbial composition of faecal microbiota of weaning piglets studied in vitro . Journal of the Science of Food and Agriculture 85, 1765–1772.
3. Bach Knudsen KE 1997. Carbohydrate and lignin contents of plant materials used in animal feeding. Animal Feed Science and Technology 67, 319–338.
4. Bach Knudsen KE 2001. The nutritional significance of 'dietary fibre' analysis. Animal Feed Science and Technology 90, 3–20.

5. Bach Knudsen KE 2014. Fiber and nonstarch polysaccharide content and variation in common crops used in broiler diets. Poultry Science 93, 2380–2393.
6. Bach Knudsen KE and Hansen I 1991. Gastrointestinal implications in pigs of wheat and oat fractions 1. Digestibility and bulking properties of polysaccharides and other major constituents. British Journal of Nutrition 65, 217–232.
7. Bach Knudsen KE and Canibe N 2000. Breakdown of plant carbohydrates in the digestive tract of pigs fed on wheat- or oat-based rolls. Journal of the Science of Food and Agriculture 80, 1253–1261.
8. Bach Knudsen KE, Jensen BB and Hansen I 1993. a. Digestion of polysaccharides and other major components in the small and large intestine of pigs fed on diets consisting of oat fractions rich in β-D-glucan.British Journal of Nutrition 70, 537–556.
9. Bach Knudsen KE, Jensen BB and Hansen I 1993. b. Oat bran but not a beta-glucan-enriched oat fraction enhances butyrate production in the large intestine of pigs. The Journal of Nutrition 123, 1235–1247.
10. Bach Knudsen KE, Laerke HN and Jorgensen H 2008. The role of fibre in nutrient utilization and animal health. In Proceedings of the 29th Western Nutrition Conference, September 23–24, Edmonton, AB, Canada, pp. 93–107.
11. Bach Knudsen KE, Hedemann MS and Laerke HN 2012. The role of carbohydrates in intestinal health of pigs. Animal Feed Science and Technology 173, 41–53.
12. Bach Knudsen KE, Jensen BB, Andersen JO and Hansen I 1991. Gastrointestinal implications in pigs of wheat and oat fractions 2. Microbial activity in the gastrointestinal tract. British Journal of Nutrition 65, 233–248.
13. Bach Knudsen KE, Serena A, Jørgensen H, Peñalvo JL and Adlercreutz H 2006. Rye and other natural cereal fiber enhance the production and plasma concentration of enterolactone and butyrate In Dietary fiber components and functions (ed. H Salovaara, F Gates and M Tenkanen), pp. 219–235. Wageningen Academic Publishers, Wageningen, The Netherlands.
14. Barthet VJ and Daun JK 2011. Seed morphology, composition, and quality In Canola: chemistry, production, processing, and utilization (ed. JK Daun, NAM Eskin and D Dickling), pp. 119–162. AOCS Press, Urbana, IL, USA.
15. Bedford MR, Patience JF, Classen HL and Inborr J 1992. The effect of dietary enzyme supplementation of rye- and barley-based diets on digestion and subsequent performance in weanling pigs. Canadian Journal of Animal Science 72, 97–105.
16. Bindelle J, Pieper R, Leterme P, Rossnagel B and Van Kessel AG 2010. Changes in intestinal microbial ecophysiology as related to the carbohydrate composition of barleys and oats cultivars in an in vitro model of the pig gastrointestinal tract. Livestock Science 133, 151–153.
17. Bouhnik Y, Raskine L, Simoneau G, Vicaut E, Neut C, Flourié B, Brouns F and Bornet FR 2004. Capacity of nondigestible carbohydrates to stimulate bifidobacteria in healthy humans. American Journal of Clinical Nutrition 80,

1658–1664.

18. Brennan CS and Cleary LJ 2005. The potential use of cereal (1/3,1/4)-β-D-glucans as functional food ingredients. Journal of Cereal Science 42, 1–13.
19. Brown I, Warhurst M, Arcot J, Playne M, Illman RJ and Topping DL 1997. Fecal numbers of bifidobacteria are higher in pigs fed Bifidobacterium longum with a high amylose cornstarch than with a low amylose cornstarch. The Journal of Nutrition 127, 1822–1827.
20. Bugaut M 1987. Occurrence, absorption and metabolism of short chain fatty acids in the digestive tract of mammals. Comparative Biochemistry and Physiology B 86, 439–472.
21. Canibe N and Bach Knudsen KE 2002. Degradation and physicochemical changes of barley and pea fibre along the gastrointestinal tract of pigs. Journal of the Science of Food and Agriculture 82, 27–39.
22. Carneiro MSC, Lordelo MM, Cunha LF and Freire JPB 2008. Effects of dietary fibre source and enzyme supplementation on faecal apparent digestibility, short chain fatty acid production and activity of bacterial enzymes in the gut of piglets. Animal Feed Science and Technology 146, 124–136.
23. Cervantes-Pahm SK, Yanhong L and Stein HH 2014. Comparative digestibility of energy and nutrients and fermentability of dietary fiber in eight cereal grains fed to pigs. Journal Science Food Agriculture 94, 841–849.
24. Chabeauti E, Noblet J and Carre B 1991. Digestion of plant cell walls from four different sources in growing pigs. Animal Feed Science and Technology 32, 207–213.
25. Choct M 1997. Feed non-starch polysaccharides: chemical structures and nutritional significance. Feed Milling International 6, 13–26.
26. Cummings JH and Stephen AM 2007. Carbohydrate terminology and classification. European Journal of Clinical Nutrition 61, 5–18.
27. Davidson MH and McDonald A 1998. Fiber: forms and functions. Nutrition Research 18, 617–624.
28. DeVries JW, Prosky L, Li B and Cho S 1999. A historical perspective on defining dietary fiber. Cereal Foods World 44, 367–369.
29. Dierick NA, Vervaeke IJ, Demeyer DI and Decuypere JA 1989. Approach to the energetic importance of fibre digestion in pigs. I. Importance of fermentation in the overall energy supply. Animal Feed Science and Technology 23, 141–167.
30. Dongowski G, Huth M, Gebhardt E and Flamme W 2002. Dietary fiber-rich barley products beneficially affect the intestinal tract of rats. The Journal of Nutrition 132, 3704–3714.
31. Drew MD, Van Kessel AG, Estrada AE, Ekpe ED and Zijlstra RT 2002. Effect of dietary cereal on intestinal bacterial populations in weaned pigs. Canadian Journal of Animal Science 82, 607–609.
32. Dritz SS, Shi J, Kielian TL, Goodband RD, Nelssen JL, Tokach MD, Chengappa MM, Smith JE and Blecha F 1995. Influence of dietary beta-glucan on growth

performance, nonspecific immunity, and resistance to Streptococcus suis infection in weanling pigs. Journal of Animal Science 73, 3341–3350.

33. Drochner W, Kerler A and Zacharias B 2004. Pectin in pig nutrition, a comparative review. Journal of Animal Physiology and Animal Nutrition 88, 367–380.

34. Duodu KG, Taylor JRN, Belton PS and Hamaker BR 2003. Mini review: factors affecting sorghum protein digestibility. Journal Cereal Science 38, 117–131.

35. Edwards SA 1996. A new look on the role of fibre in the diet of pigs. In Proceedings of the 6th European Society of Veterinary Internal Medicine, September 12–14, Veldhoven, The Netherlands, pp. 90–91.

36. Estrada A, Drew MD and Van Kessel AG 2001. Effect of dietary supplementation of fructooligosacchadies and Bifidobacterium longum to early weaned pigs on performance and fecal bacterial populations. Canadian Journal of Animal Science 81, 141–148.

37. Fairbairn SL, Patience JF, Classen HL and Zijlstra RT 1999. The energy content of barley fed to growing pigs: characterizing the nature of its variability and developing prediction equations for its estimation. Journal of Animal Science 77, 1502–1512.

38. FAO Statistical Yearbook 2013. World food and agriculture. Food and Agriculture Organization of the United Nations, pp. 172 Rome, Italy.

39. Flint HJ, Bayer EA, Rincon MT, Lamed R and White BA 2008. Polysaccharide utilization by gut bacteria: potential for new insights from genomic analysis. Nature Reviews Microbiology 6, 121–131.

40. Freire JPB, Guerreiro AJG, Cunha LF and Aumaitre A 2000. Effect of dietary fibre source on total tract digestibility, caecum volatile fatty acids and digestive transit time in the weaned piglet. Animal Feed Science and Technology 87, 71–83.

41. Gallaher DD, Wood KJ, Gallaher CM, Marquart LF and Engstrom AM 1999. Intestinal contents supernatant viscosity of rats fed oat-based muffins and cereal products. Cereal Chemical 76, 21–24.

42. Gaskins HR 2001. Intestinal bacteria and their influence on swine growth In Swine nutrition, 2nd edition (ed. AJ Lewis and LL Southern), pp. 585–608. CRC Press, Boca Raton, FL, USA.

43. Gdala J, Johansen HN, Bach Knudsen KE, Knap IH, Wagner P and Jørgensen OB 1997. The digestibility of carbohydrates, protein and fat in the small and large intestine of piglets fed non-supplemented and enzyme supplemented diets. Animal Feed Science and Technology 65, 15–33.

44. Gebreyes WA, Thakur S and Morrow WEM 2006. Comparison of prevalence, antimicrobial resistance, and occurrence of multidrug-resistant Salmonella in antimicrobial-free and conventional pig production. Journal of Food Protection 69, 743–748.

45. Gee JM, Lee-Finglas W, Wortley GW and Johnson IT 1996. Fermentable carbohydrates elevate plasma enteroglucagon but high viscosity is also necessary

to stimulate small bowel mucosal cell proliferation in rats.The Journal of Nutrition 126, 373–379.

46. Gibson GR and Roberfroid MB 1995. Dietary modulation of the human colonic microbiota: introducing the concept of prebiotics. The Journal of Nutrition 125, 1401–1412.

47. Gibson GR, Probert HM, Loo JV, Rastall RA and Roberfroid MB 2004. Dietary modulation of the human colonic microbiota: updating the concept of prebiotics. Nutrition Research Reviews 17, 259–275.

48. Gidenne T 2015. Dietary fibres in the nutrition of the growing rabbit and recommendations to preserve digestive health: a review. Animal 9, 227–242.

49. Glitsø LV, Brunsgaard G, Højsgaard S, Sandström B and Bach Knudsen KE 1998. Intestinal degradation in pigs of rye dietary fibre with different structural characteristics. British Journal of Nutrition 80, 457–468.

50. Gutierrez NA, Serão NVL, Kerr BJ, Zijlstra RT and Patience JF 2014. Relationships among dietary fiber components and the digestibility of energy, dietary fiber, and amino acids and energy content of nine corn coproducts fed to growing pigs. Journal of Animal Science 92, 4505–4517.

51. Hogberg A and Lindberg JE 2004. Influence of cereal non-starch polysaccharides on digestion site and gut environment in growing pigs. Livestock Production Science 87, 121–130.

52. Holtekjolen AK, Uhlen AK, Brathen E, Sahlstrom S and Knutsen SH 2006. Contents of starch and non-starch polysaccharides in barley varieties of different origin. Food Chemistry 94, 348–358.

53. Houdijk JGM, Verstegen MWA, Bosch MW and van Laere KJM 2002. Dietary fructooligosaccharides and transgalactooligosaccharides can affect fermentation characteristics in gut contents and portal plasma of growing pigs. Livestock Production Science 73, 175–184.

54. Imoto S and Namioka S 1978. VFA production in the pig intestine. Journal of Animal Science 47, 467–478.

55. Izydorczyk MS, Storsley J, Labossiere D, MacGregor AW and Rossnagel BG 2000. Variation in total and soluble beta-glucan content in hulless barley: effects of thermal, physical, and enzymatic treatments. Journal of Agricultural and Food Chemistry 48, 982–989.

56. Jaworski NW, Lærke HN, Bach Knudsen KE and Stein HH 2015. Carbohydrate composition and in vitro digestibility of dry matter and nonstarch polysaccharides in corn, sorghum, and wheat and coproducts from these grains. Journal of Animal Science 93, 1103–1113.

57. Jha R and Leterme P 2012. Feed ingredients differing in fermentable fibre and indigestible protein content affect fermentation metabolites and faecal nitrogen excretion in growing pigs. Animal 6, 603–612.

58. Jha R, Bindelle J, Van Kessel AG and Leterme P 2011. a. In vitro fibre fermentation of feed ingredients with varying fermentable carbohydrate and protein levels and

protein synthesis by colonic bacteria isolated from pigs. Animal Feed Science and Technology 165, 191–200.

59. Jha R, Rossnagel B, Pieper R, Van Kessel A and Leterme P 2010. Barley and oat cultivars with diverse carbohydrate composition alter ileal and total tract nutrient digestibility and fermentation metabolites in weaned piglets. Animal 4, 724–731.

60. Jha R, Bindelle J, Rossnagel B, Van Kessel AG and Leterme P 2011. b. In vitro evaluation of the fermentation characteristics of the carbohydrate fractions of hulless barley and other cereals in the gastrointestinal tract of pigs. Animal Feed Science and Technology 163, 185–193.

61. Jha R, Overend DN, Simmins PH, Hickling D and Zijlstra RT 2011. c. Chemical characteristics, feed processing quality, voluntary feed intake, growth performance, and energy digestibility among wheat classes fed in pelleted diets fed to weaned pigs. Animal Feed Science and Technology 170, 78–90.

62. Jha R, Owusu-Asiedu A, Simmins PH, Pharazyn A and Zijlstra RT 2012. Degradation and fermentation characteristics of wheat coproducts from flour milling in the pig intestine, studied in vitro. Journal of Animal Science, E-suppl. 90, 173–175.

63. Jha R, Woyengo A, Li J, Bedford MR, Vasanthan T and Zijlstra RT 2015. Enzymes enhance degradation of the fiber–starch–protein matrix of distillers dried grains with solubles as revealed by a porcine in vitro fermentation model and microscopy. Journal of Animal Science 93, 1039–1051.

64. Jensen BB and Jorgensen H 1994. Effect of dietary fiber on microbial activity and microbial gas production in various regions of the gastrointestinal tract of pigs. Applied and Environmental Microbiology 60, 1897–1904. [PMC free article]

65. Johansen HN, Bach Knudsen KE, Wood PJ and Fulcher RG 1997. Physico-chemical properties and the degradation of oat bran polysaccharides in the gut of pigs. Journal of the Science of Food and Agriculture 73, 81–92.

66. Jorgensen H, Zaho X and Eggum B 1996. The influence of dietary fibre and environmental temperature on the development of the gastrointestinal tract, digestibility, degree of fermentation in the hind-gut and energy metabolism in pigs. British Journal of Nutrition 75, 365–378.

67. Jorgensen H, Larsen T, Zhao XQ and Eggum BO 1997. The energy value of short-chain fatty acids infused into the caecum of pigs. British Journal of Nutrition 77, 745–756.

68. Just A, Fernández J and Jørgensen H 1983. The net energy value of diets for growth in pigs in relation to the fermentative processes in the digestive tract and the site of absorption of the nutrients. Livestock Production Science 10, 171–186.

69. Karr-Lilienthal LK, Kadzere CT, Grieshop CM and Fahey GC Jr 2005. Chemical and nutritional properties of soybean carbohydrates as related to nonruminants: a review. Livestock Production Science 97, 1–12.

70. Kim JC, Simmins PH, Mullan BP and Pluske JR 2005. The digestible energy value of wheat for pigs, with special reference to the post-weaned animal [Review]. Animal Feed Science and Technology 122, 257–287.

71. Konstantinov SR, Awati A, Smidt H, Williams BA, Akkermans ADL and de Vos WM 2004. Specific response of a novel and abundant Lactobacillus amylovorus-like phylotype to dietary prebiotics in the guts of weaning piglets. Applied and Environmental Microbiology 70, 3821–3830. [PMC free article]

72. Lambo AM, Oste R and MEGL Nyman 2005. Dietary fibre in fermented oat and barley beta-glucan rich concentrates. Food Chemistry 89, 283–293.

73. Le Goff G, van Milgen J and Noblet J 2002. Influence of dietary fibre on digestive utilization and rate of passage in growing pigs, finishing pigs and adult sows. Animal Science 74, 503–515.

74. Le Goff G, Noblet J and Cherbut C 2003. Intrinsic ability of the faecal microbial flora to ferment dietary fibre at different growth stages of pigs. Livestock Production Science 81, 75–87.

75. Lee CJ, Horsley RD, Manthey FA and Schwarz PB 1997. Comparisons of beta-glucan content of barley and oat. Cereal Chemistry 74, 571–575.

76. Leterme P, Froidmont E, Rossi F and Thewis A 1998. The high water-holding capacity of pea inner fibers affects the ileal flow of endogenous amino acids in pigs. Journal of Agricultural and Food Chemistry 46, 1927–1934.

77. Leterme P, Théwis A, van Leeuwen P, Monmart T and Huisman J 1996. Chemical composition of pea fibre isolates and their effect on the endogenous amino acid flow at the ileum of the pig. Journal of the Science of Food and Agriculture 72, 127–134.

78. Lindberg JE 2014. Fiber effects in nutrition and gut health in pigs. Journal of Animal Science and Biotechnology 2014, 15–22. [PMC free article]

79. Lizardo R, Peiniau J and Aumaitre A 1997. Inclusion of sugar beet pulp and change of protein source in the diet of the weaned piglet and their effects on digestive performance and enzymatic activities. Animal Feed Science Technology 66, 1–14.

80. Lunn J and Buttriss JL 2007. Carbohydrates and dietary fibre. Nutrition Bulletin 32, 21–64.

81. Louis P, Scott KP, Duncan SH and Flint HJ 2007. Understanding the effects of diet on bacterial metabolism in the large intestine. Journal of Applied Microbiology 102, 1197–1208.

82. Macfarlane GT and Macfarlane S 1993. Factors affecting fermentation reactions in the large bowel.Proceedings of the Nutrition Society 52, 367–373.

83. Macfarlane GT and Gibson GD 1995. Microbiological aspects of the production of short-chain fatty acids in the large bowel In Physiological and clinical aspects of short-chain fatty acids (ed. JH Cumming, JL Rombeau and T Sakota), pp. 87–105. Cambridge University Press, Cambridge, UK.

84. Mariscal-Landin G, Seve B, Colleaux Y and Lebreton Y 1995. Endogenous

amino nitrogen collected from pigs with end-to-end ileorectal anastomosis is affected by the method of estimation and altered by dietary fiber. Journal of Nutrition 125, 136–146.

85. Martensson O, Biorklund M, Lambo AM, Duenas-Chasco M, Irastorza A, Holst O, Norin E, Welling G, Oste R and Onning G 2005. Fermented, ropy, oat-based products reduce cholesterol levels and stimulate the bifidobacteria flora in humans. Nutrition Research 25, 429–442.

86. Mateos GG, Martin F, Latorre MA, Vicente B and Lázaro R 2006. Inclusion of oat hulls in diets for young pigs based on cooked maize or cooked rice. Animal Science 82, 57–63.

87. Molist F, van Oostruma M, Pérez JF, Mateos GG, Nyachoti CM and van der Aar PJ 2014. Relevance of functional properties of dietary fibre in diets for weanling pigs. Animal Feed Science and Technology 189, 1–10.

88. Montagne L, Pluske JR and Hampson DJ 2003. A review of interactions between dietary fibre and the intestinal mucosa, and their consequences on digestive health in young non-ruminant animals. Animal Feed Science and Technology 108, 95–117.

89. Nielsen TS, Lærke HN, Theil PK, Sørensen JF, Saarinen M, Forssten S and Bach Knudsen KE 2014. Diets high in resistant starch and arabinoxylan modulate digestion processes and SCFA pool size in the large intestine and faecal microbial composition in pigs. British Journal of Nutrition 112, 1837–1849.

90. Noblet J 2007. Recent developments in net energy research for swine. Advances in Pork Production. In Proceedings of the 2007 Banff Pork Seminar, January 16–19, University of Alberta, Edmonton, AB, Canada, pp. 149–156.

91. Noblet J and Perez JM 1993. Prediction of digestibility of nutrients and energy values of pig diets from chemical analysis. Journal of Animal Science 71, 3389–3398.

92. Oakenfull D 2001. Dietary fiber in human nutrition In Physical chemistry of dietary fiber (ed. GA Spiller), pp. 33–47. CRC Press, Boca Raton, FL, USA.

93. Owusu-Asiedu A, Patience JF, Laarveld B, Van Kessel AG, Simmins PH and Zijlstra RT 2006. Effects of guar gum and cellulose on digesta passage rate, ileal microbial populations, energy and protein digestibility, and performance of grower pigs. Journal of Animal Science 84, 843–852.

94. Pedersen C, Boersma MG and Stein HH 2007. Digestibility of energy and phosphorus in 10 samples of distillers dried grains with solubles fed to growing pigs. Journal Animal Science 85, 1168–1176.

95. Pedersen MB, Dalsgaard S, Bach Knudsen KE, Yu S and Lærke HN 2014. Compositional profile and variation of distillers dried grains with solubles from various origins with focus on non-starch polysaccharides. Animal Feed Science and Technology 197, 130–141.

96. Pedersen MB, Dalsgaard S, Arenta S, Lorentsena R, Bach Knudsen KE, Yu S and Lærke HN 2015.Xylanase and protease increase solubilization of non-starch

polysaccharides and nutrient release of corn- and wheat distillers dried grains with solubles. Biochemical Engineering Journal 98, 99–106.

97. Pieper R, Jha R, Rossnagel B, Van Kessel AG, Souffrant WB and Leterme P 2008. Effect of barley and oat cultivars with different carbohydrate compositions on the intestinal bacterial communities in weaned piglets.FEMS Microbiology Ecology 66, 556–566.

98. Pluske JR, Black B, Pethick DW, Mullan BP and Hampson DJ 2003. Effects of different sources and levels of dietary fibre in diets on performance, digesta characteristics and antibiotic treatment of pigs after weaning.Animal Feed Science and Technology 107, 129–142.

99. Queiroz-Monici KdS, Costa GEA, da Silva N, Reis SMPM and de Oliveira AC 2005. Bifidogenic effect of dietary fiber and resistant starch from leguminous on the intestinal microbiota of rats. Nutrition 21, 602–608.

100. Roberfroid M 2007. Prebiotics: the concept revisited. The Journal of Nutrition 137, 830–837.

101. Salunkhe DK, Chavan JK, Adsule RN and Kadam SS 1992. Rapeseeds. In World oilseeds: chemistry, technology, and utilization (ed. DK Salunkhe, JK Chavan, RN Adsule and SS Kadam), pp. 59–96. Van Nostrand Reinhold, New York, NY, USA.

102. Salvador V, Cherbut C, Barry JL, Bertrand D, Bonnet C and Delort-Laval J 1993. Sugar composition of dietary fibre and short-chain fatty acid production during in vitro fermentation by human bacteria. British Journal of Nutrition 70, 189–197.

103. Schiavon S, Bortolozzo A, Bailoni L and Tagliapetra F 2004. Effects of sugar beet pulp on growth and health status of weaned piglets. Italian Journal Animal Science 4, 337–351.

104. Schmidt-Willig U, Enss M, Coenen M, Gartner K and Hedrich H 1996. Response of rat colonic mucosa to a high fiber diet. Annals of Nutrition and Metabolism 40, 343–350.

105. Schneeman BO, Diane Richter B and Jacobs LR 1982. Response to dietary wheat bran in the exocrine pancreas and intestine of rats. The Journal of Nutrition 112, 283–286.

106. Schulze H, van Leeuwen P, Verstegen MWA and Van den Berg JW 1995. Dietary level and source of neutral detergent fiber and ileal endogenous nitrogen flow in pigs. Journal of Animal Science 73, 441–448.

107. Schulze H, van Leeuwen P, Verstegen MW, Huisman J, Souffrant WB and Ahrens F 1994. Effect of level of dietary neutral detergent fiber on ileal apparent digestibility and ileal nitrogen losses in pigs. Journal of Animal Science 72, 2362–2368.

108. Serena A and Knudsen KEB 2007. Chemical and physicochemical characterisation of co-products from the vegetable food and agro industries. Animal Feed Science and Technology 139, 109–124.

109. Solá-Oriol D, van Kempen T and Torrallardona D 2010. Relationships between

glycaemic index and digesta passage of cereal-based diets in pigs. Livestock Science 134, 41–43.

110. Taylor JRN and Emmambux MN 2010. Developments in our understanding of sorghum polysaccharides and their health benefits. Cereal Chemical 87, 263–271.
111. Theander O, Westerlund E, Åman P and Graham H 1989. Plant cell walls and monogastric diets. Animal Feed Science and Technology 23, 205–225.
112. Thomsen LE, Knudsen KEB, Jensen TK, Christensen AS, Møller K and Roepstorff A 2007. The effect of fermentable carbohydrates on experimental swine dysentery and whip worm infections in pigs. Veterinary Microbiology 119, 152–163.
113. Topping DL and Clifton PM 2001. Short-chain fatty acids and human colonic function: roles of resistant starch and non-starch polysaccharides. Physiological Reviews 81, 1031–1064.
114. Varel VH and Yen JT 1997. Microbial perspective on fiber utilization by swine. Journal of Animal Science75, 2715–2722.
115. Varel VH, Pond WG, Pekas JC and Yen JT 1982. Influence of high-fiber diet on bacterial populations in gastrointestinal tracts of obese- and lean-genotype pigs. Applied and Environmental Microbiology 44, 107–112. [PMC free article]
116. Verstegen MWA and Williams BA 2002. Alternatives to the use of antibiotics as growth promoters for monogastric animals. Animal Biotechnology 13, 113–127.
117. Wellock IJ, Houdijk JGM and Kyriazakis I 2007. Effect of dietary non-starch polysaccharide solubility and inclusion level on gut health and the risk of post weaning enteric disorders in newly weaned piglets.Livestock Science 108, 186–189.
118. Widmer MR, McGinnis LM and Stein HH 2007. Energy, phosphorus, and amino acid digestibility of high-protein distillers dried grains and corn germ fed to growing pigs. Journal Animal Science 85, 2994–3003.
119. Wilfart A, Montagne L, Simmins H, Noblet J and van Milgen J 2007. a. Effect of fibre content in the diet on the mean retention time in different segments of the digestive tract in growing pigs. Livestock Science 109, 27–29.
120. Wilfart A, Montagne L, Simmins PH, van Milgen J and Noblet J 2007. b. Sites of nutrient digestion in growing pigs: effect of dietary fiber. Journal Animal Science 85, 976–983.
121. Williams BA, Verstegen MWA and Tamminga S 2001. Fermentation in the large intestine of single-stomached animals and its relationship to animal health. Nutrition Research Reviews 14, 207–227.
122. Williams BA, Bosch MW, Awati A, Konstantinov SR, Smidt H, Akkermans ADL, Verstegen MWA and Tamminga S 2005. In vitro assessment of gastrointestinal tract (GIT) fermentation in pigs: Fermentable substrates and microbial activity. Animal Research 54, 191–201.

123. Wong JMW, de Souza R, Kendall CWC, Emam A and Jenkins DJA 2006. Colonic health: fermentation and short chain fatty acids. Journal of Clinical Gastroenterology 40, 235–243.

124. Yen JT 2001. Anatomy of the digestive system and nutritional physiology In Swine nutrition (ed. AJ Lewis and LL Southern), pp. 31–63. CRC Press, Boca Raton, FL, USA.

5
CHAPTER

SIMULTANEOUS INCLUSION OF SORGHUM AND COTTONSEED MEAL OR MILLET IN BROILER DIETS: EFFECTS ON PERFORMANCE AND NUTRIENT DIGESTIBILITY

D. I. Batonon-Alavo, D. Bastianelli , P. Lescoat , G. M. Weber and M. Umar Faruk

Keywords: Broiler, sorghum, millet, cottonseed meal, digestibility

IMPLICATIONS

Sorghum, millet and cottonseed meal can be used in poultry nutrition to replace corn and soybean meal. Results showed that the overall performance is not affected when high quality ingredients (low tannin, low gossypol) are used. Some effects on feed intake and feed efficiency were identified and discussed. The present study provides additional knowledge to the use of sorghum, millet and cottonseed meal in commercial poultry production.

INTRODUCTION

The worldwide competitiveness of the poultry industry requires finding alternative to major feed ingredients used in broiler feeding to reduce the cost of feeding. Sorghum (*Sorghum bicolor*) and millet (*Pennisetum glaucum,Panicum miliaceum*) have been widely investigated in broiler nutrition as substitutes to corn. Cottonseed meal (*Gossypium* spp.) has also been examined as an alternative to soybean meal. Results indicated that similar (Davis *et al.*, 2003; Hidalgo *et al.*, 2004) or improved performance (Baurhoo*et al.*, 2011) was obtained with millet-based diets compared with corn-based diets. Whereas for sorghum and cottonseed meal, a reduction in feed efficiency (Jacobs

Citation: D. I. Batonon-Alavo, D. Bastianelli, P. Lescoat, G. M. Weber and M. Umar Faruk Simultaneous inclusion of sorghum and cottonseed meal or millet in broiler diets: effects on performance and nutrient digestibility doi:10.1017/S1751731116000033.

and Parsons, 2013) or equivalent performance as the control diet were obtained (Jacob *et al.*, 1996; Azman and Yilmaz, 2005).

In a meta-analysis on the use of these feed ingredients (Batonon-Alavo *et al.*,2015), we observed that millet-based diets gave similar performance as the corn-based diets while cottonseed meal-based diets increased feed intake. Average daily gain was reduced with both sorghum- and cottonseed meal-based diets. However, there was no significant dose–response effect with millet and cottonseed meal. Growth performance decreased with the increasing level of substitution of corn with sorghum.

This reduction in growth performance obtained in sorghum and cottonseed meal-based diets may be related to their content of anti-nutritional factors. According to several authors, tannins and phytate in sorghum form complexes with protein and carbohydrates, particularly with starch, thereby reducing nitrogen and starch digestibility (Selle *et al.*, 2010; Mahmood *et al.*,2014). Birds fed high tannin diets suffered from a severe decrease of growth compared with low tannin or control fed birds (Flores *et al.*, 1994; Mahmood*et al.*, 2006). In addition, Kafirin has been examined in sorghum by Selle *et al.* (2010). Its presence in sorghum is associated with a poor amino acid digestibility. In cottonseed meal, free gossypol decreases lysine digestibility and inhibits the activity of pepsin and trypsin, thus reducing growth in broilers (Nagalakshmi *et al.*, 2007).

For more flexibility in the type of ingredients used in the least cost feed formulation, it could be hypothesized that simultaneous utilization of more than one of these alternative feed ingredients in poultry diets will become commonplace. Consequently, the question arises whether simultaneous inclusion of sorghum and cottonseed meal in broiler feeding would result in additive effect or a synergy of the anti-nutritional factors. Besides, the protein content of millet is higher than corn and the essential amino acid profile is more balanced in millet than in corn (Heuzé and Tran, 2012). The high essential amino acid concentrations, the high digestibility of these amino acids (Adeola and Orban, 1995; Yin *et al.*, 2002) and the changes in the small intestine mucosa morphology were reported to be the factors leading to a better feed efficiency with millet (Baurhoo *et al.*, 2011; Goodarzi Boroojeni *et al.*, 2011). It could therefore be assumed that bird's nutrient digestibility increases when millet is used in sorghum-based diets.

This work was, therefore, designed to evaluate the effects of partial or total substitution of corn with sorghum and millet, and partial replacement of soybean meal with cottonseed meal on performance and nutrient digestibility in broiler. It is the aim of this study to investigate the interactions that might be induced on broiler performance when sorghum and cottonseed meal are simultaneously included in the diet or when sorghum and millet totally replaced corn in the diet.

MATERIAL AND METHODS

General

Two trials were conducted: the first experiment combined sorghum and cottonseed meal and the second one used sorghum and millet. Nutritional composition of these feed ingredients are described in Table 1. The same batch of sorghum (S. bicolor) originating from France was used in the two experiments. It was a sorghum with red pericarp, but without testa layer, and therefore had low condensed tannin content (1.80 g/kg catechin equivalents) and low total phenolics (3.5 g/kg tannic acid equivalent).

Table 1: Analyzed composition of sorghum (S), cottonseed meal (CM) and millet (M)

	Sorghum	Cottonseed meal	Millet
Analyzed composition (g/kg, as fed)			
DM	864	920	869
CP	82.8	402	116
Crude fat	35.2	25.0	45.2
Starch	631	8.70	582
Total sugars	3.80	68.6	1.20
NDF	138.6	206.6	151.9
ADF	50.8	119.1	84.2
ADL	17.3	29.9	20.0
Condensed tannins	1.80	nd	nd
Free gossypol	nd	1.03	nd
ME (MJ/kg)[1]	13.33	7.69	13.19

nd = not determined; ME = metabolizable energy; DM = dry matter; EE = ether extract; CF = crude fibre; NFE = nitrogen free extract.

[1]ME content was estimated using European energy calculation (Janssen, 1989). S (tannin < 4 g) MEn = 31.02 × CP + 77.03 × EE + 37.67 × NFE; CM expeller or solvent, MEn = 21.36 × DM + 47.13 × EE − 30.55 × CF; M (corn equation), MEn = 36.21 × CP + 85.44 × EE + 37.26 × NFE; with DM, EE, CF and NFE = DM − (EE + CP + Ash + CF).

Millet (P. miliaceum) originated from France. Prepressed solvent extracted cottonseed meal (Gossypium spp., bought in France) having 402 g/kg CP and 1.03 g/kg free gossypol content was used in Experiment 1. Birds were reared in wire-floored battery cages (83 cm wide × 52 cm height × 75 cm depth) in an environmentally controlled house. The environmental temperature was maintained at 32°C during the 1st week and decreased to 23°C until the end of the experiment according to Ross guidelines (Aviagen, 2007). Experimental procedures and animal care were carried out according to French legislation at the time of the study (2013) and were approved by the local ethical committee.

Experiment 1: Simultaneous Use of Sorghum and Cottonseed Meal (Experiment 1)

Four hundred and eighty six-day-old male Ross PM3 broilers were fed ad libitum from 1 to 8 days of age with a starter diet containing 12.6 MJ/kg of metabolizable energy (ME) and 225 g/kg of CP. On day 8, birds were randomly selected and divided into nine groups. Each group contained nine replicates of six birds. Birds were assigned to replicates on the basis of BW such that homogenous BW were obtained between each replicate and treatment. The experimental period ran from 8 to 28 days of birds age. Birds were given ad libitum access to feed and water throughout the experiment. Nine different treatments were offered in a pelleted (steam pelleting with moisture addition) form (Table 2). The control treatment (C1) was based on corn and soybean meal as the main ingredients and contained 12.3 MJ/kg ME and 231 g/kg CP. The experiment was designed with dietary level of substitution and type of ingredient as the main effects. Three experimental diets were formulated by substituting 30%, 45% or 60% of the corn by sorghum (S) and two other diets were similarly formulated by replacing 15% or 40% of the soybean meal with cottonseed meal (CM).

Table 2: Dietary composition and nutrient content of experimental diets given from 8 to 28 days in Experiment 1

	Control	Sorghum-based diets			Cottonseed meal-based diets		Sorghum and cottonseed meal-based diets		
	C1	S30	S45	S60	CM15	CM40	S30/CM40	S45/CM25	S60/CM15
Ingredient (g/kg as fed)									
Wheat	60.0	60.0	73.1	80.0	60.0	60.0	54.6	71.5	99.2
Corn	513.2	385.6	304.4	230.0	513.2	514.1	385.6	314.4	222.0
S	–	165.0	243.0	320.2	–	–	165.0	230.0	290.0
Soybean meal (48%)	365.1	346.0	340.0	332.0	307.0	207.9	206.0	241.0	275.0
CM	–	–	–	–	55.0	146.0	130.0	91.2	58.0
Soybean oil	25.0	7.0	2.0	–	27.0	30.0	15.0	10.0	15.0
Limestone	4.0	4.0	4.0	4.0	4.0	3.2	4.0	4.0	4.0
Dicalcium phosphate	23.0	23.0	23.0	23.0	22.1	23.5	24.5	24.0	24.1
Salt	3.0	2.5	3.0	2.9	3.0	2.1	2.3	2.5	2.5
Trace mineral premix[1]	4.1	4.1	4.6	4.6	4.1	4.1	4.1	4.1	4.1
D,L-Methionine	1.3	1.3	1.3	1.5	1.6	2.0	1.9	1.7	1.6
L-Lysine HCl	1.1	1.3	1.4	1.6	2.2	5.0	5.0	4.1	3.3
L-Threonine	0.2	0.2	0.2	0.2	0.8	2.1	2.0	1.5	1.2
Calculated composition (g/kg unless specified)[2]									
CP	215	216	217	218	216	218	218	218	219
Crude fat	51.8	34.7	29.6	27.5	54.4	58.4	44.0	38.4	42.2
Ca	9.4	9.3	9.4	9.4	9.1	9.0	9.5	9.4	9.5
Av. P	4.8	4.6	4.6	4.5	4.5	4.5	4.5	4.5	4.6
Dig Lys	11.3	11.1	11.1	11.1	11.0	11.1	11.0	11.1	11.2
Dig Meth + Cyst	7.5	7.4	7.4	7.5	7.4	6.9	6.8	6.9	7.1
Dig Thr	7.5	7.4	7.4	7.3	7.4	7.4	7.4	7.4	7.5
Dig Trp	2.2	2.2	2.2	2.2	1.9	1.5	1.6	1.8	2.0
Dig Arg	13.4	13.0	12.9	12.7	12.2	10.3	10.1	10.8	11.5
Free gossypol (mg/kg)	–	–	–	–	5.0[3]	12.0[3]	13.0	9.0	6.0
ME (MJ/kg)[4]	12.3	12.0	11.9	11.9	12.5	12.3	11.9	11.9	12.1
Analyzed composition (g/kg)									
CP	231	224	222	227	226	217	219	215	215
Crude fat	53.4	42.3	39.0	36.6	61.6	66.6	50.0	45.9	52.0
Condensed tannins	0.47	0.60	0.63	0.77	0.66	0.81	0.89	0.78	0.68
NDF	120	118	116	120	130	131	119	126	136
ADF	33.7	37.4	39.5	41.0	39.6	49.0	44.6	46.4	48.6
ADL	5.5	6.3	7.5	7.9	8.0	5.4	1.9	2.7	13.9

C1 = control diet; S = sorghum; CM = cottonseed meal; ME = metabolizable energy.
The number represents the level of substitution of corn for S-based diets or soybean meal for CM-based diets.
[1]Premix composition (per kg): vitamin A, 1 100 000 UI; vitamin D_3, 300 000 UI; vitamin E, 4000 UI; vitamin B_1, 250 mg; vitamin B_2, 800 mg; vitamin B_6, 500 mg; vitamin B_{12}, 2.5 mg; vitamin PP, 5000 mg; vitamin C, 10 000 mg; vitamin K_3, 300 mg; Zn, 5400 mg; Cu, 3000 mg; Fe, 6000 mg; I, 124 mg; Se, 29.7 mg; Mn, 8000 mg; Ca, 0.17 mg; Co, 60 mg; Mg, 0.08 g; folic acid, 150 mg; choline chloride, 50 004 mg; biotin, 15 mcg. Premix also supplied titanium 1.0 g/kg of diet and Avatec, 0.6 g/kg of diet.
[2]Nutritional values were calculated, based on Sauvant *et al.* (2004).
[3]Analyzed composition.
[4]ME content estimated using European energy equation (Fisher and McNab, 1987).

To test the effect of simultaneous inclusion of both S and CM, three other diets combined the level of substitution of sorghum and cottonseed meal as follows: 30% S and 40% CM, 45% S and 25% CM; 60% S and 15% CM. Diets were formulated to meet Ross broiler nutrient specifications (Aviagen, 2007) and included 1 g/kg titanium dioxide (TiO2) as digestibility marker. All diets were formulated to be isocaloric and isonitrogenous by adjusting soybean oil, wheat and amino acids supply to avoid any essential amino acids as limiting factor.

Experiment 2: Total Corn Substitution by Millet and Sorghum in Broiler Diets (Experiment 2)

Four hundred and eighty six-day-old male Ross PM3 broilers were fed ad libitum with the same starter feed as in Experiment 1 from 1 to 8 days. On day 8, nine replicates (of six chicks per cage) of similar BW were constituted per treatment. The experimental period ran from 8 to 27 days of age. Birds were given one of the 10 treatments: a control (C2) complete feed and nine experimental diets designed as a factorial arrangement (3×3) with dietary level of substitution and type of ingredient as the main effects (Table 3). All diets were fed as pellets. The aim of this experiment was to establish whether it would be possible to substitute corn with sorghum beyond the levels of substitution used in Experiment 1. Therefore, three other experimental diets were formulated by replacing 60%, 80% or 100% of the corn with sorghum (S60, S80 and S100). Three millet-based diets (M60, M80 and M100) were formulated in the same way.

Table 3: Dietary composition and nutrient content of experimental diets given from 8 to 27 days in Experiment 2

	Control	Sorghum-based diets			Millet-based diets			Sorghum and millet-based diets		
	C2	S60	S80	S100	M60	M80	M100	S30/M30	S40/M40	S50/M50
Ingredient (g/kg as fed)										
Wheat	61.8	50.0	60.4	71.0	70.1	75.9	105.2	68.8	66.2	106.7
Corn	500.7	213.6	120.0	–	220.0	120.0	–	220.0	129.2	–
S	–	310.0	400.7	514.0	–	–	–	150.0	200.5	250.2
Soybean meal (48%)	375.0	357.9	354.4	349.5	334.0	318.0	300.0	347.7	338.3	325.9
M	–	–	–	–	310.7	420.1	528.2	150.0	200.5	250.2
Soybean oil	28.0	30.0	30.0	30.0	30.0	30.0	30.0	29.0	30.0	30.8
Limestone	3.7	3.3	3.1	3.5	3.7	5.2	5.4	3.9	4.0	4.2
Dicalcium phosphate	21.3	22.7	22.9	23.2	22.2	20.8	20.5	21.7	22.0	22.0
Salt	3.3	2.7	2.5	2.5	3.0	2.9	2.8	2.8	2.9	2.8
Trace mineral premix[1]	4.6	4.1	4.1	4.1	4.1	4.1	4.1	4.1	4.1	4.1
D,L-Methionine	1.2	4.0	1.1	1.2	1.1	1.2	1.3	1.2	1.2	1.3
L-Lysine HCl	0.4	0.7	0.8	1.0	1.1	1.6	2.1	0.8	1.1	1.6
L-Threonine	–	1.0	–	–	–	0.2	0.4	–	–	0.2
Calculated composition (g/kg unless specified)[2]										
CP	219	219	217	218	218	218	218	218	218	218
Crude fat	63.2	61.3	59.7	57.6	64.1	63.6	62.0	61.5	61.8	59.8
Ca	9.0	9.0	9.0	9.1	9.0	9.2	9.1	9.0	9.1	9.1
Av. P	4.5	4.5	4.5	4.50	4.70	4.5	4.5	4.50	4.50	4.5
Dig Lys	11.5	11.1	11.0	11.0	11.1	11.1	11.0	11.0	11.0	11.1
Dig Meth + Cyst	7.9	10.0	7.3	7.3	7.3	7.1	7.0	7.5	7.4	7.3
Dig Thr	7.9	8.5	7.5	7.4	7.4	7.4	7.4	7.5	7.4	7.4
Dig Trp	2.4	2.3	2.4	2.4	2.5	2.6	2.6	2.4	2.5	2.5
Dig Arg	14.2	13.1	12.6	12.2	13.5	13.4	13.2	13.1	13.3	12.7
ME (MJ/kg)[3]	12.5	12.6	12.5	12.6	12.8	12.8	12.8	13.0	12.5	12.8
Analyzed composition (g/kg)										
CP	224	226	219	224	226	221	220	224	223	218
Crude fat	57.7	60.7	60.1	61.2	66.4	70.4	72.7	72.8	62.9	74.5
Condensed tannins	–	0.60	0.54	0.78	–	–	–	0.31	0.40	0.43
NDF	76.8	90.0	81.7	81.4	91.3	101.0	111.8	91.9	95.3	95.2
ADF	30.2	32.1	32.1	33.7	38.3	44.6	50.0	36.7	39.5	39.8
NDL	4.7	5.1	5.6	6.1	4.8	5.8	7.7	5.4	5.1	3.4

C2 = control diet; S = sorghum; M = millet; ME = metabolizable energy.
The number represents the level of substitution of corn.
[1]Premix composition (per kg): vitamin A, 1 100 000 UI; vitamin D_3, 300 000 UI; vitamin E, 4000 UI; vitamin B_1, 250 mg; vitamin B_2, 800 mg; vitamin B_6, 500 mg; vitamin B_{12}, 2.5 mg; vitamin PP, 5000 mg; vitamin C, 10 000 mg; vitamin K_3, 300 mg; Zn, 5400 mg; Cu, 3000 mg; Fe, 6000 mg; I, 124 mg; Se, 29.7 mg; Mn, 8000 mg; Ca, 0.17 mg; Co, 60 mg; Mg, 0.08 g; folic acid, 150 mg; choline chloride, 50 004 mg; biotin, 15 mcg. Premix also supplied titanium 1.0 g/kg of diet and Avatec, 0.6 g/kg of diet.
[2]Nutritional values were calculated, based on Sauvant *et al.* (2004).
[3]ME content estimated using European energy equation (Fisher and McNab, 1987).

In order to assess the interactions that might exist when these ingredients are used simultaneously, three other diets were formulated by substituting different proportions of the corn with both S and M (S30/M30, S40/M40, S50/M50). Birds were given ad libitum access to feed and water throughout the experiment. All diets were formulated to be isocaloric (12.69 ± 0.05 MJ/kg) and isonitrogenous (222.5 ± 0.9 g/kg) and formulated to meet Ross broiler nutrient specifications (Aviagen, 2007) as in Experiment1. TiO_2 was included at 1 g/kg in all diets as digestibility marker.

Measurements

Performance (Experiments 1 and 2)

Chicken BW was measured at days 8, 15, 21 and 28 in Experiment 1. In Experiment 2, birds were weighed at days 8, 14 and 27. Feed intake was measured every week by weighing feed refusals in all experiments. Due to practical reasons, feed intake and BW were not measured at day 20 in Experiment 2. Feed conversion ratio (FCR) was calculated as the ratio between total feed intake and BW gain of all birds in each cage. Feed intake, BW gain and FCR were calculated after correcting for mortalities.

Apparent Ileal Digestibility (AID) (Experiments 1 and 2)

At the end of each experimental period, all birds were slaughtered by cervical dislocation. The contents of the terminal part of the ileum, defined as the region between 17 and 2 cm proximal to the ileocaecal junction were collected for ileal digestibility determinations. Digesta of birds were pooled per cage, freeze-dried and ground before analysis. All digesta samples were analyzed for dry matter (DM), nitrogen and concentration of titanium as indigestible marker. The AID coefficient of nutrients in the two experiments was calculated according to the following equation:

$$\text{AID}\ (\%) = \left\{1 - \left(\frac{Ti_{Diet}}{Ti_{ileum}} \times \frac{Nutrient_{ileum}}{Nutrient_{Diet}}\right)\right\} \times 100$$

where Ti_{Diet} and Ti_{ileum} represent concentration of titanium in feed and ileal digesta, respectively. $Nutrient_{Diet}$ and $Nutrien_{tileum}$ are the nutrients (protein and energy) in feed and ileal digesta.

Total Tract Digestibility (Experiment 2)

In the second experiment, plastic trays were placed under cages for total excreta collection. Representative samples of excreta were then collected from six cages per treatment over a period of 3 × 24 h from day 22 to 24. All excreta samples were freeze-dried and then ground to pass through 0.5 mm sieve. Samples of diets and excreta were analyzed for DM, gross energy (GE), starch, fat, total N and protein N. Apparent metabolizable energy (AME) was calculated as the difference between energy intake and excretion, relative to feed intake. AME was then corrected for the energy of N retention (AMEn) according to Hill and Anderson (1958).

Chemical Analysis

Chemical composition of feeds and faeces was determined following standard procedures (AOAC International, 1995). DM was determined with oven drying at 103°C. Crude fat contents were determined after a saponification with determination by gas chromatography (Büchi, B820). GE was determined using a bomb calorimeter (C2000 Basic; IKA, Germany). Total nitrogen analysis was carried out with a Leco N analyzer (LECO, FP528) and CP was then calculated as N×6.25. Starch was measured using a polarimeter (ADP 410; Bellingham and Stanley, UK). Total sugars were determined according to Luff-Schoorl method. NDF, ADF and ADL in feed ingredients and diets were measured on Fibertech according to Van Soest and Wine (1967). Titanium concentrations were determined by Induction Coupled Plasma after HNO_3/NH_4F mineralization in microwaves. Total tract protein nitrogen (Terpstra and de Hart, 1974) and total tract starch were determined by Near Infrared Spectroscopy (NIRS) according to Bastianelli et al. (2010). Condensed tannins were determined by vanillin assay and expressed as catechin equivalent. Gossypol contents were determined by colorimetry following the standard ISO6866.

Statistical Analysis

Data analyses were performed using R version 3.0.2 (R Core Team, 2013). The mortality recorded for all treatments in each experiment was submitted to a one-way ANOVA. In Experiment 1, data collected was analyzed based on three periods: (1) 8 to 15 days; (2) 15 to 21 days (3) 21 to 27 days. In Experiment 2, analyses were performed based on two periods: (1) 8 to 14 days and (2) 14 to 27 days. A one-way ANOVA was performed in each period of age to test the treatment effect (nine levels in Experiment 1; 10 levels in Experiment 2) on feed intake, weight gain and FCR. Data were also submitted to a one-way ANOVA to evaluate the effect of the type of ingredient irrespective of its level: C1 v. S v. CM v. S/CM in Experiment 1 and C2 v. S v. M v. S/M in Experiment 2. To assess the effect of the level of substitution, treatments of each ingredient were submitted to a Kruskal–Wallis test at each period of age. These analyses were also realized on the data collected for AID and total tract digestibility. Tested factors were considered significant if P0.10. A Bonferroni–Dunnet pairwise comparison was used to compare differences between mean after each ANOVA test being performed. A multiple comparisons post hoc test (Siegel and Castellan, 1988) was realized to compare differences between means for the level of substitution effect in each ingredient-based diets.

RESULTS

Experiment 1: simultaneous use of sorghum and cottonseed meal Performance

The overall mortality from 8 to 28 days was 4.17 ± 0.91% and was not influenced by the treatments. Feed intake was not affected by treatments throughout the experimental period (Table 4). However, from 21 to 28 days and 8 to 28 days, it tended to be influenced by the type of ingredient with the lowest values in CM-based diets and the highest in S-based diets (P = 0.07). Feed intake of the control diet and S/CM-based diets were intermediate. There was no significant effect of the level of substitution of each ingredient on feed intake. Weight gain was similar among treatments and was not affected by the inclusion of sorghum or cottonseed meal. A high weight gain value (82.7 g/b per day) was recorded for S30 diet at days 15 to 21. This induced a trend for a level effect in sorghum at this period. CM-based diets resulted in a similar FCR to control diet (1.49 v. 1.50). Higher FCR were obtained with sorghum- (1.55) and S/ CM-based diets (1.54). The lowest FCR value was observed with CM15 (1.47) while the highest was obtained with S45/ CM25 (1.58). However, only a few significant differences of FCR were obtained in the first two periods regarding the dose– response effect of each ingredient. FCR increased with the level of substitution of S and CM. When the two ingredients were included in the same diet from 8 to 15 days, FCR was affected due to a low value in S60/CM15.

AID of Protein and Energy

Table 5 shows the effect of treatments on AID of protein and energy. AID of protein of all experimental treatments was lower to control diet (82.93±0.70%). This difference was not significant for diets based on sorghum (80.69±0.34%), but it was significant for CM-based diets (78.81±0.59%). AID of protein of S/CM-based diets was the lowest (76.71±0.55%), with values significantly lower from control and S-based diets. No effect of the level of substitution was found on AID of protein in S- and CM-based diets. However, the combination of these two ingredients affected AID of protein in relation to the low value in S30/CM40.

Table 4: Effect of diets with sorghum (S), cottonseed meal (CM) or their combination (S/CM) on the performance of broilers fed from 8 to 28 days old (Experiment 1)

	Feed intake (g/b per day)				Weight gain (g/b per day)				FCR			
Age (days)	8 to –15 days	15 to 21 days	21 to 28 days	8 to 28 days	8 to 15 days	15 to 21 days	21 to 28 days	8 to 28 days	8 to 15 days	15 to 21 days	21 to 28 days	8 to 28 days
C1	64.1	115	151	111	50.3	80.8	89.2	74.2	1.27^{a}	1.43^{a}	1.70	1.50^{ba}
S30	67.0	121	156	115	49.8	82.7	90.0	74.1	1.35^{ab}	1.46^{a}	1.73	1.55^{bc}
S45	65.1	118	161	115	48.7	78.2	95.1	74.0	1.33^{ab}	1.50^{a}	1.70	1.55^{b}
S60	66.8	124	160	117	49.8	79.1	94.3	74.4	1.34^{ab}	1.57^{b}	1.69	1.57^{b}
CM15	65.0	113	148	109	50.2	81.1	90.0	73.8	1.29^{ab}	1.40^{a}	1.65	1.47^{a}
CM40	65.2	119	153	112	49.5	79.6	94.6	74.6	1.32^{ab}	1.49^{ab}	1.62	1.51^{ba}
S30/CM40	67.3	114	157	112	49.2	78.8	92.5	73.5	1.37^{ab}	1.44^{a}	1.72	1.53^{b}
S45/CM25	67.4	115	159	114	48.6	79.6	89.9	72.6	1.38^{b}	1.51^{a}	1.78	1.58^{c}
S60/CM15	65.9	118	154	113	50.1	80.2	92.7	74.3	1.31^{ab}	1.48^{a}	1.67	1.52^{ba}
SEM	1.3	3	4	2	0.7	2.1	2.7	1.2	0.02	0.03	0.03	0.01
Treatment effect[1]	ns	ns	ns	ns	ns	ns	ns	ns	*	**	ns	***
Type of ingredient[2]	ns	ns	†	†	ns	ns	ns	ns	**	*	ns	***
Level of substitution[3]												
S	ns	ns	ns	ns	ns	†	ns	ns	ns	*	ns	ns
CM	ns	ns	ns	ns	ns	ns	ns	ns	*	**	ns	ns
S and CM	ns	ns	ns	ns	ns	ns	ns	ns	**	ns	ns	ns

C1 = control diet; FCR = feed conversion ratio.
The number represents the level of substitution of corn for S-based diets or soybean meal for CM-based diets.
[1]Effect of treatments: ANOVA on nine diets.
[2]Effect of the type of ingredient: ANOVA on four treatments (control *v.* S-based diets *v.* CM-based diets *v.* S/CM-based diets).
[3]Effect of the level of substitution (dose–response effect) within each type of ingredient.
[a,b,c]Values with the same superscript are not significantly different at $P < 0.05$.
***$P < 0.001$; **$P < 0.01$; *$P < 0.05$; †$P \leq 0.10$. ns: not significant at $P > 0.10$.

AID of energy was different among treatments with the highest values in control (78.11±0.66%) and the lowest in S/CM-based diets (72.05±0.52%). AID of energy obtained with S-based diets (76.12±0.31%) was not significantly different from the control whereas in CM-based diets (74.19±0.51%) it was lower than the control. The level of substitution of sorghum tended to affect AID of energy (P=0.09) while in CM-based diets, AID of energy decreased with the increasing level of substitution. Combination of sorghum and cottonseed meal affected AID of energy with the lowest values in S30/CM40 diet and the highest in S45/CM25 diet.

Table 5: Apparent ileal digestibility of protein and energy of broilers fed sorghum- (S) and/or cottonseed meal (CM) based diets measured at 28 days (Experiment 1)

	AID of protein (%)	AID of energy (%)
C1	82.9^{a}	78.1^{a}
S30	81.3^{a}	77.1^{ac}
S45	80.5^{ab}	76.2^{ac}
S60	80.3^{ab}	75.1^{bcd}
CM15	79.6^{ab}	75.7^{ab}
CM40	78.0^{bc}	72.7^{de}
S30/CM40	74.7^{c}	70.2^{e}
S45/CM25	78.0^{b}	73.3^{bd}
S60/CM15	77.4^{bc}	72.7^{de}
SEM	0.77	0.60
General treatment effect[1]	***	***
C *v.* S *v.* CM *v.* S/CM[2]	***	***
Level of substitution[3]		
S	ns	†
CM	ns	***
S and CM	*	*

C1 = control diet; AID = apparent ileal digestibility.
The number represents the level of substitution of corn for S-based diets or soybean meal for CM-based diets.
[1]Effect of treatments: ANOVA on nine diets.
[2]Effect of the type of ingredient: ANOVA on four treatments (control *v.* S-based diets *v.* CM-based diets *v.* S/CM-based diets).
[3]Effect of the level of substitution (dose–response effect) within each type of ingredient.
[a,b,c,d,e]Values with the same superscript are not significantly different at $P<0.05$.
***$P<0.001$; **$P<0.01$; *$P<0.05$; †$P\leqslant 0.10$. ns: not significant at $P>0.10$.

Experiment 2: Total Corn Substitution by Millet and Sorghum in Broiler Diets Performance

The overall mortality recorded in this experiment was 2.41±0.62% and was not affected by the treatments.

Simultaneous utilization of sorghum and millet resulted in similar feed intake and growth performance as the control diet (Table 6). As main effects, sorghum and millet did not have a significant impact on feed intake, weight gain and FCR. No difference of level of substitution of the feed ingredient was observed for M-based diets and S/M-based diets. However, there was a significant difference between the three S-based diets from 8 to 14 days of age with no significant post hoc effect.

Table 6: Effect of diets with sorghum (S), millet (M) or their combination (S/M) on the performance of broilers fed from 8 to 27 days old (Experiment 2)

	Feed intake (g/b per day)			Weight gain (g/b per day)			FCR		
	8 to 14 days	15 to 20 days	8 to 27 days	8 to 14 days	15 to 27 days	8 to 27 days	8 to 14 days	15 to 20 days	8 to 27 days
C2	62.7	134	109	47.3	84.6	71.6	1.33	1.58	1.52
S60	61.6	131	107	45.9	85.3	71.5	1.34	1.54	1.50
S80	61.9	132	108	45.8	85.8	71.8	1.35	1.54	1.50
S100	63.9	132	108	47.6	82.8	70.4	1.34	1.60	1.54
M60	63.6	130	107	46.4	85.9	72.1	1.37	1.51	1.48
M80	62.1	134	109	44.8	86.5	71.9	1.38	1.55	1.51
M100	62.4	132	108	46.4	85.0	71.5	1.35	1.55	1.51
S30/M30	59.8	128	104	45.3	81.7	68.9	1.32	1.56	1.51
S40/M40	63.1	132	108	45.0	84.6	71.2	1.42	1.55	1.52
S50/M50	63.3	133	109	48.2	84.2	71.6	1.34	1.58	1.52
SEM	1.5	2.0	1.5	1.3	1.6	1.2	0.02	0.02	0.02
Treatment effect[1]	ns	ns	ns	ns	ns	ns	ns	ns	ns
Type of ingredient[2]	ns	ns	ns	ns	ns	ns	ns	ns	ns
Level of substitution[3]									
S	ns	ns	ns	ns	ns	ns	*	ns	ns
M	ns	ns	ns	ns	ns	ns	ns	ns	ns
S and M	ns	ns	ns	ns	ns	ns	ns	ns	ns

C2 = control diet; FCR = feed conversion ratio.
The number represents the level of substitution of corn.
[1]Effect of treatments: ANOVA on 10 diets.
[2]Effect of the type of ingredient: ANOVA on four treatments (control *v.* S-based diets *v.* M-based diets *v.* S/M-based diets).
[3]Effect of the level of substitution (dose–response effect) within each type of ingredient.
*$P<0.05$; ns: not significant at $P>0.10$.

AID of Protein and Energy

The effects of the type of cereal on AID of protein and energy are shown in Table 7. AID of protein of all experimental treatments was higher than the control diet (80.40±1.44%). This difference was not significant for S- (82.17±0.44%) and S/M-based diets (82.15±0.60%). However, it was significant for millet-based diets (83.73±0.23%). AID of protein was affected with the increasing level of millet when it was used separately or combined with sorghum. However, no significant difference was found between these diets with the post hoc test. No dose–response effect of sorghum was found on AID of protein.

Similar results were obtained for the AID of energy with the highest values in M-, followed by S/M-, S-based and control diets in descending order. The level of substitution tended to affect AID of energy in S-based diets (P=0.09) while it significantly affected AID of energy in M-based diets and S/M-based diets.

Table 7: Apparent ileal digestibility of protein and energy of broilers fed sorghum- (S) and/or millet- (M) based diets measured at 27 days (Experiment 2)

	AID of protein (%)	AID of energy (%)
C2	80.4^{ac}	74.6^{a}
S60	82.8^{abc}	75.9^{ab}
S80	81.3^{abc}	74.6^{a}
S100	82.4^{abc}	76.3^{ab}
M60	84.3^{b}	78.4^{b}
M80	84.2^{b}	77.9^{b}
M100	82.6^{abc}	75.9^{ab}
S30/M30	83.2^{abc}	77.0^{ab}
S40/M40	83.7^{ab}	77.7^{b}
S50/M50	79.5^{c}	74.6^{a}
SEM	0.69	0.54
Treatment effect[1]	***	***
Type of ingredient[2]	**	**
Level of substitution[3]		
S	ns	†
M	**	**
S and M	**	*

C2 = control diet; AID = apparent ileal digestibility.
The number represents the level of substitution of corn.
[1]Effect of treatments: ANOVA on 10 diets.
[2]Effect of the type of ingredient: ANOVA on four treatments (control *v.* S-based diets *v.* M-based diets *v.* S/M-based diets).
[3]Effect of the level of substitution (dose–response effect) within each type of ingredient.
[a,b,c]Values with the same superscript are not significantly different at $P<0.05$.
***$P<0.001$; **$P<0.01$; *$P<0.05$; †$P\leqslant 0.10$. ns: not significant at $P>0.10$.

Total tract digestibility of DM, protein, fat and starch and AMEn values

Total tract digestibility of nutrients is shown in Table 8. There were significant differences in DM digestibility, although none of the experimental treatments was significantly different from the control diet (68.81±0.46%). Lower DM digestibility was observed with S-based diets (67.52±0.17%) compared with M-based diets (69.15±0.30%). S/M-based diets were intermediate (68.57±0.25%). In S and S/M, there was a significant reduction of DM digestibility when the level of substitution increased.

Table 8: Total tract nutrient digestibility and nitrogen-corrected apparent metabolizable energy (AMEn; MJ/kg DM) of S- and/or M-based diets of broilers measured on faeces collected from 22 to 24 days (Experiment 2)

	Nutrient digestibility (%)				
	DM	Protein	Fat	Starch	AMEn (MJ/kg DM)
C2	68.8^{abc}	81.5^{a}	82.2^{abc}	97.2^{ab}	13.3^{ab}
S60	68.1^{abc}	80.5^{ab}	79.2^{a}	96.4^{abc}	13.2^{ab}
S80	67.5^{bc}	78.9^{ab}	80.1^{ab}	95.7^{cd}	13.1^{ab}
S100	66.9^{c}	78.7^{abc}	80.2^{ab}	94.9^{d}	13.0^{b}
M60	69.3^{ab}	78.7^{ab}	82.7^{abc}	97.6^{a}	13.4^{a}
M80	69.3^{ab}	77.6^{abc}	83.2^{abc}	96.9^{ae}	13.4^{a}
M100	68.8^{abc}	75.0^{c}	84.0^{bc}	97.0^{abe}	13.3^{ab}
S30/M30	69.5^{a}	79.4^{ab}	86.1^{c}	96.8^{ae}	13.4^{a}
S40/M40	68.4^{abc}	77.3^{bc}	82.1^{abc}	96.1^{bce}	13.3^{ab}
S50/M50	67.8^{abc}	77.0^{bc}	83.4^{abc}	95.8^{de}	13.2^{ab}
SEM	0.38	0.75	0.87	0.24	0.07
Treatment effect[1]	***	***	***	***	***
Type of ingredient[2]	***	***	***	***	**
Level of substitution[3]					
S	*	†	ns	*	†
M	ns	†	ns	†	ns
S and M	*	ns	**	*	ns

C2 = control diet; S = sorghum; M = millet; DM = dry matter.
The number represents the level of substitution of corn.
[1]Effect of treatments: ANOVA on 10 diets.
[2]Effect of the type of ingredient: ANOVA on four treatments (control *v.* S-based diets *v.* M-based diets *v.* S/M-based diets).
[3]Effect of the level of substitution (dose–response effect) within each type of ingredient.
[a,b,c,d,e]Values with the same superscript are not significantly different at $P<0.05$.
***$P<0.001$; **$P<0.01$; *$P<0.05$; †$P\leqslant 0.10$. ns: not significant at $P>0.10$.

Total tract digestibility of protein was higher in the control diet (81.50±0.85%) compared with all other treatments. M- (77.09±0.62%) and S/M-based diets (77.89±0.54%) were significantly lower than the control diet whereas S-based diets (79.35±0.32%) were not significantly different. Total tract digestibility of protein in S- and M-based diets tended to decrease linearly with the increasing level of substitution of corn. However, there was no effect of the level of substitution on total tract digestibility of protein when both sorghum and millet are simultaneously used in the diet.

No significant differences were observed between the control diet (82.23±1.64%) and the experimental treatments. However, S-based diets (79.82±0.46%) showed lower total tract digestibility of fat compared with M- (83.30±0.42%) and S/M-based diets (83.83±0.62%). The level of substitution of sorghum or millet did not impact the total tract digestibility of fat. However, the combination of these two ingredients in broiler diet significantly affected fat digestibility with the highest level in the S30/M30 diet and the lowest level with the S50/M50 diet.

Higher total tract digestibility of starch was obtained with the control diet (97.21±0.17%) and M-based diets (97.17±0.14%) compared with the S-based diets (95.66±0.22%). Total tract digestibility of starch in S/M-based diets was intermediate (96.28±0.16%) and not different from the control diet. The level of substitution of millet tended to affect total tract digestibility of starch whereas a significant reduction was observed when the level of substitution increased in S and S/M-based diets.

The treatment significantly affected AMEn with the highest values obtained with M-based diets (13.38±0.05 MJ/kg DM). Results observed with the control diet (13.26±0.07 MJ/kg DM) were similar to S- (13.11±0.03 MJ/kg DM) and S/M-based diets (13.29±0.05 MJ/kg DM). AMEn tended to decrease (P=0.06) with the level of substitution in sorghum-based diets. No dose–response effect was found in millet or when this latter was mixed with sorghum.

DISCUSSION

Partial or total substitution of corn by sorghum and millet, and soybean meal replacement by cottonseed meal had no significant effect on broiler's feed intake and growth performance. This is in accordance with reports of Jacob et al. (1996), Hidalgo et al. (2004) and Azman and Yilmaz (2005). However, these results are not in agreement with those reported by Jacobs and Parsons (2013) who observed a reduction in feed efficiency with sorghum-based diets, and related it to the feed particle size, which induces a higher feed intake. However, the present findings confirm the results obtained with the meta-analysis (Batonon-Alavo et al., 2015) in which no significant difference in feed intake was observed between sorghum- or millet-based diets and corn-based diets, and between cottonseed meal and control diets. However, sorghum-based diets increased FCR in Experiment 1 while no difference was found between the control- and sorghum-based diets in Experiment 2. In both experiments, there were differences between formulated and calculated ME and CP contents in all treatments (Tables 2 and 3). This is because estimated ME and CP contents of major feed ingredients were used in formulating diets and these differ from determined CP contents and recalculated ME contents based on the laboratory determination of starch, sugar and fat. The trend of higher feed consumption in sorghum-based diets in Experiment 1 could then be related to their lower recalculated ME content (11.9 MJ/kg) compared with the control diet (12.3 MJ/kg). In addition, sorghum-based diets were lower in analyzed fat content (39.3±1.65 g/kg) than the control diet (53.4 g/kg) in Experiment 1. Since bird tend to decrease its consumption with the level of energy in the diet (Pérez-Bonilla et al., 2012) it is possible that ME and fat contents contributed to explain the increased FCR observed in sorghum-based diets in Experiment 1 instead of the sorghum effect per se. In addition, there was no difference in lysine intake and Met+Cyst intake. Therefore, no deficiency in these amino acids seems to exist with young birds.

Another aim of this study was to evaluate the effects of simultaneous inclusion of sorghum and cottonseed meal or millet on broiler performance and nutrient digestibility. Most substitution studies involved only one ingredient and interactions are therefore not studied (Hidalgo et al., 2004; Azman and Yilmaz, 2005). The present results showed that a combination of cottonseed meal and sorghum in the same diet did not decrease bird performance. Simultaneous replacement of corn by sorghum and millet also resulted in equivalent performance as the control diet. However, significant differences were observed in AID of protein and energy between feed ingredients, with the cottonseed meal- and sorghum/cottonseed meal-based diets having lower protein and energy digestibility than corn-based diets (Experiment 1). These results are consistent with those reported by Li et al. (2012) and Gonzalez-Vega and Stein (2012) who observed a decline in nitrogen and energy digestibility of growing and finishing pigs fed with various sources of cottonseed meal. Free gossypol in cottonseed meal is known to have an inhibitory action on certain enzymes in bird gastrointestinal tract like pepsinogen, pepsin and trypsin by binding with their free epsilon amino groups of lysine, thus reducing protein digestibility (Sharma et al., 1978). However, in this study free gossypol levels were too low (5 to 13 ppm) to induce any detrimental effect on growth performance (Nagalakshmi et al., 2007); but the effect of free gossypol could not be excluded from the reduction of digestibility with cottonseed meal-based diets. Similarly, lower total tract digestibility of fat and starch and lower AMEn values were obtained with sorghum-based diets consistently with previous reports (Selle et al., 2010; Mahmood et al.,2014) where digestibility coefficients were found inferior in sorghum than in other cereals due to the presence of phytate and phenolic compounds. Despite the lower fat and starch digestibility, our results suggest that birds fed low tannin sorghum-based diets were able to demonstrate equivalent performance as the control group since protein was equivalently available for both groups. Birds in millet-based diets have the same ability as those in corn-based diets to metabolize the energy and have a good feed efficiency. Higher coefficients of digestibility were obtained in millet-based diets consistently with Yin et al. (2002) and Baurhoo et al. (2011) and might be the factors leading to equivalent performance to the control diets.

Slaughtering method can have an effect on the results when digestibility is estimated by collection in the digestive tract. It should be noted that in the current study, birds were killed by cervical dislocation. Despite a debate on the best method of slaughtering the birds (i.e. cervical dislocation v. CO_2 asphyxiation) in ileal collection, Poureslami et al. (2012) suggest that cervical dislocation, by the means of cutting off the nervous stimulation from the brain to the gut, might actually slow or stop peristalsis sooner and therefore is more advantageous than asphyxiation slaughtering method.

The results obtained in the two experiments also suggest that the level of substitution of the experimental feed ingredients does not affect broiler performance, thus confirming the meta-analytic findings (Batonon-Alavo et al., 2015). Corn can be partially replaced by sorghum (Experiment 1) or totally substituted by sorghum or millet without any damages on bird performance consistently with several authors (Davis et al., 2003; Manwar and Mandal, 2009). Likewise, soybean meal substitution at various levels by cottonseed meal did not impact bird's response. However, the present study demonstrated that nutrient digestion can be compromised with these feed ingredients. The inclusion of exogenous enzymes in poultry diets is an established practice to improve performance and proteases are of particular relevance since they can be used to reduce the levels of trypsin inhibitors and lectins, thus improving protein digestibility (Bedford and Partridge, 2010). Other feed processing technologies like hydrothermal processes combined with reducing agents may enhance the solubility and digestibility of sorghum protein by either cleaving disulphide linkages or preventing their formation (Liu et al., 2013). Consequently, further investigations are required on the best strategies to enhance the nutritional value of these feed ingredients and lead to an increasing utilization in poultry.

ACKNOWLEDGEMENTS

The authors are grateful to the poultry research team of the Research Center for Animal Nutrition and Health (CRNA), DSM Nutritional Products France for their technical assistance as well as their full commitment in the data collection. The authors also thank CIRAD SELMET laboratory staff, Laurent Bonnal and Elodie Baby for the chemical analyses.

REFERENCES

1. Adeola, O and Orban, JI 1995. Chemical composition and nutrient digestibility of pearl millet (*Pennisetum glaucum*) fed to growing pigs.Journal of Cereal Science 22, 177–184.
2. AOAC International 1995. Official methods of analysis vol. 2, 16th edition. Association of Analytical Communities, Arlington, VA, USA.
3. Aviagen 2007. Broiler Ross 308: nutrition specifications. Aviagen, Scotland, UK, 8pp.
4. Azman, MA and Yilmaz, M 2005. The growth performance of broiler chicks fed with diets containing cottonseed meal supplemented with lysine. Revue de Médecine Vétérinaire 156, 104–106.
5. Bastianelli, D, Bonnal, L, Juin, H, Mignon-Grasteau, S, Davrieux, F andCarré, B 2010. Prediction of the chemical composition of poultry excreta by near infrared spectroscopy. Journal of Near Infrared Spectroscopy 18, 69–77.
6. Batonon-Alavo, DI, Umar Faruk, M, Lescoat, P, Weber, GM andBastianelli, D 2015. Inclusion of sorghum, millet and cottonseed meal in broiler diets: a meta-analysis of effects on performance. Animal 9, 1120–1130.
7. Baurhoo, N, Baurhoo, B, Mustafa, AF and Zhao, X 2011. Comparison of corn-based and Canadian pearl millet-based diets on performance, digestibility, villus morphology, and digestive microbial populations in broiler chickens. Poultry Science 90, 579–586.
8. Bedford, MR and Partridge, GG 2010. Enzymes in farm animal nutrition. CAB International, London, UK.

9. Davis, AJ, Dale, NM and Ferreira, FJ 2003. Pearl millet as an alternative feed ingredient in broiler diets. Journal of Applied Poultry Research 12, 137–144.

10. Fisher, C and McNab, JM 1987. Techniques for determining the ME content of poultry feeds. In Recent advances in animal nutrition (ed. W Haresign and DJA Cole), pp. 3–17. Butterworths, London.

11. Flores, MP, Castañon, JIR and McNab, JM 1994. Effect of tannins on starch digestibility and TME of triticale and semi purified starches from triticale and field beans. British Poultry Science 35, 281–286.

12. Gonzalez-Vega, JC and Stein, HH 2012. Amino acid digestibility in canola-, cottonseed- and sunflower-products fed to finishing pigs. Journal of Animal Science 90, 4391–4400.

13. Goodarzi Boroojeni, F, Samie, AH, Edriss, MA, Khorvash, M, Sadeghi, G,Van Kessel, A and Zentek, J 2011. Replacement of corn in the diet of broiler chickens using foxtail millet produced by 2 different cultivation strategies. Poultry Science 90, 2817–2827.

14. Heuzé, V and Tran, G 2012. Pearl millet (*Pennisetum glaucum*) grain. Feedipedia.org. A programme by INRA, CIRAD, AFZ and FAO. Retrieved August 8, 2014, from http://www.feedipedia.org/node/724.

15. Hidalgo, MA, Davis, AJ, Dale, NM and Dozier III, WA 2004. Use of whole pearl millet in broiler diets. Journal of Applied Poultry Research 13, 229–234.

16. Hill, FW and Anderson, DL 1958. Comparison of metabolizable energy and productive energy determinations with growing chicks. Journal of Nutrition 64, 587–603.

17. Jacob, JP, Mitaru, BN, Mbugua, PN and Blair, R 1996. The effect of substituting Kenyan Serena sorghum for maize in broiler starter diets with different dietary crude protein and methionine levels. Animal Feed Science and Technology 61, 27–39.

18. Jacobs, C and Parsons, CM 2013. The effects of coarse ground corn, whole sorghum, and a prebiotic on growth performance, nutrient digestibility, and cecal microbial populations in broilers fed diets with and without corn distillers dried grains with solubles. Poultry Science 92, 2347–2357.

19. Janssen, WMMA 1989. European table of energy values for poultry feedstuffs, 3rd edition. Spelderholt Center for Poultry Research and Information Services, Beekbergen, The Netherlands.

20. Li, JT, Li, DF, Zang, JJ, Yang, WJ, Zhang, WJ and Zhang, LY 2012. Evaluation of energy digestibility and prediction of digestible and metabolizable energy from chemical composition of different cottonseed meal sources fed to growing pigs. Asian-Australian Journal of Animal Science 25, 1430–1438.

21. Liu, SY, Selle, PH and Cowieson, AJ 2013. Strategies to enhance the performance of pigs and poultry on sorghum-based diets. Animal Feed Science and Technology 181, 1–14.

22. Mahmood, S, Ali, H, Ahmad, F and Iqbal, Z 2014. Estimation of tannins

in different sorghum varieties and their effects on nutrient digestibility and absorption of some minerals in caged white leghorn layers. International Journal of Agriculture and Biology 16, 217–221.

23. Mahmood, S, Khan, MA, Sarwar, M and Nisa, M 2006. Chemical treatments to reduce antinutritional factors in salseed (*Shorea robusta*) meal: effect on nutrient digestibility in colostomized hens and intact broilers. Poultry Science 85, 2207–2215.
24. Manwar, SJ and Mandal, AB 2009. Effect of high moisture storage of pearl millet (*Pennisetum typhoides*) with or without feed enzymes on growth and nutrient utilization in broiler chickens. Animal Science Journal 80, 438–445.
25. Nagalakshmi, D, Rama Rao, SV, Panda, AK and Sastry, VRB 2007. Cottonseed meal in poultry diets: a review. The Journal of Poultry Science 44, 119–134.
26. Pérez-Bonilla, A, Novoa, S, García, J, Mohiti-Asli, M, Frikha, M and Mateos, GG 2012. Effects of energy concentration of the diet on productive performance and egg quality of brown egg-laying hens differing in initial body weight. Poultry Science 91, 3156–3166.
27. Poureslami, R, Batal, AB and Jung, B 2012. Effect of ileal sub-section and the method of collection of digesta on the determination of apparent ileal digestibility of amino acids in broilers. Animal Feed Science and Technology 177, 130–133.
28. R Core Team 2013. R: a language and environment for statistical computing. R Foundation for Statistical Computing, Vienna, Austria.
29. Sauvant, D, Perez, JM and Tran, G 2004. Tables of composition and nutritional value of feed materials, INRA Editions and AFZ. Wageningen Academic Publishers, Paris, France.
30. Selle, PH, Cadogan, DJ, Li, X and Bryden, WL 2010. Implications of sorghum in broiler chicken nutrition. Animal Feed Science and Technology 156, 57–74.
31. Sharma, NK, Lodhi, GN and Ichhponani, JS 1978. Comparative feeding value of expeller processed undecorticated and decorticated cottonseed cakes for growing chicks. Journal of Agricultural Science 91, 531–541.
32. Siegel, S and Castellan, NJ 1988. Non parametric statistics for the behavioral sciences. MacGraw Hill Int., New York, NY, USA.
33. Terpstra, K and de Hart, N 1974. The estimation of urinary nitrogen and faecal nitrogen in poultry excreta. Zeitschrift für Tierphysiologie Tierernährung und Futtermittelkunde 32, 306–320.
34. Van Soest, PJ and Wine, RH 1967. Use of detergents in the analysis of fibrous feeds. IV. Determination of plant cell walls constituents. Journal of the Association of Official Analytical Chemists 50, 50–55.
35. Yin, YL, Gurung, NK, Jeaurond, EA, Sharpe, PH and de Lange, CFM 2002. Digestible energy and amino acid contents in Canadian varieties of sorghum, pearl millet, high-oil corn, high-oil-high-protein corn and regular corn samples for growing pigs. Canadian Journal of Animal Science 82, 385–391.

6

CHAPTER

DIFFERENTIAL EFFECTS OF SHORT-TERM β AGONIST AND GROWTH HORMONE TREATMENTS ON EXPRESSION OF MYOSIN HEAVY CHAIN IIB AND ASSOCIATED METABOLIC GENES IN SHEEP MUSCLE

K. M. Hemmingsa , Z. C. T. R. Daniel, P. J. Buttery, T. Parr and J. M. Brameld

Keywords: β agonist, growth hormone, muscle fibre type, myosin heavy chain, sheep

IMPLICATIONS

This study provides mechanistic information on the anabolic responses to growth hormone and β-adrenergic agonist in sheep muscle. It therefore has implications for the use of these agents in farm animals, which is licensed in some parts of the world (e.g. United States, Australia) but not others (e.g. EU). Regarding cross-species comparisons, there are clear differences in responses between sheep and rodents, with sheep appearing to respond in a manner more similar to humans than rats. Indeed, the fibre type composition of sheep muscle is more similar to humans than rats, suggesting that sheep may be a better model organism for conducting skeletal muscle studies.

INTRODUCTION

Administration of growth hormone (GH) or β-adrenergic agonist (BA) results in the repartitioning of nutrients, increasing skeletal muscle mass and decreasing fat mass (Buttery and Dawson, 1990). The effectiveness of these agents is thought to be reduced in young, rapidly growing animals compared with those approaching mature BW (Bell et al., 1998). In addition, the increase in skeletal muscle mass may be dependent upon muscle fibre type composition or involve fibre type transitions (Lefaucheur and

Citation: Hemmings KM, Daniel ZCTR, Buttery PJ, Parr T, Brameld JM. Differential effects of short-term β agonist and growth hormone treatments on expression of myosin heavy chain IIB and associated metabolic genes in sheep muscle. Animal. 2015;9(2):285-294. doi:10.1017/S175173111400233X.

Gerrard,2000). In recent years, measurement of myosin heavy chain (MyHC) isoforms has been used to indicate the fibre type composition of skeletal muscles (Sazili et al., 2005). In sheep and cattle, the predominant MyHC isoforms expressed in adult animals are types I, IIA and IIX, with the 'fastest, most glycolytic' isoform, MyHCIIB, being rarely detected in these species (Hemmings et al., 2009; Picard and Cassar-Malek, 2009).

The factors involved in regulating fibre type transitions are poorly understood, particularly those relating to fast glycolytic fibre types. Rodent studies have identified a number of potential regulatory signalling molecules, including the myogenic regulatory factors (MRFs; Hughes et al., 1993) and the signalling enzymes, calcineurin (CaN) and calmodulin kinase II (CaMKII; Chin, 2005; Bassel-Duby and Olson, 2006). The Six1/Eya1 protein complex has been shown to be important in the regulation of the fast glycolytic fibre type in rodents (Grifone et al., 2004), but their role in other species is not known. In addition, the peroxisome proliferator-activated receptor (PPAR) transcription factors, along with the PPAR-γ coactivator-1 (PGC-1) family of transcriptional coactivators and the transcriptional co-repressor, receptor-interacting protein 140 (RIP140), are all known to influence muscle metabolic gene expression in rodents (Lin, 2009), but again their role in other species is not known.

The objective of the current study was therefore to compare the effects of short-term (6 day) treatment with GH or BA on enzyme activities and expression of genes associated with skeletal muscle metabolism and fibre type in young (rapidly growing) and mature (fattening) lambs. BA had previously been shown to alter MyHC isoform expression after 1 week of treatment in pigs (Gunawan et al., 2007) and therefore this short-term (6 day) treatment was chosen in order to investigate the initial (inductive) response to these agents and potential regulatory mechanisms involved.

MATERIAL AND METHODS

Sheep studies were approved by the University of Nottingham Local Ethical Review Committee and were carried out in accordance with the UK Home Office guidelines (Animals (Scientific Procedures) Act, 1986).

Sample Collection and Preparation

Male (Mule×Charolais) twin lambs were group housed with their mothers until 53±4 days of age, when they were weaned and pairs were split into day 60 (D60) and day 120 (D120) groups. These ages were chosen based on a previous (unpublished) study at the University of Nottingham, showing that D60 was within the rapid growth phase and D120 represented mature lambs with reduced growth rates and increased backfat deposition. Creep feed (48.48% barley, 27% soya, 15% nutritionally improved straw, 5% molassed feed meal, 2.5% vitamins and minerals, 1% limestone, 0.5% salt, 0.5% vegetable oil and 0.02% Minsal g109 E50) was introduced ~28 days before weaning and lambs were given free access from weaning onwards. Lambs were individually penned with ad-libitum access to creep feed for 7 days before the commencement of treatments. The D60 and D120 groups were split into three treatment groups, ensuring that the mean initial BWs for each group were not significantly different. The control group (CO, n=11) were fed the creep dietad libitum; the GH group (n=10) were fed the same diet ad libitum and administered a prolonged release bovine GH formulation (POSILAC, Monsanto, St Louis, Missouri, USA) on the 1st day of treatment by a single subcutaneous injection ~5 cm from the nape of the neck, at a dose of 3.75 mg/kg BW, as described by McLaughlin et al. (1994). The BA group (n=10) were fed the same creep diet ad libitum, but containing theβ2-adrenergic agonist, cimaterol, at 10 ppm (10 mg/kg). The treatments were for 6 days, starting at 60±4 or 120±4 days of age. This short-treatment period was chosen in order to investigate the initial (inductive) response to these agents, particularly in terms of intracellular signalling. At slaughter, blood was collected and plasma prepared, while samples of the Longissimus dorsi (LD, from the thoracic region at the 10th rib),Supraspinatus (SS) and Semitendinosus (ST) muscles from the left side of the carcass, were snap frozen in liquid nitrogen and stored at −80°C; these were subsequently crushed and mixed before further analyses. LD and ST muscles were chosen for their commercial importance and as representative 'fast' muscles, whereas SS was selected as a 'slower' muscle type (Hemmings et al., 2009). The whole LD, ST, SS, Vastus lateralis(Vlat) and Vastus intermedius (Vint) muscles were dissected from the right side of the carcass and weighed along with the liver, all within 1 h of slaughter.

Plasma IGF-I and Glycerol Concentrations and LD Skeletal Muscle Glycogen Content

Plasma IGF-I concentrations were determined using an enzyme-linked immunosorbent assay kit (Diagnostic Systems Laboratories, Oxford, UK) and plasma glycerol concentrations were measured using the Serum Triglyceride determination kit (Sigma Aldrich, Poole, UK), both procedures were carried out according to the manufacturers' instructions. Acid-soluble muscle glycogen was determined using a protocol adapted from Dreiling et al. (1987). Briefly, 1 g of muscle was homogenised in 4 ml of 8% (vol/vol) perchloric acid (PCA), then 1 ml was removed and centrifuged at 14 000×g at 4°C for 10 min. To neutralise the supernatant, 133 µl was mixed with 150 µl of saturated sodium bicarbonate (5 M) solution and 167 µl of 200 mM sodium acetate buffer, pH 4.8. A range of glycogen standards were prepared in neutralised PCA containing 0 to 300 µg of bovine liver glycogen (Sigma Aldrich) in a 200 µl volume. Amyloglucosidase (5 µl of 80 U/ml, Sigma Aldrich) was added to 200 µl of each standard or neutralised sample supernatant and incubated at 37°C for 30 min to release the glucose from the glycogen. Samples were heated at 100°C for 5 min to stop the reaction and then centrifuged at 14 000×g for 10 min at room temperature. Glucose concentrations were determined using Enzymatic Glucose Reagent (Thermo Electron Corporation, Fife, UK), by measuring the absorbance at 550 nm on a 680 XR 96-well microplate reader (Bio-Rad, Hemel Hempstead, UK).

Skeletal muscle total RNA Extraction and cDNA Synthesis

Total RNA was extracted from LD, SS and ST muscle samples using TRIZOL (Invitrogen, Paisley, UK), according to the manufacturer's instructions, and then treated with DNase (Promega, Southampton, UK). Following DNase treatment, RNA was quantified using the Nanodrop spectrophotometer ND-1000 (Thermo Scientific, Wilmington, DE, USA) and RNA integrity was checked by running samples on a 1% agarose gel and visualising the 28S and 18S ribosomal RNA bands. First strand complementary DNA (cDNA) was synthesised from 1 μg of total RNA using random hexamer primers and Moloney murine leukemia virus reverse transcriptase (Promega) in a final volume of 25 μl (Hemmings et al., 2009).

MyHC Isoform Expression

The relative expression of the adult MyHC isoforms (MyHCI, IIA, IIX and IIB) were determined at both the mRNA and protein level, as described previously (Hemmings et al., 2009). Briefly, SDS–PAGE separation of MyHC isoforms was carried out to determine percentage MyHC protein expression, comparing samples to a bovine positive control (Longissimus thoracis) sample (kindly provided by Dr Brigitte Picard, INRA, France) previously shown to express all four adult MyHC isoforms (Picard and Cassar-Malek, 2009). Quantitative real-time PCR was carried out on muscle cDNA to determine percentage MyHC mRNA expression, using ovine specific primers and probes for MyHCI, IIA and IIX (Supplementary Table S1). The relative percentage of each isoform was calculated as described previously (Hemmings et al., 2009). Primers and probes designed to the available sequence for the ovine MyHCIIX isoform would be expected to generate amplicons from any MyHCIIB transcripts present as well, as this region is identical in the porcine sequences for MyHCIIX and IIB, therefore expression is stated as MyHCIIX/IIB.

The mRNA expression of the MyHCIIB isoform was analysed by semi-quantitative RT-PCR using published primers previously described as detecting only ovine MyHCIIB (Vuocolo et al., 2007). MyHCIIB expression was compared with the semi-quantitative RT-PCR analysis of MyHCIIX+IIB combined (using the primers in Supplementary Table S1).

Metabolic Enzyme Activities

Isocitrate dehydrogenase (ICDH) and lactate dehydrogenase (LDH) activities were determined in LD homogenates using the methods described by Brandstetter et al. (1998) as an indication of oxidative and glycolytic metabolism, respectively.

CaN and Calmodulin Kinase Activities

Protein content was determined in LD homogenates using the Bradford assay (Bradford, 1976) and samples adjusted to constant protein before assay. CaN activity was determined in skeletal muscle homogenates using the BIOMOL CaN cellular assay kit plus (Enzo Life Sciences Ltd, Exeter, UK). CaMKII activity was determined using the SignaTECT calcium/calmodulin-dependent protein kinase (CaMKII) Assay kit (Promega).

Real-time PCR Analysis of Signalling Molecule Gene Expression

Ovine specific primers and probes were designed using Primer Express software (Applied Biosystems, Warrington, UK) and are shown in Supplementary Table S1. LD cDNA was diluted 1 in 4 and a pool of samples generated and serially diluted to generate a standard curve. Each sample cDNA was further diluted 1 in 8 and 5 μl was used for real-time PCR reactions (in triplicate), using 384 well plates and 1X Lightcycler 480 probes master PCR mastermix (Roche, Burgess Hill, UK) in 15 μl final reaction volume (containing 0.3 μM of each primer and 0.2 μM of probe). Reactions were run on the Lightcycler 480 PCR machine (Roche) with the following conditions: 95°C for 10 min and then 45 cycles of 95°C for 10 s, 60°C for 50 s and 72°C for 1 s, during which fluorescence was measured. Crossing point (Cp) values were determined using the second derivative maximum method and quantified relative to the standard curve (Brown et al., 2012). Supplementary Table S1 includes the primers used for PCR. After being unable to find a suitable housekeeping gene for normalising the data, it was decided to express the real-time data relative to the equivalent quantity of RNA in the reaction, given that the total RNA concentrations had been normalised before cDNA synthesis.

Statistical Analysis

Data are presented as means or as a percentage value. All data had a normal distribution. Data for each age group was analysed separately by one-way ANOVA (Genstat release 11.1. (Lawes Agricultural Trust, Hertfordshire, UK)), blocking for gel where appropriate. s.e.d. values are included for the ANOVA. When significant differences were observed (with ANOVA), a post hoc Dunnett's test was then used to compare each treatment group to the control. If not significant (i.e. $P>0.05$), the P-values shown relate to the ANOVA, but in the text significant differences are indicated using P-values for the Dunnett's test. Tables include the P-values for the ANOVA, with * used to indicate significant differences for the Dunnett's test.

RESULTS

Lamb Responsiveness to BA and GH Treatments

The effects of a 6-day treatment with BA or GH on lamb carcass characteristics are shown in Table 1. Neither treatment significantly altered BWs ($P>0.05$) and there were no significant differences in any muscle weights at D60 or in LD and VInt muscle weights at D120 ($P>0.05$). However, BA treatment resulted in significant increases in SS and ST muscle weights at D120 ($P<0.05$), with VLat weights also tending to be higher in the BA treated older lambs ($P=0.058$). As expected, GH significantly increased liver weights ($P<0.05$; Table 1) and plasma IGF-1 levels ($P<0.05$; Table 2) at both ages; while BA significantly reduced LD glycogen content at both ages ($P<0.05$; Table 2). Surprisingly, plasma glycerol levels were unaffected ($P>0.05$; Table 2) by treatment at both ages.

Table 1: The effects of short-term (6 day) BA or GH treatment of lambs at 60 or 120 days of age on lamb carcass characteristics

	D60					D120				
Trait	CO	BA	GH	s.e.d.	*P*-value	CO	BA	GH	s.e.d.	*P*-value
n	11	10	10			11	10	10		
BW (kg)	23.9	23.9	24.5	1.4	0.863	43.0	46.4	44.3	1.6	0.122
Liver (g)	475	460	567[a]	32	0.005	928	926	1101[a]	55	0.004
Muscle weights (g)										
Longissimus dorsi	305	306	327	27	0.655	592	667	632	35	0.108
Supraspinatus	53.5	55.2	55.9	3.9	0.812	98.5	116.4[a]	98.0	6.3	0.009
Semitendinosus	63.3	67.1	65.0	4.3	0.667	112.4	131.2[a]	120.4	6.6	0.023
Vastus lateralis	85.4	88.0	89.3	6.3	0.815	164.5	180.3	163.8	7.5	0.058
Vastus intermedius	28.5	27.5	30.9	2.3	0.331	51.7	55.7	52.4	3.6	0.487

BA = β agonist; GH = growth hormone; CO = control.
[a]Significantly different to control (Dunnett's test; $P<0.05$).

Table 2: The effects of short-term (6 day) BA or GH treatment of lambs at 60 or 120 days of age on plasma IGF-1 and glycerol concentrations and muscle glycogen content

	D60					D120				
Trait	CO	BA	GH	s.e.d.	*P*-value	CO	BA	GH	s.e.d.	*P*-value
n	11	10	10			11	10	10		
IGF-1 (ng/ml)	375	373	928[a]	129	<0.001	972	964	2092[a]	127	<0.001
Glycerol, (μg/ml)	6.3	8.2	7.3	1.1	0.191	8.2	9.4	9.5	2.4	0.826
LD Glycogen (mg/g)	25.5	19.0[a]	23.8	1.6	<0.001	21.6	18.3[a]	24.2	1.3	<0.001

BA = β agonist; GH = growth hormone; CO = control; LD = *Longissimus dorsi*.
[a]Significantly different to control (Dunnett's test; $P<0.05$).

Effects of BA and GH on MyHC Expression in LD, SS and ST Muscles

As an indication of the effects of treatment on the fibre type composition of the LD, SS and ST muscles, MyHC expression was determined at both the mRNA and protein levels. GH treatment had no effects (P>0.05) on MyHC isoform mRNA (Table 3) or protein (Table 4) expression in any of the muscles studied. In contrast, there were significant effects of BA treatment on percentage MyHC isoform mRNA expression in all three muscles at both ages (Table 3). At D60, BA treatment decreased the %MyHCI mRNA in the LD (P<0.05), but there were no effects in the SS and ST muscles (P>0.05) and there were no effects on %MyHCI mRNA expression in any muscle at D120 (P>0.05). BA treatment consistently reduced %MyHCIIA mRNA expression in all three muscles at both ages (P<0.05). Interestingly, BA treatment increased the %MyHCIIX/IIB mRNA expression in the two fast muscles, LD and ST, at both ages (P<0.05). The response observed in the SS muscle was similar, but either not significant at D120 or appeared to only increase relative to the GH group at D60.

Table 3: The effects of short-term (6 day) BA or GH treatment of lambs at 60 or 120 days of age on MyHC isoform mRNA expression in different muscles

		D60					D120				
Muscle	Isoform (%)	CO	BA	GH	s.e.d.	P-value	CO	BA	GH	s.e.d.	P-value
	n	10 to 11	10	10			10 to 11	10	9 to 10		
Longissimus dorsi	MyHCI	7.2	3.9[a]	6.6	0.8	<0.001	6.8	5.0	8.1	1.9	0.277
	MyHCIIA	26.9	7.0[a]	34.1	3.4	<0.001	21.1	5.2[a]	25.4	3.6	<0.001
	MyHCIIX/IIB	65.9	89.1[a]	59.3	3.6	<0.001	72.1	89.8[a]	66.5	4.2	<0.001
Supraspinatus	MyHCI	34.7	37.7	42.3	8.8	0.693	56.6	51.6	45.5	9.6	0.520
	MyHCIIA	21.2	9.1[a]	27.9	3.8	<0.001	15.8	5.3[a]	17.2	2.1	<0.001
	MyHCIIX/IIB	44.1	53.2	29.8	7.3	0.012	27.6	43.1	37.3	9.1	0.243
Semitendinosus	MyHCI	13.5	8.9	11.7	2.4	0.161	11.2	14.6	13.5	2.2	0.259
	MyHCIIA	20.6	4.3[a]	25.5	2.8	<0.001	17.1	2.7[a]	20.4	2.6	<0.001
	MyHCIIX/IIB	66.0	86.8[a]	62.7	4.0	<0.001	71.7	82.7[a]	66.2	3.8	<0.001

BA = β agonist; GH = growth hormone; MyHC = myosin heavy chain; CO = control.
[a]Significantly different to control (Dunnett's test; $P < 0.05$).

Table 4: The effects of short-term (6 day) BA or GH treatment of lambs at 60 or 120 days of age on MyHC isoform protein expression in different muscles

		D60					D120				
Muscle	Isoform (%)	CO	BA	GH	s.e.d.	P-value	CO	BA	GH	s.e.d.	P-value
	n	11	10	10			11	10	10		
Longissimus dorsi	MyHCI	10.2	9.4	9.4	1.1	0.680	8.1	8.9	8.2	1.0	0.802
	MyHCIIA	33.3	29.9	35.0	1.8	0.028	31.9	30.7	33.3	1.5	0.236
	MyHCIIX/IIB	56.5	60.7	55.6	2.3	0.060	60.0	60.4	58.4	1.8	0.533
Supraspinatus	MyHCI	42.1	45.4	42.0	8.2	0.852	48.4	34.2	47.2	8.5	0.166
	MyHCIIA	20.6	19.2	22.0	2.0	0.387	20.3	18.2	17.7	1.6	0.429
	MyHCIIX/IIB	37.3	35.3	36.0	6.7	0.916	31.3	47.6	35.1	7.6	0.086
Semitendinosus	MyHCI	11.0	9.0	8.7	1.8	0.308	7.3	9.2	6.9	1.7	0.347
	MyHCIIA	26.3	26.2	29.1	1.5	0.142	25.0	24.0	24.8	2.0	0.920
	MyHCIIX/IIB	62.7	64.8	62.2	2.8	0.578	67.8	66.8	68.2	3.3	0.870

BA = β agonist; GH = growth hormone; MyHC = myosin heavy chain; CO = control.

The effects of BA treatment on MyHC isoform mRNA expression were not observed at the protein level (Table 4). However, BA treatment did appear to reduce %MyHCIIA protein in the LD at D60, but only relative to the GH group, and BA tended to increase %MyHCIIX protein in the LD at D60 and SS at D120 (Table 4).

As BA treatment significantly increased %MyHCIIB/IIX mRNA in the LD and ST muscles, expression of MyHCIIB mRNA was determined by semi-quantitative RT-PCR and compared with the expression of MyHCIIX and IIB mRNA combined (Figure 1). MyHCIIB mRNA was induced by BA treatment in the LD of lambs at both ages, and expression of the MyHCIIB isoform was increased by BA in the SS and ST muscles. However, MyHCIIB was not detected at the protein level in any of the muscles studied (Figure 2).

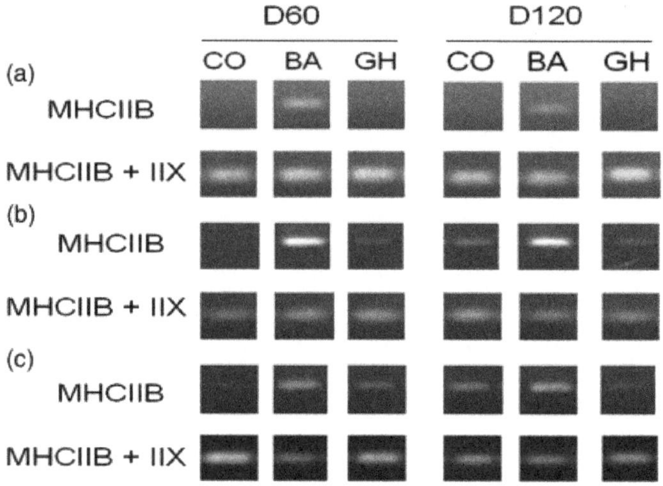

Figure 1: Representative semi-quantitative RT-PCR for myosin heavy chain (MyHC) IIB and IIB+IIX mRNAs in (a) Longissimus dorsi (LD), (b)Supraspinatus (SS) and (c) Semitendinosus (ST) muscles from D60 and D120 lambs. The 6-day administration of β agonist .

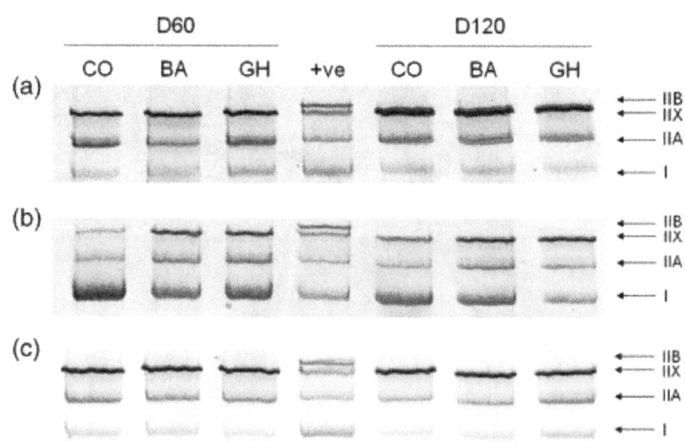

Figure 2: Representative SDS–PAGE separation of myosin heavy chain (MyHC) isoforms in (a) Longissimus dorsi (LD), (b) Supraspinatus(SS) and (c) Semitendinosus (ST) muscles from D60 and D120 lambs. +ve represents a bovine Longissimus thoracis sample containing .

Overall, the changes in MyHC isoform expression indicate that muscles from BA-treated lambs were in transition towards a faster, more glycolytic fibre type composition, with MyHCIIA fibres switching to MyHCIIX and/or IIB fibres. Because of the observed induction of de novo mRNA expression of the MyHCIIB isoform in the LD in response to BA treatment, further characterisation of the metabolic response and postulated intracellular signalling mechanisms were conducted in that muscle.

Effects of BA and GH on metabolic enzyme activities of LD

ICDH and LDH activities were determined in the LD as an indication of oxidative and glycolytic metabolic activities, respectively (Table 5). ICDH activity was significantly decreased in BA treated lambs at both ages (P<0.05); but there were no effects of treatment on LDH activities (P>0.05, Table 5). Overall, the LD appears to be changing to a less oxidative metabolism, which would agree with the changes in MyHC isoform mRNA expression described previously.

Table 5: The effects of short-term (6 day) BA or GH treatment of lambs at 60 or 120 days of age on metabolic and signalling enzyme activities in the Longissimus dorsi

	D60					D120				
Trait	CO	BA	GH	s.e.d.	*P*-value	CO	BA	GH	s.e.d.	*P*-value
n	10 to 11	7 to 9	10			10 to 11	9 to 10	10		
ICDH activity (mOD/min per mg tissue)	49.3	33.9[a]	49.4	4.9	0.004	37.3	28.2[a]	33.1	3.4	0.041
LDH activity (mOD/min per mg tissue)	510	489	519	86	0.928	637	618	652	114	0.957
CaMKII (pmol/min per μg protein)	11.5	14.3	10.9	2.1	0.246	13.8	14.9	11.5	1.8	0.160
CaN ($\times 10^{-3}$) arbitrary units	49.5	43.0	44.4	15.5	0.901	34.1	27.1	22.5	11.1	0.562

BA = *β* agonist; GH = growth hormone; CO = control; ICDH = isocitrate dehydrogenase; LDH = lactate dehydrogenase; CaMKII = calmodulin kinase II; CaN = calcineurin.
[a]Significantly different to control (Dunnett's test; $P < 0.05$).

Effects of BA and GH on Signalling Molecules and Regulatory Factors in LD

The enzyme activity (Table 5) or mRNA expression (Table 6) of selected mediators thought to influence muscle contractile and/or metabolic properties were determined in LD. Surprisingly, the activities of the calcium dependent signalling enzymes, CaN and CaMKII, were unaffected by treatment at either age (P>0.05 for all, Table 5). There were no significant effects of treatment on PGC-1α mRNA expression at either age (P>0.05, Table 6), whereas PGC-1β mRNA expression was significantly reduced by both BA and GH treatment at D60 (P<0.05, Table 6), but only by BA treatment at D120 (P<0.05, Table 6). In contrast, RIP140 mRNA expression was significantly increased by BA treatment at both ages (P<0.05, Table 6). GH treatment reduced PPARα mRNA expression in the D120 lambs (P<0.05, Table 6), but there were no effects at D60 (P>0.05, Table 6). Similarly, BA increased PPARδ mRNA expression in the D120 lambs (P<0.05, Table 6), but there were no effects in the D60 lambs (P>0.05, Table 6). PPARγ mRNA expression was unaffected by treatment (P>0.05 for all, Table 6) at either age. Surprisingly, Six1 and Eya1 mRNA expression were significantly decreased by BA treatment in both age groups (P<0.05, Table 6). There were no significant treatment effects (P>0.05, Table 6) on Myf-5 or MRF4 mRNA expression at either age, nor on MyoD mRNA expression in the D60 lambs, but BA treatment reduced MyoD mRNA expression in the D120 lambs (P<0.05, Table 6). In contrast, BA treatment increased myogenin mRNA expression at both ages (P<0.05), suggesting a possible effect on myogenic/satellite cell differentiation.

Table 6: The effects of short-term (6 day) BA or GH treatment of lambs at 60 or 120 days of age, on mRNA expression of regulators of muscle contractile and metabolic properties in the Longissimus dorsi

	D60					D120				
Gene mRNA target[a]	CO	BA	GH	s.e.d.	*P*-value	CO	BA	GH	s.e.d.	*P*-value
n	11	10	10			11	10	10		
PPARα	4.46	3.82	3.94	0.31	0.096	4.53	3.74	2.91[b]	0.35	<0.001
PPARδ	4.30	5.75	4.08	0.86	0.124	3.82	5.14[b]	4.22	0.35	0.002
PPARγ	4.85	2.91	4.39	1.76	0.516	4.46	3.45	4.14	0.48	0.111
PGC1α	4.53	4.36	3.70	0.77	0.521	4.28	4.67	3.85	0.58	0.376
PGC1β	5.11	2.87[b]	3.85[b]	0.45	<0.001	4.16	2.68[b]	3.42	0.44	0.007
RIP140	3.59	5.58[b]	4.30	0.66	0.015	3.86	5.63[b]	3.98	0.55	0.005
Six1	4.31	2.58[b]	4.11	0.41	<0.001	4.66	2.24[b]	4.24	0.27	<0.001
Eya1	4.38	2.73[b]	4.29	0.53	0.006	5.35	2.93[b]	5.04	0.69	0.002
MyoD	2.95	2.66	2.73	0.40	0.735	3.00	2.01[b]	3.28	0.39	0.008
Myogenin	3.34	4.65[b]	3.65	047	0.023	3.15	4.54[b]	3.23	0.40	0.002
MRF4	3.02	4.47	3.59	1.36	0.552	4.08	4.14	5.06	0.63	0.230
Myf-5	2.71	3.69	3.22	0.40	0.059	4.33	3.71	3.71	0.57	0.436

BA = β agonist; GH = growth hormone; CO = control; PPAR = peroxisome proliferator-activated receptor; PGC = PPAR-γ coactivator; RIP = receptor-interacting protein; MRF = myogenic regulatory factor.
[a]Values for gene targets expressed as arbitrary units per unit total RNA.
[b]Significantly different to control (Dunnett's test; $P < 0.05$).

DISCUSSION

The major findings from this sheep study relate to BA administration which, at both ages, induced changes that are indicative of muscle fibre type transitions towards a fast glycolytic phenotype, with de novoexpression of MyHCIIB mRNA being observed. These changes in MyHC isoform mRNA expression corresponded with a decrease in oxidative metabolism, which are likely induced by the increase in RIP140 and corresponding decrease in PGC-1β mRNA expression.

Responsiveness to Agents

Both GH and BA have previously been shown to induce skeletal muscle hypertrophy and have the effect of repartitioning nutrients from adipose tissue towards skeletal muscle; but they are thought to do this via different mechanisms (National Research Council, 1994). The majority of anabolic effects exerted by GH are thought to be mediated indirectly via the stimulation of IGF-1 production by the liver and other tissues (Bellet al., 1998), whereas the effects of GH on adipose tissue are thought to be direct. In contrast, BAs are thought to act directly on tissues (particularly muscle) by binding to β2-adrenergic receptors (β2-ARs) expressed by those tissues (National Research Council, 1994). In the present study, very different effects of GH and BA treatment were observed, particularly on factors indicative of fibre type transitions and potential regulatory factors.

Despite only a short-term (6 day) treatment, the lambs responded to both agents at both ages, although the magnitude of the responses did vary with age. At both ages, plasma IGF-1 levels and liver weights were increased by GH, which are commonly observed responses to GH administration (Brameld et al., 1996). As expected, BA decreased LD glycogen content at both ages. Surprisingly, neither agent altered plasma glycerol concentrations, which is commonly used as an indicator of adipose tissue lipolysis. Previous studies have shown that free fatty acid levels remain elevated during chronic BA administration, but plasma glycerol concentrations return to a basal level (O'Connor et al., 1991).

At D60, there were no effects of either treatment on muscle weights, whereas at D120, BA treatment significantly increased SS and ST weights. Therefore, there was an age differential in terms of muscle weight responses to BA administration, with older animals responding more. The reason for the lack of an effect on muscle weights in the younger lambs is unclear, but may be due to lower receptor numbers, a decreased binding affinity for the agonist or a more rapid desensitisation to the BA (National Research Council, 1994). It has previously been suggested that fast muscle fibres are more responsive to BA administration and that selective hypertrophy occurs in these fibres (Kim et al., 1987). However, there was a greater muscle weight response in the slower SS compared with the LD, which has the higher proportion of fast fibres. Of interest is the fact that a significant increase in muscle weight was observed in the ST but not the LD, which have very similar MyHC compositions. This suggests that MyHC composition may not be a major factor in determining muscle responses to BA administration. Further histochemical analysis (data not included) showed that the increased ST weights were associated with increased muscle fibre diameters (i.e. hypertrophy), with the fast fibres being mainly affected. However, we did not determine whether similar increases in fibre diameter were present in the LD as well.

Changes in MyHC Isoforms

There were no effects of GH treatment for 6 days on the relative expression of the MyHC isoforms at either the mRNA or protein levels. This may simply reflect the short (6 day) administration period, as increases in MyHCIIX protein have been reported in humans after a more chronic (12 week) period of GH treatment (Lange et al., 2002). In contrast, MyHC isoform transitions were observed at the mRNA level in response to BA treatment. BA induced a significant decrease in the %MyHCI and/or IIA mRNA in all three muscles studied, as well as an increase in the %MyHCIIX/IIB mRNA in the LD and ST. Importantly, BA induced de novo MyHCIIB mRNA expression in the LD (where it is not normally expressed) and increased MyHCIIB expression in the SS and ST. These differences were not observed at the protein level, but this may be due to the MyHC proteins having a slower rate of turnover compared with the corresponding mRNA, as the half-life of the MyHC proteins is up to 30 days (Kay, 1978). Therefore, it appears that BA is initiating changes in gene expression that are indicative of fibre type transitions, but the 6-day treatment may be insufficient to see a significant response at the protein level. Initially it was thought that MyHCIIB was not expressed in large mammals (Hämäläinen and Pette., 1995); however, it is expressed at both the mRNA and protein level in pigs (Lefaucheur et al., 2004). More recently MyHCIIB protein expression has been observed in muscles of the Blonde d'Aquitaine breed of bull (Picard and Cassar-Malek, 2009), but not other breeds of cattle. As in our previous paper (Hemmings et al., 2009), MyHCIIB was not expressed at the protein level in any of the three sheep muscles studied. It does, however, appear to have a low level of expression at the mRNA level in the SS and ST muscles, which was subsequently increased by BA treatment.

Interestingly, our data is similar to that of Vuocolo et al. (2007), who showed that MyHCIIB is expressed at the mRNA level in muscles from callipyge sheep. Like BA treatment, the callipyge phenotype results in muscle hypertrophy and a switch in fibre type, with a decrease in type I fibres and a shift towards type IIX and IIB fibres. When compared with normal (wild type) sheep, the authors reported 13- and 6-fold increases in MyHCIIB expression at birth and 12 weeks of age in the callipyge sheep. In that study (Vuocolo et al.,2007), four adult MyHC isoforms were detected at the mRNA level in sheep LD skeletal muscle, with MyHCIIX being the predominant isoform and MyHCIIB being expressed at extremely low levels (`1% of total adult MyHC expression). The effects of BA treatment are therefore very similar to the callipyge sheep phenotype. Longer term administration of BA might therefore result in the de novo expression of MyHCIIB protein, the isoform not normally expressed at the protein level in sheep muscles (Hemmings et al., 2009). The observed decrease in oxidative capacity (ICDH activities) suggests that BA treatment is driving muscle fibres towards a faster, less oxidative fibre type, consistent with previous observations in rats (Polla et al.,2001), cattle (Cassar-Malek et al., 1998) and pigs (Gunawan et al., 2007).

It has been suggested that the switch in fibre type contributes to the muscle hypertrophy associated with BA administration, as glycolytic fibres have a larger cross-sectional area (Lefaucheur and Gerrard, 2000). An alternative theory is that the switch in fibre type might be secondary to the hypertrophy that occurs as a result of BA treatment (Gunawan et al., 2007). On the basis of the current study, it seems unlikely that the changes in MyHC are secondary to the hypertrophy, as changes in mRNA expression have already been observed in the LD at both ages, before significant changes in muscle weight. Our alternative hypothesis is that the fibre type transitions are simply a mechanism for the muscle to adapt to changes in availability of energy substrates, with changes in metabolism preceding changes in contractile properties. BA treatment not only stimulates glycogenolysis in skeletal muscles (Sensky et al., 2006), it also increases plasma glucose concentrations (Eisemann et al., 1988). Therefore, the muscle fibres might simply be adapting to the increased glucose availability, which may have a particularly pronounced effect in ruminants, given that they are more dependent on acetate and other volatile fatty acids for energy metabolism. The significant effects of BA on MyHC expression in the LD muscle, with a strong induction of MyHC IIB expression, provide a good model for characterising the signalling events involved in fibre type switching towards a faster phenotype.

Effects Of Treatment on Signalling Molecules and Regulatory Factors

CaN and CaMKII Activity

CaN and CaMKII are calcium dependent signalling molecules implicated in both muscle hypertrophy (Dunn et al., 1999; Chin, 2005) and muscle fibre type transitions (Chin, 2005; Bassel-Duby and Olson, 2006), as they are thought to regulate the slow oxidative phenotype (Naya et al., 2000; Chin, 2005). Both are activated by calcium, being dependent on the frequency of the intracellular fluctuations in calcium concentration (Chin et al., 1998; Chin, 2005). CaN has been shown to increase MyHCIIA expression (Allen and Leinwand, 2010) and potentially decrease MyHCIIX and IIB expression (Da Costa et al., 2007) in mouse C2C12 muscle cells. Neither GH nor BA treatment affected the activities of either of these enzymes and therefore we conclude that the observed changes in MyHC expression and metabolism were independent of changes in activity of the CaN/ calmodulin kinase pathway in this study. Obviously we cannot discount that effects of these agents on this pathway may occur before or after 6 days.

Metabolic Regulatory Factors

Transcriptional co-activators of the PGC-1 family, as well as the co-repressor RIP140, have previously been shown to influence the metabolic capacity of muscle fibres, with PGC-1 increasing oxidative metabolism and RIP140 decreasing it (Lin, 2009). Indeed, they appear to have opposite effects and actually counter each other in their transcriptional regulatory roles. In the current study, there was no effect of BA treatment on PGC-1α mRNA expression, but PGC-1β mRNA was significantly decreased by BA at both ages, coincident with the observed decrease in oxidative metabolism. This might suggest that PGC-1β, rather than PGC-1α, plays an important role in regulating the BA-induced changes in oxidative metabolism in sheep muscle. The lack of an effect on PGC-1α mRNA expression was unexpected as it is thought to mediate the effects of exercise on muscle, which involves β2-AR stimulation (Miura et al., 2007). Indeed, administration of the β2-adrenergic agonist, clenbuterol, has previously been shown to decrease PGC1α mRNA and protein expression, as well as oxidative metabolism, in rat muscles (Hoshino et al.,2011). However, administration of the endogenous β-adrenergic agonist, adrenaline, had no effect on the expression of PGC-1α mRNA in human muscle (Viguerie et al., 2004) and the authors suggested that the effects of adrenaline on PGC-1α expression may be species dependent. Our data would appear to agree with this suggestion.

RIP140 mRNA expression was increased by BA administration, which agrees with a recent study in rats (Hoshino et al., 2011). This increase, together with the decrease in PGC-1β mRNA, coincides with the decrease in oxidative metabolism, suggesting that these two factors are involved in the BA-induced metabolic response.

The PPAR nuclear receptors are targets of PGC-1 and RIP140 co-activation/co-repression and are thought to be major players in regulating lipid metabolism (Hock and Kralli, 2009). In the current study, BA increased PPARδ mRNA expression, which may be involved in the increased energy expenditure observed previously in response to short-term BA treatment (MacRae et al., 1988). PPARα expression was decreased in response to GH treatment, but what role/affect this has on muscle function or metabolism is still unclear. As the short-term administration of GH did not alter MyHC expression or any of the metabolic measurements in the LD, the lack of effects on the various signalling molecules is unsurprising. However, GH did have significant effects on PGC-1β and PPARα mRNA expression, both of which are associated with metabolic changes (Hock and Kralli, 2009) and may therefore indicate that changes in muscle metabolism might be observed if GH were administered for longer. The altered expression of these metabolic regulatory genes in response to BA treatment gives additional support to the hypothesis that changes in MyHC isoform expression (and presumably fibre type transitions) are metabolically driven. The fact that changes in oxidative metabolism (ICDH activity) appear to precede any changes in protein levels of the MyHC isoforms, would support this. These key metabolic regulators may represent alternative signalling mechanisms for BA to the well-characterised signalling through β2-ARs and the cAMP activated second messenger pathway.

Six1 and Eya1

In recent years, a number of factors have been identified that regulate the slow oxidative muscle phenotype, particularly the stimulatory effects of the PGC-1 and PPAR pathways on oxidative metabolism and mitochondrial biogenesis and function. In comparison, relatively little is known about the factors that regulate the fast glycolytic phenotype. The transcription factor Six1 and cofactor Eya1, have been identified as initiating a shift in expression towards a faster muscle phenotype in mice (Grifone et al., 2004). However, our data would appear to contradict this as both factors were downregulated when the shift towards a faster, less oxidative phenotype was observed in LD in response to BA treatment. It has been reported that Six1 mRNA expression was downregulated in human skeletal muscle in response to lengthening contractions as early as 3 h after exercise (Kostek et al., 2007), whereas other genes associated with muscle hypertrophy (FHC1 and NEXN) were upregulated. Hence, Six1 may actually be a negative regulator of muscle hypertrophy that responds to BA treatment in sheep as well as to exercise in humans.

MRFs

The development and differentiation of muscle is dependent on the MRFs, MyoD, Myf-5, myogenin and MRF4. Muscle fibre number is thought to be fixed at birth in most large mammals and therefore postnatal muscle growth relates to fibre hypertrophy. This is thought to be dependent upon the differentiation and fusion of satellite cells to existing muscle fibres in order to increase the DNA content of fibres (Brameldet al., 1998). There are also suggestions that the MRFs may influence muscle fibre type composition, as MyoD mRNA expression is highest in fast glycolytic muscles of rats, whereas myogenin expression is highest in slow oxidative muscles (Hughes et al., 1993). BA administration significantly increased myogenin and decreased MyoD mRNA expression. As these changes coincided with the decrease in oxidative metabolism, the increase in MyHCIIX/IIB mRNA and induction of MyHCIIB mRNA, this would appear to contradict previous studies linking MyoD and myogenin mRNA expression with fibre type transitions in rats. However, as indicated above, the MRFs have other roles in skeletal muscle development, particularly satellite cell proliferation and fusion with muscle fibres during muscle regeneration and hypertrophy. One possible explanation for the described effects of BA in sheep muscle, is that BA increases satellite cell differentiation/fusion, which would be associated with increased myogenin and reduced MyoD expression (Brown et al., 2012). In addition, the activities/functions of these transcription factors may be dependent upon their phosphorylation status rather than simply the relative mRNA or protein expression (Puri and Sartorelli, 2000).

CONCLUSION

Over the 6-day time-course, BA had greater effects on muscle growth and phenotype than GH, especially in the older lambs. The BA-induced muscle growth was associated with changes in phenotype indicative of transitions towards a fast glycolytic fibre type in all muscles studied. The BA response included the induction of MyHCIIB mRNA expression, an isoform not normally expressed in sheep LD. These changes in MyHC expression corresponded with a decrease in oxidative metabolism (ICDH activity), which was likely induced by the combination of an increase in RIP140 and a decrease in PGC-1β mRNA expression. However, it remains unclear what (if any) factors are involved in driving towards the fast glycolytic fibre type, as expression of the previously suggested regulatory factors, Six1 and Eya1, was reduced by BA treatment. Our interpretation of the observations in this study are that changes in energy metabolism or substrate availability precede the changes in contractile proteins, but identification of the key factors involved in driving towards the fast glycolytic fibre type remains elusive.

ACKNOWLEDGEMENTS

The authors gratefully acknowledge D. Surgay (University of Nottingham) for technical assistance and J. Craigon (University of Nottingham) for statistical advice. This work was supported by a project grant from Pfizer Ltd (now Zoetis) and a Biotechnology and Biological Sciences Research Council (UK) Industrial CASE studentship (with Pfizer Ltd).

REFERENCES

1. Allen DL and Leinwand LA 2010. Intracellular calcium and myosin isoform transitions. Journal of Biological Chemistry 47, 45323–45330.
2. Bassel-Duby R and Olson E 2006. Signaling pathways in skeletal muscle remodelling. Annual Review of Biochemistry 75, 19–37.
3. Bell AW, Bauman DE, Beerman DH and Harrell RJ 1998. Nutrition, development and efficacy of growth modifiers in livestock species. Journal of Nutrition 128, 360S–363S.
4. Bradford MM 1976. Rapid and sensitive method for quantitation of microgram quantities of protein utilizing principle of protein-dye binding. Analytical Biochemistry 72, 248–254.
5. Brameld JM, Buttery PJ, Dawson JM and Harper JMM 1998. Nutritional and hormonal control of skeletal-muscle cell growth and differentiation. Proceedings of the Nutrition Society 57, 207–217.
6. Brameld JM, Atkinson JL, Saunders JC, Pell JM, Buttery PJ and Gilmour RS 1996. Effects of growth hormone administration and dietary protein intake on

insulin-like growth factor I and growth hormone receptor mRNA expression in porcine liver, skeletal muscle, and adipose tissue. Journal of Animal Science 74, 1832–1841.

7. Brandstetter AM, Picard B and Geay Y 1998. Muscle fibre characteristics in four muscles of growing bulls I. Postnatal differentiation. Livestock Production Science 53, 15–23.

8. Brown DM, Parr T and Brameld JM 2012. Myosin heavy chain mRNA isoforms are expressed in two distinct cohorts during C2C12 myogenesis. Journal of Muscle Research and Cell Motility 32, 383–390.

9. Buttery PJ and Dawson JM 1990. Growth promotion in farm animals. Proceedings of the Nutrition Society 49, 459–466.

10. Cassar-Malek I, Listrat A and Picard B 1998. Contrôle hormonal des caractéristiques des fibres musculaires après la naissance. INRA Productions Animales 11, 365–377.

11. Chin ER 2005. Role of Ca2+/calmodulin-dependent kinases in skeletal muscle plasticity. Journal of Applied Physiology 99, 414–423.

12. Chin ER, Olson EN, Richardson JA, Yang Q, Humphries C, Shelton JM, Wu H, Zhu W, Bassel-Duby R and Williams RS 1998. A calcineurin-dependent transcriptional pathway controls skeletal muscle fiber type. Genes and Development 12, 2499–2509. [PMC free article]

13. Da Costa N, Edgar J, Ooi P-T, Su Y, Meissner JD and Chang K-C 2007. Calcineurin differentially regulates fast myosin heavy chain genes in oxidative muscle fibre type conversion. Cell and Tissue Research 329, 515–527.

14. Dreiling CE, Brown DE, Casale L and Kelly L 1987. Muscle glycogen: comparison of iodine binding and enzyme digestion assays and application to meat samples. Meat Science 20, 167–177.

15. Dunn SE, Burns JL and Michel RN 1999. Calcineurin is required for skeletal muscle hypertrophy. Journal of Biological Chemistry 274, 21908–21912.

16. Eisemann JH, Huntington GB and Ferrell CL 1988. Effects of dietary clenbuterol on metabolism of the hindquarters in steers. Journal of Animal Science 66, 342–353.

17. Grifone R, Laclef C, Spitz F, Lopez S, Demignon J, Guidotti J-E, Kawakami K, Xu P-X, Kelly R, Petrof BJ, Daegelen D, Concordet J-P and Maire P 2004. Six1 and Eya1 expression can reprogram adult muscle from the slow-twitch phenotype into the fast-twitch phenotype. Molecular and Cellular Biology 24, 6253–6267. [PMC free article]

18. Gunawan AM, Richert BT, Schinckel AP, Grant AL and Gerrard DE 2007. Ractopamine induces differential gene expression in porcine skeletal muscles. Journal of Animal Science 85, 2115–2124.

19. Hämäläinen N and Pette D 1995. Patterns of myosin isoforms in mammalian skeletal muscle fibres. Microscopy Research and Technique 30, 381–389.

20. Hemmings KM, Parr T, ZCTR Daniel, Picard B, Buttery PJ and Brameld JM

2009. Examination of myosin heavy chain isoform expression in ovine skeletal muscles. Journal of Animal Science 87, 3915–3922.

21. Hock MB and Kralli A 2009. Transcriptional control of mitochondrial biogenesis and function. Annual Review of Physiology 71, 177–203.
22. Hoshino D, Yoshida Y, Holloway GP, Lally J, Hatta H and Bonen A 2011. Clenbuterol, a β2-adrenergic agonist, reciprocally alters PGC-1 alpha and RIP140 and reduces fatty acid and pyruvate oxidation in rat skeletal muscle. American Journal of Physiology – Regulatory, Integrative and Comparative Physiology 302, R373–R384.
23. Hughes SM, Taylor JM, Tapscott SJ, Gurley CM, Carter WJ and Peterson CA 1993. Selective accumulation of MyoD and myogenin mRNAs in fast and slow adult skeletal muscle is controlled by innervation and hormones. Development 118, 1137–1147.
24. Kay J 1978. Intracellular protein degradation. Biochemical Society Transactions 6, 789–797.
25. Kim YS, Lee YB and Dalrymple RH 1987. Effect of the repartitioning agent cimaterol on growth, carcass and skeletal muscle characteristics in lambs. Journal of Animal Science 65, 1392–1399.
26. Kostek MC, Chen Y-W, Cuthbertson DJ, Shi R, Fedele MJ, Esser KA and Rennie MJ 2007. Gene expression responses over 24 h to lengthening and shortening contractions in human muscle: major changes in CSRP3, MUSTN1, SIX1, and FBXO32. Physiological Genomics 31, 42–52.
27. Lange KHW, Andersen JL, Beyer N, Isaksson F, Larsson B, Rasmussen MH, Juul A, Bulow J and Kjaer M 2002. GH administration changes myosin heavy chain isoforms in skeletal muscle but does not augment muscle strength or hypertrophy, either alone or combined with resistance exercise training in healthy elderly men. The Journal of Clinical Endocrinology and Metabolism 87, 513–523.
28. Lefaucheur L and Gerrard D 2000. Muscle fiber plasticity in farm mammals. Journal of Animal Science 77 (E Suppl.), 1–19.
29. Lefaucheur L, Milan D, Ecolan P and Le Callennec C 2004. Myosin heavy chain composition of different skeletal muscles in Large White and Meishan pigs. Journal of Animal Science 82, 1931–1941.
30. Lin JD 2009. The PGC-1 coactivator networks: chromatin-remodeling and mitochondrial energy metabolism. Molecular Endocrinology 23, 2–10. [PMC free article]
31. MacRae JC, Skene PA, Connell A, Buchan V and Lobley GE 1988. The action of the β-agonist clenbuterol on protein and energy metabolism in fattening wether lambs. British Journal of Nutrition 59, 457–465.
32. McLaughlin CL, Hedrick HB, Veenhuizen JJ, Hintz RL, Munyakazi L, Kasser TR and Baile CA 1994. Perfomance, clinical chemistry, and carcass responses of finishing lambs to formulated sometribove (methionyl bovine somatotropin). Journal of Animal Science 72, 2544–2551.

33. Miura S, Kawanaka K, Kai Y, Tamura M, Goto M, Shiuchi T, Minokoshi Y and Ezaki O 2007. An increase in murine peroxisome proliferators-activated receptor-γ coactivator-1α (PGC-1α) mRNA in response to exercise is mediated by β-adrenergic receptor activation. Endocrinology 148, 3441–3448.

34. National Research Council 1994. Metabolic modifiers: effects on the nutrient requirements of food-producing animals. National Academy Press, Washington, DC, USA.

35. Naya FJ, Mercer B, Shelton J, Richardson JA, Williams RS and Olson EN 2000. Stimulation of slow skeletal muscle fiber gene expression by calcineurin in vivo. The Journal of Biological Chemistry 275, 4545–4548.

36. O'Connor RM, Butler WR, Finnerty KD, Hogue DE and Beermann DH 1991. Acute and chronic hormone and metabolite changes in lambs fed the beta-agonist, cimaterol. Domestic Animal Endocrinology 8, 537–548.

37. Picard B and Cassar-Malek I 2009. Evidence for expression of IIb myosin heavy chain isoform in some skeletal muscles of Blonde d'Aquitaine bulls. Meat Science 82, 30–36.

38. Polla B, Cappelli V, Morello F, Pellegrino MA, Boschi F, Pastoris O and Reggiani C 2001. Effects of the β2-agonist clenbuterol on respiratory and limb muscles of weaning rats. American Journal of Physiology – Regulatory, Integrative and Comparative Physiology 280, R862–R869.

39. Puri PL and Sartorelli V 2000. Regulation of muscle regulatory factors by DNA-binding, interacting proteins, and post-transcriptional modifications. Journal of Cellular Physiology 185, 155–173.

40. Sazili AQ, Parr T, Sensky PL, Jones SW, Bardsley RG and Buttery PJ 2005. The relationship between slow and fast myosin heavy chain content, calpastatin and meat tenderness in different ovine skeletal muscles. Meat Science 69, 17–25.

41. Sensky PL, Jewell KK, Ryan KJP, Parr T, Bardsley RG and Buttery PJ 2006. Effect of anabolic agents on calpastatin promoters in porcine skeletal muscle and their responsiveness to cyclic adenosine monophosphate- and calcium-related stimuli. Journal of Animal Science 84, 2973–2982.

42. Viguerie N, Clement K, Barbe P, Courtine M, Benis A, Larrouy D, Hanczar B, Pelloux V, Poitou C, Khalfallah Y, Barsh GS, Thalamas C, Zucker J-D and Langin D 2004. In vivo epinephrine-mediated regulation of gene expression in human skeletal muscle. The Journal of Clinical Endocrinology and Metabolism 89, 2000–2014.

43. Vuocolo T, Byrne K, White J, McWilliam S, Reverter A, Cockett NE and Tellam RL 2007. Identification of a gene network contributing to hypertrophy in callipyge skeletal muscle. Physiological Genomics 28, 253–272.

7
CHAPTER

PREVALENCE OF LAMENESS AND CLAW LESIONS DURING DIFFERENT STAGES IN THE REPRODUCTIVE CYCLE OF SOWS AND THE IMPACT ON REPRODUCTION RESULTS

L. M. Pluym, A. Van Nuffel, S. Van Weyenberg and D. Maes

Keywords: Sow, lameness, reproduction, culling, claw lesions

IMPLICATIONS

Knowledge on the economic impact of claw lesions and lameness helps to understand how both the diseases influence economic losses, and indicates how these problems should best be addressed. The fact that the prevalence of lameness varies throughout the reproductive cycle and peaks after sows were housed in the insemination stable is crucial for risk factor analysis. Furthermore, it implicates that focus on housing in the insemination stable is important to improve the prevention strategy.

INTRODUCTION

Lameness is a growing concern in swine breeding herds. Owing to the pain, suffering and limited freedom of movement, lameness is recognized as an important welfare concern (Whay *et al.*, 2003). Nowadays, the disease condition is implemented in the Welfare Quality® Assessment Protocol for pigs (Welfare Quality®,2009).

In addition to the impaired welfare, lameness is allied to financial loss estimated at €37 per lame sow in Germany (Grandjot, 2007), $180 per lame sow in the United States (Deen *et al.*, 2008) and in a Dutch study, €20 to 30 per sow present in the farm (Schuttert, 2008). Economic losses can be attributed to increased work load, higher

---

Citation: Pluym LM, Van Nuffel A, Van Weyenberg S, Maes D. Prevalence of lameness and claw lesions during different stages in the reproductive cycle of sows and the impact on reproduction results. Animal. 2013;7(7):1174-1181. doi:10.1017/S1751731113000232.

veterinary costs because of treatment or euthanasia, higher risk for total or partial condemnation resulting in less slaughter revenue and to an impact on reproduction (Rowles, 2001; Schuttert, 2008). The impact on reproduction can be divided into a direct and an indirect effect. Literature on the indirect effect of lameness on reproduction is more widespread and much more consistent than the direct impact on reproduction.

Indirectly, lameness may affect reproduction of sows by influencing sows' longevity, behaviour and feed intake. Lameness is the most important reason for euthanasia and the second most important reason for involuntary culling (Engblom *et al.*, 2007; Jensen *et al.*, 2010), leading to a detrimental impact on sow longevity. According to Bonde *et al.*, (2004), lameness leads to uncontrolled lying-down behaviour and, as a consequence, it may augment the risk of crushing piglets. The higher level of acute-phase proteins in lame sows indicates the presence of inflammatory processes (Heinonen *et al.*, 2006). Cytokines released by the inflammatory processes predispose to anorexia and lethargic behaviour (Johnson, 1997). In addition, Fitzgerald *et al.*, (2012) reported a negative effect of overgrown toes on feed consumption during lactation. Early culling of sows, reduced feed intake and impaired locomotion indirectly decrease the mean number of litters per sow per year and the mean number of weaned pigs per sow per year, increasing the cost per weaned piglet.

However, the direct effect of lameness on reproduction (i.e. direct improvement or deterioration of the breeding or farrowing performance of a sow), is less clear. Several studies did not find any effect (Kronemanet *al.*, 1993; Andersen and Bøe, 1999; Heinonen *et al.*, 2006; Willgert, 2011), whereas other studies described a negative relationship between lameness and farrowing performance (Grandjot, 2007; Anil *et al.*,2009). Moreover, comparison between these studies is difficult as different methods and different stages in the reproductive cycle were used to assess lameness. As sows can recover over time, it may be recommended to assess lameness in the same sows at several times. In addition, knowledge on the stage in the reproductive cycle at which lameness is most prevalent may have a key value regarding risk factor analysis.

In breeding herds, sows are commonly housed in three main areas: the insemination stable (shortly after weaning), the gestation stable and the farrowing crates. Each of these areas is related to a different physiological status of the sow and is characterized by its own specific floor, feeding strategy, environment and management. Several studies already referred to floor characteristics (Newton *et al.*, 1980; Mouttotou *et al.*, 1999) and feed (Simmins and Brooks, 1988) as important risk factors for claw lesions and lameness. When sows are moved to a new area within the farm, the possible risk factors for development of lameness to which sows may be exposed will also change.

The aim of the present study was to evaluate the short-term effect of lameness and claw lesions on (re)production at the three main stages of the reproductive cycle in sows. In addition, the prevalence of lameness and claw lesions was investigated at each of these three stages.

MATERIAL AND METHODS

Animals and Housing

A total of 491 sows from five randomly selected farms were included in the study. General information about every farm is given in Table 1. All farms had at least 750 sows and all sows were housed in groups during gestation. On all farms, the gestation stable was provided with a partly slatted, concrete floor without bedding material. Farms 1 to 4 used free access stalls. On farm 5, sows were housed in small groups of 15 sows per pen and were fed by use of trough feeding. Space per animal during gestation ranged between 2.0 and 2.6 m^2. At the insemination unit, sows were housed individually with solid concrete flooring in the front half and concrete slats in the rear half. Only on the farm 5 metal slats were used in the rear half. The width of the concrete slats varied between 8 and 11 cm whereas the metal slats measured 1 cm. Slots of the floors on all farms did not exceed 2 cm. Farrowing stalls contained cast iron (partly), slatted flooring without bedding. During lactation, sows were provided feed and water *ad libitum*. As farm 2 had recently started with new sows, all sows had farrowed only once at the beginning of the study.

Table 1: Descriptive data of the farms (n = 5) and the investigated group of sows on each farm in the study

	Farm 1	Farm 2	Farm 3	Farm 4	Farm 5
Farms in study					
Farm size (no. of sows)	1100	1000	1700	750	750
Breed	French hybrid	PIC	Danbred	Danbred	Topigs-40
Batch farrowing system	Weekly	2 weekly	2 weekly	2 weekly	3 weekly
Type of group housing	FAS	FAS	FAS	FAS	Small pen
Litters/sow/year	2.2	2.5	na	2.2	2.3
Piglets weaned/sow/year	23.6	29.2	na	26.0	28.2
Piglets born alive/litter	11.6	12.8	na	11.7	13.9
% stillborn piglets	10.8	4.4	na	4.2	9.1
Sows in study					
Total no. of sows	54	109	141	86	101
Parity (mean (s.d.))	2.3 (1.5)	1.0 (0.0)	3.3 (1.3)	2.1 (1.4)	3.8 (1.9)
Parity of sows that remained in the group (mean (s.d.))	2.3 (1.7)	1.0 (0.0)	3.3 (1.4)	2.1 (1.5)	3.5 (1.7)
Parity of sows that left the group (mean (s.d.))	2.3 (1.2)	1.0 (0.0)	3.1 (0.8)	2.2 (1.2)	4.4 (2.1)

FAS = free access stalls; na = data not available.
Part of the sows left the group during the study due to culling, euthanasia, death, anoestrus, rebreeding or abortion. The mean parity of the sows that left the group as well as of the sows that remained in the group is also given.

Measurements

Within each farm, one randomly selected group of sows was followed up for a period of one reproductive cycle (from weaning to weaning). Every group was investigated five times, that is two claw lesion scorings and three lameness assessments, over a period of 5 months (Figure 1). Visual lameness assessment was performed on the day the sows were moved from one stage on the farm to another (L1, L2 and L3) and was based on the Welfare Quality Protocol. In this protocol, a lame sow is defined as a sow unable to use one or more limbs in a normal manner varying in severity from reduced ability to bear weight, to total recumbancy. However, in the present study, only distinction was made between lame and healthy sows. Claw lesion scoring of sows was carried out twice in the farrowing stable: 1 week before the sows were weaned (C1) and 1 week after farrowing (C2) (Figure 1). Seven claw parameters were scored: length of toes, length of dew claws, wall cracks, heel cracks and/or overgrowth, skin lesions above the claw, heel–sole cracks and white line cracks. The first five parameters were based on the Dutch scoring method 'Nederlandse Zeugenklauwencheck' (Hoofs, 2006). The last two parameters were scored following the Zinpro®Feet First method (Feet First®Team, 2010). Each of the four claws of the hind legs were scored from one to four, with score 1 meaning no lesion and score 4 meaning a severe lesion. The sum of the score of the four claws was made and was called the 'total score' (minimum 4 to maximum 16). However, when performing statistical analysis, total score was included as a dichotomous variable. This means that sows without lesions on any of the four claws (total score = 4) were coded 0 and sows with a total score of 5 to 16 were coded 1. By summing the seven 'total scores', a 'global score' was obtained for each sow (minimum 28 to maximum 112). For the data and statistical analyses, global scores were regarded as continuous variables.

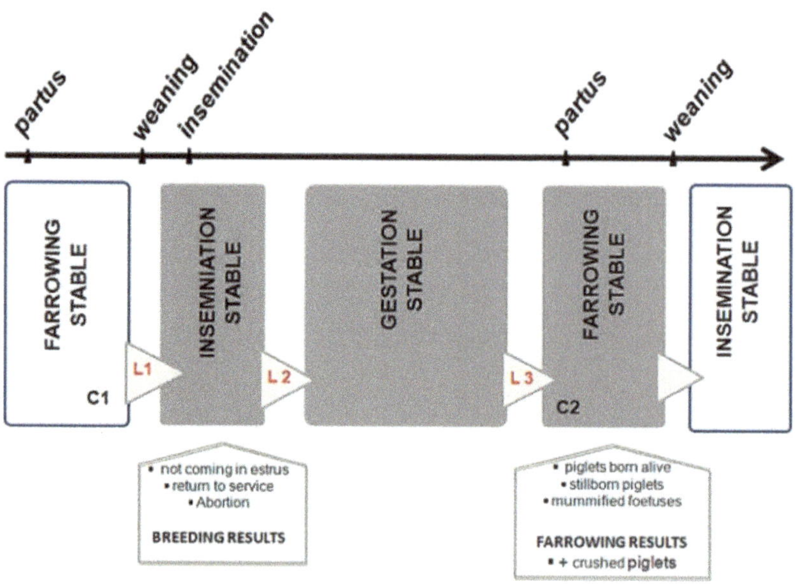

Figure 1: Measurements during the reproductive cycle of the sows: Lameness assessment was performed when sows were moved between different stables (L1 to L3). Claw lesion scoring was performed in the farrowing stable: 1 week before weaning and 1 week after farrowing Parity of sows was categorized into four groups: 1, 2, 3 to 5 and ≥ 6 parity.

Reproduction results in this study are defined as a combination of the farrowing performance, crushed piglets and the breeding performance. Information on farrowing performance and crushed piglets that was collected for the investigated cycle included the number of piglets born alive (continuous variable) and the absence or presence of stillborn piglets, mummified foetuses and crushed piglets (dichotomous variables). The breeding performance of sows included 'showing oestrus post-weaning', 'weaning-to-oestrus interval', 'returning to service' and 'abortion' (defined as expulsion of foetuses between day 35 and day 108 of gestation). The weaning-to-oestrus interval was regarded as a continuous variable, the other three as dichotomous variables. Coming into oestrus within 10 days after weaning, positive pregnancy diagnosis by ultrasound scanning 23 to 35 days after breeding and absence of abortion were coded as normal (0).

Sows that left the group during the study were recorded, as well as the time and the reason (not coming in oestrus, returning to service, abortion, death, euthanasia or culling). As it is essentially different, distinction was made between sows that left the group because they were removed from the farm (sows removed from the farm) and sows that left the group but that remained on the farm (sows removed from the group). The reasons for removing sows from the farm were culling, euthanasia or death. The reason why sows left the group but remained in the farm was because they did not show oestrus, they returned to service or they aborted.

An overview of the reproduction results across all farms is given in Table 2.

Table 2: Overview of the farrowing and breeding performance of the sows in the study

Farrowing performance	Mean	s.e.	Minimum	Maximum
Piglets born alive	14	0.18	0	23
Stillborn piglets	1.2	0.08	0	11
Mummified foetuses	0.5	0.05	0	7
Crushed piglets	0.6	0.06	0	11
Breeding performance	**Mean**	**s.e.**	**Mean %**	
Weaning-to-oestrus interval	6.0	0.33	–	
Number of sows				
Not showing oestrus	24	–	6	
Returning to service	39	–	10	
Abortion	4	–	1	

Descriptive data on the number of piglets born alive, stillborn piglets, mummified foetuses and crushed piglets over all 381 sows as well as descriptive data on breeding performance including the mean weaning-to-oestrus interval and the mean number and percentage of sows that did not show oestrus, did return to service or aborted during the study.

Statistical Analysis

Normality of continuous variables was evaluated. Linear and logistic regression models were built with farrowing and breeding-related variables as dependent variables. To account for the clustering of sow in the farm, a fixed farm effect was included. In addition, total and global claw lesion scores, as well as lameness and parity, were included as fixed effects. A stepwise forward model-building procedure was followed, and all factors with a *P*-value <0.05 were retained in the final multivariable model.

Descriptive statistics were used to illustrate the prevalence of claw lesions and lameness throughout the productive cycle. To evaluate whether parity and global score differ between healthy and lame sows (0/1), between sows that were (1) or were not (0) removed from the farm, and between sows removed because of lameness (1) or because of other reasons (0), univariable linear regressions, with either parity or global score as dependent variable and farm as fixed effect, were performed. To evaluate the difference in global score between the first and the second claw lesion scoring, mixed model linear regression was performed, with global score as dependent variable, sow as random effect, to correct for repeated measurements, and time of claw lesion scoring, farm and interaction time × farm as fixed effects.

All statistical analyses were performed using IBM SPSS Statistics 19 (IBM, New York, USA).

To analyse the effect of the three measuring periods (L1, L2 and L3) on lameness, a multilevel logistic regression model was fit using MLwiN 2.02 (Centre for Multilevel Modeling, Bristol, UK). Farm (1 to 5), measuring period and the interaction farm × measuring period were included as fixed effects.

RESULTS

Impact on (re)Production Results

After stepwise forward model building, significant associations were found between the independent variables (parity, lameness, claw score and farm) and the dependent variables (stillborn piglets, mummified foetuses, crushed piglets and number of live born piglets) (Table 3). In the final model, no significant associations between risk factors and breeding performance (showing oestrus post weaning, weaning-to-oestrus interval, returning to service and abortion) were found. However, several claw lesions were related to farrowing performance. The presence of skin lesions above the claw ($P = 0.021$) and white line lesions ($P = 0.036$) significantly increased the odds of stillborn piglets. Heel lesions were significantly associated with the presence of crushed piglets ($P = 0.017$). The odds of mummified foetuses increased when wall cracks were present ($P = 0.044$). An influence of lameness could only be found for the presence of mummified foetuses. The odds of having mummified foetuses were twice as high in lame sows compared with healthy sows. A significant farm effect was found for all of the farrowing results, whereas parity was significantly associated with the presence of stillborn piglets and crushed piglets (Table 3).

Table 3: Final multivariable logistic regression models related to the risk factors for farrowing performance and crushed piglets with farm, parity, claw score and lameness as independent variables and presence of stillborn piglets, mummified foetuses and crushed piglets as dependent variables

Variable	*n*	b	s.e.	Odds ratio	95% CI	*P*-value
Stillborn piglets (0/1)[e]						
Intercept		0.24	0.47	-		
Farm	5					0.025
Parity	4					0.012
Parity 1		Reference	-			
Parity 2		1.61	0.51	5.00	13.59−1.84	0.002
Parity 3 to 5		1.14	0.43	3.14	7.26−1.35	0.008
Parity ≥6		1.27	0.73	3.56	14.89−0.85	0.080
White line 1[a]	306	0.65	0.31	1.91	1.04−3.51	0.036
Skin 1[b]	306	1.42	0.62	4.13	1.23−13.79	0.021
Mummified fetuses (0/1)[e]						
Intercept		−1.08	0.43	-		
Farm	5					0.052
Wall cracks 1[c]	306	−0.75	0.37	0.47	0.23−0.98	0.044
Lameness	306	0.87	0.35	2.38	1.19−4.75	0.014
Crushed piglets (0/1)[e]						
Intercept		1.58	0.97	-		
Farm	4					<0.001
Parity	4					0.029
Parity 1		Reference	-			
Parity 2		0.20	0.52	1.23	3.38−0.44	0.696
Parity 3 to 5		1.18	0.47	3.24	8.17−1.29	0.013
Parity ≥6		1.29	0.73	3.63	15.19−0.87	0.077
Heel lesions 1[d]	236	−2.23	0.93	0.11	0.02−0.67	0.017

Regression analyses were based on the results of 306 sows.
[a]Total claw score for white line lesions.
[b]Total claw score for skin lesions above the claw.
[c]Total claw score for wall cracks.
[d]Total claw score for heel cracks and overgrowth. The number behind the total score refers to the first time (at weaning) or second time (at farrowing) claw lesion scoring was performed.
[e]Absence (0) or presence (1) of stillborn, mummified and crushed piglets, respectively.
b, regression coefficient.

Prevalence of Claw Lesions

The percentage of sows with and without lesions for each of the seven claw parameters that were scored the first and second time claw lesion scoring was performed, are represented in Figure 2. Heel cracks were found to be the most frequent lesions (96% sows with lesions) whereas skin lesions above the claw were only occasionally found (8% of sows with lesions). No significant difference in global score, the sum of seven total scores, was present between the sows that remained in the group and the sows that left the group during the study. In general, the mean global score slightly improved from first (mean (s.d.) score of 35.8 (3.9)) to second time (mean (s.d.) score of 34.9 (3.2)). However, a significant farm × time interaction ($P = 0.012$) was found. The change between the first and second global score was found to be significant only at farms 1 and 4.

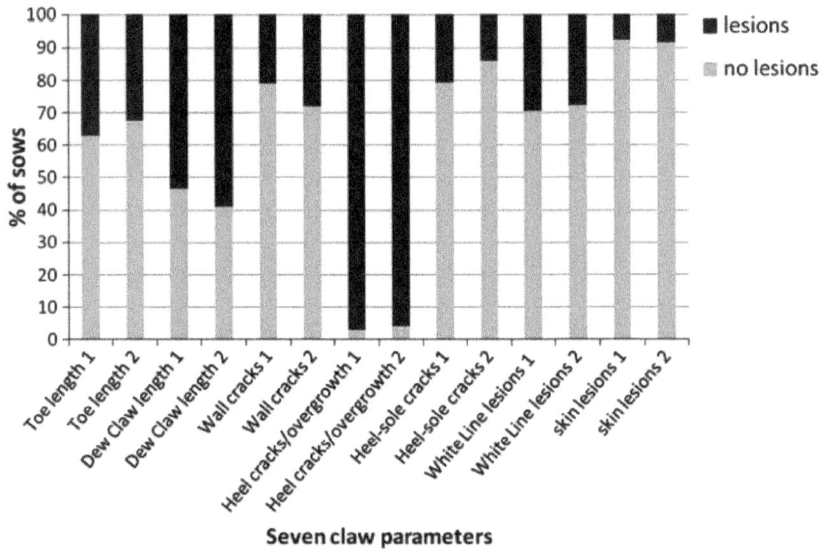

Figure 2: Percentage of sows (total of 381) with and without lesions for each of the seven claw parameters that were scored at the first (1) and second (2) time claw lesion scoring was carried out. 1, performed at the end of the first lactation period, at the beginning of the study; 2, performed 5 days after parturition in the second lactation period, at the end of the study.

Prevalence of Lameness

In total, 106 sows (21.6%) left the group during the study. Of these, 54 sows (51.0%) were removed from the group with returning to service as the most important reason. The remaining 52 sows (49.0%) were removed from the farm because of culling, euthanasia or death. Udder health problems (33%) and feet and leg problems (15%) were the most important reasons for culling. Sows that were removed from the farm had a mean parity of 4.0 ± 1.8, whereas sows specifically culled because of feet and leg problems had a mean parity of 2.6 ± 1.3.

Figure 3 shows the percentage of sows that were scored lame at the three visual lameness assessments (L1-L2-L3). The mean percentage of lame sows for all farms was 5.5 ± 3.3%, 8.1 ± 6.1% and 4.1 ± 4.4% at L1, L2 and L3, respectively. The difference between L1, L2 and L3 was found to be significant ($P = 0.027$) and the prevalence of lameness significantly varied according to the farm ($P = 0.002$). The interaction between farm and measuring period was not significant. On every farm, the percentage of lame sows peaked at the moment of the second lameness investigation, that is, when sows were moved from the insemination stable to the gestation stable. At L1, 27 sows were assessed to be lame. Of these, nine sows left the group before L2 could be performed. At L2, a total of 32 lame sows were found. Only two sows that were lame at L1 were still lame at L2 and four sows that were lame at L2 were also lame at L3. There was no significant difference in parity between sows that were lame (mean parity of 2.6 ± 1.7) and sows that were not lame (mean parity of 2.5 ± 1.7).

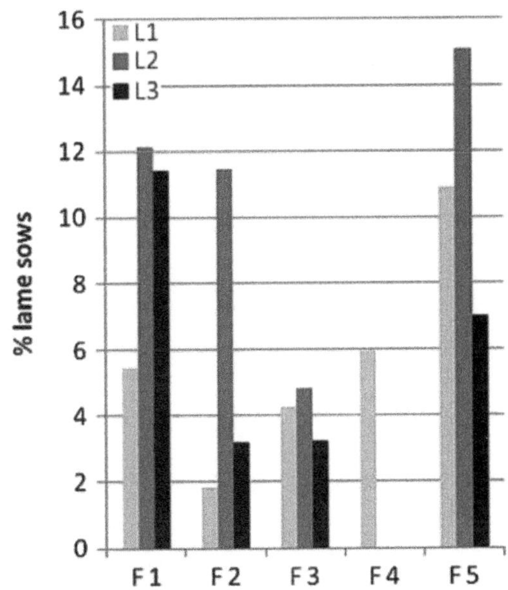

Figure 3: Percentage of sows that were lame at the three visual lameness assessments (L1 to L3) for each farm (F1 to F5).

DISCUSSION

Direct Impact

On the basis of the models, some claw lesions seem to have a direct impact on farrowing performance of sows. An effect of claw lesions on reproductive performance was also reported by De Pita (2010), Anil (2011) and Fitzgerald *et al.* (2012) but could not be shown by Enokida *et al.* (2011). In the present study, white line lesions and skin lesions above the claw increased the risk for stillborn piglets. Both skin lesions and white line lesions may facilitate entry of bacteria causing infection, pain and clinical lameness. An association between white line lesions and lameness in breeding sows was described by Anil *et al.* (2007). De Pita (2010) reported a positive association between white line lesions and the likelihood of having $\leq$ 10 piglets born alive. In the same study, the likelihood of having $\leq$ 10 piglets born alive was positively associated with the number of stillborn piglets. The reason why sows with wall cracks have lower odds of having mummified foetuses remains unknown.

Contrary to claw lesions, a clear direct impact of lameness on breeding and farrowing performance could not be shown with the exception of the effect on mummified foetuses. Lameness is a clinical sign that shows that movement is painful. The higher serum concentrations of acute-phase proteins in lame sows reported by Heinonen *et al.* (2006) indicate the presence of inflammatory processes. Pain and inflammation may cause reduction in feed intake and consequently may lead to a poor body condition of the sow. Poor body condition has been associated with the presence of mummified foetuses in sows (Knauer *et al.*, 2006).

Apart from the effect on mummified foetuses, no effect of lameness on reproduction was found in the present study. This is in contrast with the lower number of piglets born alive and higher piglet losses before weaning reported by Grandjot (2007) and Anil *et al.* (2009). The absence of a significant association between lameness and reproductive performance may be explained by the fact that reproduction results may strongly differ between individual sows. Therefore, it can be suggested to evaluate the impact on reproduction in longitudinal studies covering several reproductive cycles and using a sow as her own reference. However, in this study, parity and farm, as indicators for breed effect, were already taken into account when performing linear and logistic regression.

Indirect Impact

In addition to the direct impact, lameness and claw lesions are assumed to have an indirect impact on reproduction results. Indeed, a relationship with the risk of removal from the farm was found as well as a significant association between heel lesions and crushed piglets.

With a prevalence of 15%, feet- and leg problems appeared to be the second most important reason for sow removal in this study. Moreover, sows removed from the farm because of feet and leg problems were significantly younger compared with sows removed because of other reasons. The finding that feet and leg problems are important reasons of early culling is in agreement with the results of other studies (Stalder *et al.*,2003; Engblom *et al.*, 2007; Jensen *et al.*, 2010). Sows reach peak production between the third and sixth parity. Hence, removal of sows at a mean parity of 2.6 implies that part of the sows did not reach their most productive parities, nor repaid their investment costs yet (Ritter *et al.*, 1999; Stalder *et al.*, 2003; Anil *et al.*,2009). A higher removal rate also increases replacement costs and the frequency of adding new gilts, which in turn implicates a higher health risk to the animals currently in the farm.

In this study, a negative association was found between the presence of heel lesions and presence of crushed piglets. Heel lesions are very common among sows. The high percentage of sows with heel lesions (96%) in this study is in agreement with the results of previous studies (Gjein and Larssen, 1995; Anil *et al.*, 2007; Pluym *et al.*, 2011). Only a few sows did not have any heel lesion. Owing to the low number of sows without lesions, the association found between heel lesions and crushed piglets may not be an effect but just coincidence.

Prevalence of Claw Lesions and Lameness

To be able to perform thorough risk factor analysis, and to start prevention at the right moment, knowledge on the changes in the prevalence of lameness and claw lesions throughout the reproductive cycle is important. The mean global score in general remained unchanged but significantly improved from first to second claw lesion scoring on farms 1 and 4. Trimming of long toes and healing of lesions may have contributed to this improvement. The high prevalence of sows with heel cracks and overgrowth was also observed in other studies (Gjein and Larssen, 1995; Anil *et al.*, 2007; Pluym *et al.*, 2011). The mean prevalence of lameness namely 5.6%, 8.7% and 5.1% at L1, L2 and L3, respectively, were slightly lower compared with the prevalence of 9.7% that was found in an earlier study conducted on Belgian farms (Pluym*et al.*, 2011). High differences between farms, as seen in other studies (Heinonen *et al.*, 2006; Pluym *et al.*,2011), were also noticed in this study. Group housing is mentioned as an important risk factor for lameness (Anil *et al.*, 2005; Chapinal *et al.*, 2010). Notwithstanding, in this study, the prevalence of lameness was lowest at the end of the gestation period (L3) and highest at the moment the sows were moved from the insemination stable to the gestation stable (L2). On farm 2, lameness assessment could not be done when sows were moved from the insemination to the gestation stable, and therefore had to be done 2 weeks after the sows were introduced in group housing (stable groups). However, even without this farm, the prevalence remained highest at L2 (data not shown). Group housing probably can influence the development of lameness, especially shortly after introduction when sows fight to establish a dominance hierarchy. When this equilibrium is found, sows may have the time to heal from their injuries (Arey, 1998). Nevertheless, results of this study suggest that even housing in the insemination stable may have an impact on the development of lameness. The increase in prevalence at L2 is caused by a higher number of lame sows and not by a lower number of total sows owing to the sows leaving the group between L1 and L2. In the insemination stable, sows were housed in conventional stalls. Owing to immobility, muscle conformation and bone strength can deteriorate (Marchant and Broom, 1994). However, in the present study, sows were housed in stalls for at most 5 weeks, whereas impact of restricted mobility on the development of muscle and bone strength (Marchant and Broom, 1994), as well as on lameness (Karlen *et al.*, 2007), could only be shown after housing sows in stalls during almost the entire gestation period.

In addition to the differences between L1, L2 and L3, the prevalence of lameness significantly varied between farms. Differences in management (e.g. handling of animals, moving the sows at the time of oestrus detection) might explain the variation between farms. These results indicate that the prevalence of lameness differs throughout the reproductive cycle and that changes vary according to the farm. This substantiates our suggestion that lameness should be assessed several times during the reproductive cycle.

CONCLUSION

A direct effect of certain claw lesions on the farrowing performance of sows was shown, whereas for lameness a clear indirect impact on (re)production results was found. The prevalence of lameness differed between farms and varied throughout the reproductive cycle. Housing in the insemination stable seems to induce more lameness than housing in the gestation stable in the farms studied. Therefore, further investigation on risk factors, especially related to the insemination stable and introduction of sows in group housing is needed.

ACKNOWLEDGEMENTS

This research was granted by 'The Institute for Innovation through Science and Technology (IWT) Flanders, Belgium' (Innovation study SB-091420). The authors are grateful to the farmers for their willingness to cooperate in this study.

REFERENCES

1. Andersen IL, Bøe KE 1999. Straw bedding or concrete floor for loose-housed pregnant sows: consequences for aggression, production and physical health. Acta Agriculturae Scandinavica, Section A – Animal Science 49, 190–195
2. Anil SS 2011. Epidemiology of lameness in breeding female pigs. PhD University of Minnesota.
3. Anil SS, Anil L, Deen J 2009. Effect of lameness on sow longevity. Journal of the American Veterinary Medical Association 235, 734–738
4. Anil L, Anil SS, Deen J, Baidoo SK, Wheaton JE 2005. Evaluation of well-being, productivity and longevity of pregnant sows housed in groups in pens with an electronic sow feeder or separately in gestation stalls. American Journal of Veterinary Research 66, 1630–1638
5. Anil SS, Anil L, Deen J, Baidoo SK, Walker RD 2007. Factors associated with claw lesions in gestating sows. Journal of Swine Health and Production 15, 78–83
6. Arey DS 1998. Time course for the formation and disruption of social organization in group-housed sows. Applied Animal Behaviour Science 62, 199–207
7. Bonde M, Rousing T, Badsberg JH, Sørensen JT 2004. Associations between lying-down behaviour problems and body condition, limb disorders and skin lesions of lactating sows housed in farrowing crates in commercial sow farms. Livestock Production Science 87, 179–187
8. Chapinal N, Ruiz de la Torre JL, Cerisuelo A, Gasa J, Baucells MD, Coma J, Vidal A, Manteca X 2010. Evaluation of welfare and productivity in pregnant sows kept in stalls or in 2 different group housing systems. Journal of Veterinary Behavior 5, 82–93
9. Deen J, Anil SS, Anil L, Baidoo S 2008. Lameness overview and awareness: implications for welfare, housing, performance and economics. FeetFirstTM Symposium on Sow Lameness, Asten/Sterksel, The Netherlands, 4pp.
10. De Pita ADS 2010. Sow lameness and claw lesions evaluation. Master Thesis, University of Minnesota.
11. Engblom L, Lundeheim N, Dalin AM, Andersson K 2007. Sow removal in Swedish commercial farms. Livestock Science 106, 76–86
12. Enokida M, Sasaki Y, Hoshino Y, Saito H, Koketsu Y 2011. Claw lesions in lactating sows on commercial farms were associated with postural behavior but not with suboptimal reproductive performance or culling risk. Livestock Science 136, 256–261

13. Feet First® Team 2010. Feet First® swine claw lesion identification. Retrieved December 23, 2010, from http://www.zinpro.com/lameness/swine/lesion-identification

14. Fitzgerald RF, Stalder KJ, Karriker LA, Sadler LJ, Hill HT, Kaisand J, Johnson AK 2012. The effect of hoof abnormalities on sow behavior and performance. Livestock Science 145, 230–238

15. Gjein H, Larssen RB 1995. Housing of pregnant sows in loose and confined systems – a field study. 2. Claw lesions: morphology, prevalence, location and relation to age. Acta Veterinaria Scandinavica 36, 433–442

16. Grandjot G 2007. Claw problems cost money SUS-Schweinezucht und Schweinemast Landwirtschaftsverlag GmbH. Munster-Hiltrup, Germany: 5, 28–31

17. Heinonen M, Oravainen J, Orro T, Seppä-Lassila L, Ala-Kurikka E, Virolainen J, Tast A, Peltoniemi OAT 2006. Lameness and fertility of sows and gilts in randomly selected loose-housed farms in Finland. The Veterinary Record 159, 383–387

18. Hoofs AIJ, Netwerk 'perfecte zeugenklauw' 2006. De Nederlandse Zeugenklauwencheck (in Dutch). Retrieved December 23, 2010, from http://www.verantwoordeveehouderij.nl/netwerken/enquete/zeugenklauwencheck/Klauwencheck.pdf

19. Jensen TB, Bonde MK, Kongsted AG, Toft N, Sørensen JT 2010. The interrelationships between clinical signs and their effect on involuntary culling among pregnant sows in group-housing systems. Animal 4, 1922–1928

20. Johnson RW 1997. Inhibition of growth by pro-inflammatory cytokines: an integrated view. Journal of Animal Science 75, 1244–1255

21. Karlen GAM, Hemsworth PH, Gonyou HW, Fabrega E, Strom AD, Smits RJ 2007. The welfare of gestating sows in conventional stalls and large groups on deep litter. Applied Animal Behaviour Science 105, 87–101

22. Knauer MT, Stalder KJ, Karriker L, Johnson C, Layman L 2006. Factors influencing sow culling. Retrieved November 2, 2012, from http://www.nsif.com/Conferences/2006/pdf/FactorsInfluencingSowCulling.pdf

23. Kroneman A, Vellenga L, van der Wilt FJ, Vermeer HM 1993. Field research on veterinary problems in group-housed sows – a survey of lameness. Journal of Veterinary Medicine A 40, 704–712

24. Marchant JN, Broom DM 1994. Effects of housing system on movement and leg strength in sows. Applied Animal Behaviour Science 41, 275–276

25. Mouttotou N, Hatchell FM, Green LE 1999. Foot lesions in finishing pigs and their associations with the type of floor. The Veterinary Record 144, 629–632

26. Newton GL, Booram CV, Hale OM, Mullinix G Jr 1980. Effects of four types of floor slats on certain feet characteristics and performance of swine. Journal of Animal Science 50, 7–20

27. Pluym L, Van Nuffel A, Dewulf J, Cools A, Vangroenweghe F, Van Hoorebeke S, Maes D 2011. Prevalence and risk factors of claw lesions and lameness in pregnant sows in two types of group housing. Veterinarni Medicina 56, 101–109
28. Ritter LA, Xue J, Dial GD, Morrison RB, Marsh WE 1999. Prevalence of lesions and body condition scores among female swine at slaughter. Journal of the American Veterinary Medical Association 214, 525–528
29. Rowles C 2001. Sow lameness. Journal of Swine Health and Production 9, 130–131
30. Simmins PH, Brooks PH 1988. Supplementary biotin for sows: effect on claw integrity. Veterinary Record 122, 431–435
31. Schuttert M, 2008. The economical impact of lameness in sows. FeetFirstTM Symposium on Sow Lameness, Asten/Sterksel, The Netherlands, 4pp.
32. Stalder KJ, Lacy CR, Cross TL, Conatser GE 2003. Financial impact of average parity of culled females in a breed-to-wean swine operation using replacement gilt net present value analysis. Journal of Swine Health and Production 11, 69–74
33. Welfare Quality ® 2009. Welfare Quality® assessment protocol for pigs (sows and piglets, growing and finishing pigs). Welfare Quality® Consortium, Lelystad, The Netherlands
34. Whay HR, Main DCJ, Green LE, Webster AJF 2003. Animal-based measures for the assessment of welfare state of dairy cattle, pigs and laying hens: consensus of expert opinion. Animal Welfare 12, 205–217
35. Willgert K, 2011. The economic and welfare impact of lameness in sows in England. Retrieved January 10, 2012, from http://www.fao.org/fileadmin/user_upload/animalwelfare/TheeconomicandwelfareimpactoflamenessinsowsinEngland.pdf

8
CHAPTER

ASSESSMENT OF COMPLETENESS OF REPORTING IN INTERVENTION STUDIES USING LIVESTOCK: AN EXAMPLE FROM PAIN MITIGATION INTERVENTIONS IN NEONATAL PIGLETS

A. O'Connor, R. Anthony, L. Bergamasco, J. F. Coetzee, R. S. Dzikamunhenga, A. K. Johnson, L. A. Karriker, J. N. Marchant-Forde, G. P. Martineau, S. T. Millman, E. A. Pajor, K. Rutherford, M. Sprague, M. A. Sutherland, E. von Borell and S. R. Webb

Keywords: Animal welfare, data collection, piglets, pain, reviews

IMPLICATIONS

Authors have an ethical responsibility to report the study design and results in a manner that enables reproduction of results and assessment of bias. In this paper we discuss approaches for comprehensive reporting in animal welfare studies. Checklists such as the Reporting guide lines for randomized controlled trials for livestock and food safety statement provide guidance for reporting studies. Such standards represent the current minimum for reported standards.

INTRODUCTION

Complete reporting of study conduct and results has always been an important part of the scientific process, however, in recent years there has been a renewed focus on the importance of complete and accurate reporting. Driving forces behind this focus include (1) an increased scrutiny of scientific findings, (2) the manner in which scientific

Citation: A. O'Connor, R. Anthony, L. Bergamasco, J. F. Coetzee, R. S. Dzikamunhenga, A. K. Johnson, L. A. Karriker, J. N. Marchant-Forde, G. P. Martineau, S. T. Millman, E. A. Pajor, K. Rutherford, M. Sprague, M. A. Sutherland, E. von Borell and S. R. Webb (2016). Assessment of completeness of reporting in intervention studies using livestock: an example from pain mitigation interventions in neonatal piglets. animal, 10, pp 660-670. doi:10.1017/S1751731115002323.

information is applied to the decision-making process and (3) concerns over wastage of animals and resources used in research endeavors (O'Connor et al., 2010; Sargeant and O'Connor, 2013; Ioannidis et al., 2014). The increased use of formal research synthesis techniques, such as risk assessment, systematic reviews and meta-analysis, in the decision-making process of public policy makers and for regulatory purposes also places greater importance on the incorporation of primary research into these methods. These explicit uses of research data have led to efforts that ensure accurate estimates of the magnitude of the effect and that potential for biases are incorporated into research synthesis techniques. If studies are incompletely reported, then the results may not be useable for secondary purposes, and the financial resources are wasted and the ethical value of the animals is unappreciated. In order to avoid waste of recourses and to appropriately recognize the ethical value of animal research subjects, authors have an ethical obligation to provide as complete and as accurate a report as possible and editors and peer-reviewers have an obligation to ensure that the authors do so. A common research question used for policy development is the assessment of interventions designed to mitigate an adverse outcome. Recently developed guidelines exist for identifying what a complete account of an intervention assessment study in animal studies represents (Kilkenny et al., 2010; O'Connor et al., 2010). We are unaware of other studies that have assessed the completeness of reporting in studies focused on interventions for animal welfare outcomes. The primary objective was to assess completeness of reporting interventions designed to mitigate pain in neonatal piglets undergoing routine management procedures. Our second objective was to illustrate how authors can report the items recommended by a single/uniform reporting guideline framework using examples from existing animal welfare science literature (Dawkins, 2006). We sought to identify aspects of study design, analysis and results that were inadequately reported and provide examples so that education of animal welfare science researchers could be targeted to improve reporting in the future.

MATERIAL AND METHODS

Study Population

This project used literature identified for a systematic review to identify research gaps and develop recommendations related to pain mitigation in the neonatal piglet undergoing castration, tail docking or ear notching (National Research Council (US) Committee on Recognition and Alleviation of Pain in Laboratory Animals, 2009; Dzikamunhenga et al., 2014; O'Connor et al., 2014). Details about the protocol, search, screening process to identify relevant studies and resulting review are available elsewhere (Dzikamunhenga et al., 2014; O'Connor et al., 2014). For the assessment of comprehensive reporting, we used the studies relevant to the original review. The unit of concern for reporting was a study/trial. Two or more studies/trial were occasionally reported in a single article. An intervention study/trial must have at least two arms (treatment groups).

Reporting consistent with Reflect (Reporting guide lines for randomized controlled trials for livestock and food safety) guidelines

The REFLECT statement is a reporting guideline for randomized controlled trials that assess interventions for food-producing animals such a swine and is therefore suitable for this topic area (http://www.REFLECT-statement. org) (O'Connor et al., 2010; Sargeant et al., 2010). The REFLECT statement comprises 22 checklist items (Table 1), of which we assessed the reporting of 17. The rationale for including these items in a publication is provided by Sargeant et al. (2010). The reporting of five REFLECT checklist items was not assessed (Table 1). We did not assess study flow (REFLECT checklist item 13) because we expected that studies relevant to the interventions were of such short duration that it was unlikely any loss-to-follow-up would occur, that is, few piglets would leave the study because the outcome could be assessed. We did not assess REFLECT checklist items 2 (Introduction and Background), 20 (Discussion and Interpretation), 21 (Generalizability) and 22 (Overall Evidence) because they are more prone to subjective assessment.

For REFLECT checklist item 3 (Methods and Participants), we extracted the country in which a study was conducted if it was explicitly reported in the article. Otherwise, the reviewer scored location as 'not reported' and the item was 'partially reported.' For REFLECT checklist item 5 (Objectives) to be considered 'completely reported,' the objectives had to be associated with a hypothesis that related to the outcomes. For REFLECT checklist item 6 (Outcomes), we considered for studies that assessed only one outcome that this was the primary outcome. Otherwise, we expected the authors to designate a primary outcome or this checklist item was considered 'incompletely reported.' We also added one item to assess if the studies reported random allocation to group. This was necessary because the REFLECT statement makes the a priori assumption that studies are randomized. Based on the assumption that the study is randomized, the REFLECT asks for information about the steps in the randomization approach for assessment of its validity. That is, sequence generation, allocation concealment and implementation. If a study does not randomize to group, then the steps of randomization will not be reported and listed as missing from the report.

We assessed the reporting of statistical analyses (REFLECT checklist item 12) using the guidelines by Lang and Altman (2014). We considered statistical analyses fully reported if all of the following were provided:

- A full description of the main methods for analyzing the primary and/or secondary objectives of the study;
- Clear methodology used for each analysis, rather than just listing in one place all the statistical methods used;
- Confirmation that data conformed to assumptions of the test used to analyze them. In particular, if the analyses specified that (1) skewed data were analyzed with non-parametric tests, (2) paired data were analyzed with paired tests and (3) the underlying relationship analyzed with linear regression models was linear;
- Whether and how any allowance or adjustments were made for multiple comparisons (performing multiple hypotheses tests on the same data) when the

reported results suggested such adjustment was necessary. For example, when studies reported comparison of multiple time points or trials with 3 + trial arms in the results we expected a report of the approach to adjusting for such pairwise comparisons, that is, Tukey's, Bonferroni's, etc. If authors did not report the approach, but did report that adjustment was conducted, this was considered 'complete reporting';

- For t-tests only, whether tests were one- or two-tailed and justification for the use of one-tailed tests;
- Description of the α level (e.g., 0.05) that defined statistical significance;
- The name of the statistical package or program used in the analyses. In this situation we considered reporting complete even if only the program, rather than the package, was reported, that is, both SAS® and SAS® PROC MIXED were considered 'complete reporting.'

Table 1: Checklist for REFLECT statement and the frequency of reporting of REFLECT checklist items

Location in paper	Item number	REFLECT checklist descriptor	Reported	Partially reported	Not reported
Title and abstract	1	How study units were allocated to interventions (i.e., 'random allocation,' 'randomized,' or 'randomly assigned' or 'weight matched')	5	0	47
Methods participants	3	Eligibility criteria for owner/managers and study units at each level of the organizational structure, and the settings and locations where the data were collected	0	10	42
Interventions	4a	Precise details of the interventions intended for each group, the level at which the intervention was allocated, and how and when interventions were actually administered	21	31	0
Objectives	5	Specific objectives and hypotheses	7	8	37
Outcomes	6	Clearly defined primary and secondary outcome measures and the levels at which they were measured and, when applicable, any methods used to enhance the quality of measurements (e.g., multiple observations, training of assessors)	6	5	41
Sample size	7	How sample size was determined and, when applicable, explanation of any interim analyses and stopping rules	0	0	52
Randomization – sequence generation	8	Method used to generate the random allocation sequence at the relevant level of the organizational structure, including details of any restrictions (e.g., blocking, stratification)	0	0	52
Randomization – allocation concealment	9	Method used to implement the random allocation sequence at the relevant level of the organizational structure (e.g., numbered containers or central telephone), clarifying whether the sequence was concealed until interventions were assigned	0	0	52
Randomization – implementation	10	Who generated the allocation sequence, who enrolled study units and who assigned study units to their groups at the relevant level of the organizational structure?	0	0	52
Blinding (masking)	11	Whether or not participants administering the interventions, caregivers and those assessing the outcomes were blinded to group assignment. If done, how the success of blinding was evaluated. Provide justification for not using blinding if it was not used	0	18	34
Statistical methods	12	Statistical methods used to compare groups for all outcome(s); clearly state the level of statistical analysis and methods used to account for the organizational structure, where applicable; methods for additional analyses, such as subgroup analyses and adjusted analyses	0	44	8
Results recruitment	14	Dates defining the periods of recruitment and follow-up	6	0	46
Baseline data	15	Baseline demographic and clinical characteristics of each group	0	5	47
Numbers analyzed	16	Number of study units (denominator) in each group included in each analysis and whether the analysis was by 'intention-to-treat.' State the results in absolute numbers when feasible (e.g., 10/20, not 50%)	1	7	44
Outcomes and estimation	17	For each primary and secondary outcome, a summary of results for each group, accounting for each relevant level of the organizational structure, and the estimated effect size and its precision (e.g., 95% confidence interval)	0	11	41
Ancillary analyses	18	Address multiplicity by reporting any other analyses performed, including subgroup analyses and adjusted analyses, indicating those pre-specified and those exploratory	0	0	52
Adverse events	19	All important adverse events or side effects in each intervention group	0	22	30

REFLECT = Reporting guidElines For randomized controLled trials for livEstoCk and food safety.
The following were not assessed, Introduction (REFLECT item 2), Study flow (REFLECT item 13) Discussion (REFLECT item 20), Generalizability (REFLECT Item 21) and Overall evidence (REFLECT item 22).

If at least one but not all of the above were reported, then we considered statistical analyses 'partially reported.' The presence or absence of each REFLECT checklist item was independently evaluated by two reviewers. Disagreements were initially resolved by one of the reviewers. Where there was disagreement between reviewers about the presence of a checklist item, one reviewer would re-evaluate the article. If this approach did not resolve the conflict, then the item was discussed with a third reviewer. As with any assessment of comprehensive reporting, quality assessments were not made. For example, we did not assess if the method used to allocate piglets to treatment groups reduced bias, rather we assessed if the approach to allocation was reported.

Reporting of Procedures, Trial Characteristics, Study Design Features and Summary Measures

REFLECT checklist items are very general, and as some sources of heterogeneity are domain specific, we also determined if specific aspects of some checklist items were reported. We specifically assessed if the following were reported: type of production system (i.e., all in/all out or continuous flow or not reported), and facility types where the research was conducted (i.e., university-owned farm or laboratory/research facility or privately owned/commercial operation or not reported). We extracted specifics about the reporting of the interventions. We also evaluated reporting of descriptors of the study design: number of animals enrolled in the trial, and number of animals enrolled in trial arms. The inclusion in the report of statistical descriptions of the outcomes, including effect sizes and measures of precision were also evaluated.

RESULTS

A total of 622 articles were identified by original search and of those, 52 studies from 40 articles met the eligibly criteria for the review and were eligible for assessment of the approach to reporting (Dzikamunhenga et al., 2014; O'Connor et al., 2014). All the studies were experimental and therefore should have been randomized trials; no relevant cohort studies were identified. The characteristics of the studies assessed are provided in Supplementary Table S1. A summary of the completeness of reporting of items from the REFLECT checklist is shown in Table 1. No single study reported all of the REFLECT checklist items evaluated in this analysis. None of the studies assessed, reported the selection criteria for farms or animals, the approach to allocation to group, the sample size rationale, complete description of statistical methods, baseline data by group for animals enrolled, complete description of the results, information about ancillary analyses or the occurrence of adverse events by group. Other checklist items were only reported by some of the studies (Table 1). The reporting of the information that would enable end-users to understand the relevance of the study population to a target population was poor. Often, eligibility criteria for the farms and animals used were missing. The frequency of reporting country of conduct and study setting is shown in Supplementary Table S1. Specific intervention information (REFLECT checklist item 4) was reasonably well reported; all studies provided at least some information about the interventions assessed. Supplementary Table S2 provides reporting examples for the studies that assessed non-steroidal anti-inflammatory drug interventions. In the interest of space the other interventions are not included. In Table 2 we provide a simple summary of basic outcome measures: means (or proportions) and measures of precision and trial arm sample size; frequently this information was not reported. In Table 3, Table 4 and Supplementary Table S2 we provide examples where the REFLECT items were well reported from the studies included in the review. In a few situations, no examples could be found in the 52 studies and examples were drawn from other animal studies. Table 3 focuses on the description of the methods and materials, while Table 4 focuses on presentation of the results. The material in the Supplementary Table S3 relates to the introduction and discussion in a manuscript. The three tables should be used together when preparing a manuscript.

Supplementary Table S1: Characteristics of 52 studies evaluated in 40 articles

Study characteristics	Castration	Tail Docking	Ear Notch-ing	Teeth Clipping
Experiments	42	10	2	2
Technique	0	0	0	0
Castration – cut	21	0	0	0
Castration – tear	5	0	0	0
Castration – not reported	16	0	0	0
Tail docking – side cutters	0	7	0	0
Tail docking – blade	0	2	0	0
Tail docking – surgical cutters	0	1	0	0
Ear ID – notching	0	0	2	0
Teeth trimming – clip	0	0	0	2
Language				
English	24	8	2	2
German	17	0	0	0
French	2	2	0	0
Danish	1	0	0	0
Type of Publication				
Published article	38	7	2	2
Conference proceeding	2	0	0	0
Thesis	2	3	0	0
Report	1	0	0	0
Not Reported	1	0	0	0
Country				
Australia	0	2	2	2
Belgium	1	0	0	0
Brazil	1	0	0	0

Canada	0	0	0	0
France	1	2	0	0
Germany	8	0	0	0
Sweden	1	0	0	0
Switzerland	2	0	0	0
United States	2	1	0	0
Not reported	28	5	0	0
Setting				
University farm	3	2	2	2
Private farm	4	3	0	0
Laboratory facility	4	2	0	0
Not reported	33	3	0	0
Production system				
All in/all out	3	3	0	0
Not reported	41	7	2	2
Management system				
Farrowing crates	10	6	0	0
Farrowing pens	1	0	0	0
Not reported	32	4	2	2
Number of animals	9766	5065	115	117
Not reported	2	0	0	0
Age				
Range	1–28 days	18 hours–7 days	0	0
Min	1 day	18 hours	0	0
Max	28 days	7 days	3 days	3 days
Not reported	1	0	0	0
Weight (kg)				
Range	1–10.6	0	0	0
Min	1	0	0	0
Max	10.6	0	0	0
Mean		2.2		
Not reported	31	8	2	2
Breed				
Meat breed	31	9	2	2
Not reported	12	1	0	0
Other	1	0	0	0

Number of relevant arms in study				
2	23	8	2	2
3	11	0	0	0
4	7	0	0	0
5	2	2	0	0
6	1	0	0	0
Intervention				
Pharmacological	39	5	0	0
Non-pharmacological	6 (1 says both)	7 (2 say both)	2	2

Supplementary Table S2: Reporting of compound, dose and time of administration for studies that used non-steroidal anti-inflammatory drug (NSAID)

Study	**Drugs/intervention**	**Dose reported**	**Time of administration**
(Courboulay *et al.*, 2010)2010	1% Ketoprofen	0.75 ml per piglet	Not reported
(Hansson *et al.*, 2011)	Meloxicam	0.2 ml of 5 mg/kg BW	Not reported
(Kluivers-Poodt *et al.*, 2012)	Meloxicam	0.4 mg/kg	15 minutes before castration
(Langhoff *et al.*, 2009)	Meloxicam	0.4 mg/kg BW	15–30 minutes before castration
	Flunixin meglumine	2.2 mg/kg BW	15–30 minutes before castration
	Carprofen	1.4 mg/kg BW	15–30 minutes before castration
(Reiner *et al.*, 2012)	Flunixin	5 mg	30 minutes before castration
	Flunixin	5 mg	Immediately before castration

	Meloxicam	2 mg or 2 mg/kg	Immediately before castration
(Schmidt *et al.*, 2012)	Meloxicam	0.4 mg/kg	10 minutes before castration
(Schwab *et al.*, 2012)	Ketoprofen	0.03 mg/kg	10–30 minutes before castration
(Sutherland *et al.*, 2012)	Flunixin meglumine	Not reported	At the time of castration
(Sutherland *et al.*, 2012)	Flunixin meglumine	Not reported	At the time of castration
(Tenbergen, 2012)	Meloxicam	0.4 mg/kg	30 min prior to procedure
(Tenbergen, 2012)	Ketoprofen	3 mg/kg	30 min before procedure
(Wavreille *et al.*, 2012)	Tolfenamic acid	2 mg/kg (0.08–0.18 ml)	1 h prior to castration
	Meloxicam	0.4 mg/kg (0.13–0.29 ml)	1 h prior to castration
(Zoels *et al.*, 2006)	Meloxicam 20 mg/ml	0.4 mg/kg	15 min before castration

Supplementary Table S3: Examples of check items from the REFLECT statement sections related to the introduction and discussion from the reviewed trials reported.

Paper section and topic	Item	Example from review studies
Title & Abstract	1	"Piglets were *randomly* assigned to receive 1.0 ml of a 12% sucrose solution (treatment group) or a placebo (1.0 ml of air) administered via syringe in the mouth, 60 s before commencement of one of the management procedures" (Rand *et al.*, 2002).

Introduction Back-ground	2	"Carbon dioxide gas is currently being used in several countries to stun pigs before exsanguination during slaughter. Carbon dioxide gas is usually administered at concentrations greater than 80% to minimize the aversion period experienced by the animal and reduce the risk of inadequate anesthesia (Nowak et al., 2007). Kohler et al. (1998) concluded that CO_2 anesthesia with a concentration of 80% can be induced safely and rapidly in pigs and that castration can be performed without any reaction, but stress induced by handling and manipulation before castration is not reduced. Administering CO_2 can cause behavioral signs of aversion in pigs. Svendsen (2006), on the other hand, stated that aversion before losing consciousness is compensated for by the fact that piglets experience complete anesthesia and analgesia during castration. Gerritzen et al. (2008) observed that heavy breathing was the only typical behavior that piglets exhibited when exposed to a mixture of 70% CO_2 and 30% O_2 . Gerritzen et al. (2008) also concluded that the period of unconsciousness and analgesia achieved by this gas mixture was long enough to castrate pigs without them experiencing pain. Trembling behavior has been observed in piglets, lambs, and dogs for several days after castration, which is indicative that castration causes pain (Morton and Griffiths, 1985; Wemelsfelder and van Putten, 1985; Molony et al., 1997). Other pain-specific behaviors observed in piglets after castration include spasms, tail wagging, scratching, and huddling up, as well as changes in lying and suckling behavior (McGlone et al., 1993; Taylor et al., 2001; Hay et al., 2003; Llamas Moya et al., 2008). However, the effect of CO_2 anesthesia before castration on pain-related behaviors displayed by piglets after castration has not yet been studied. The hypothesis of this study is that piglets will experience less pain and discomfort after castration when anesthetized with CO_2 before castration, thus improving their overall welfare." (Beirendonck *et al.*, 2011)
Discussion -- Interpretation	20	"In our study, it was shown that the required current of 1.25 A stunned a vast majority (92%) of the animals correctly. When using a lower current (0.6 A), however, stun quality results were poor. Furthermore, both trials reported here generally support earlier findings that tongs' position influences stun quality, where a frontal position (as required by Swedish animal welfare legislation) is more efficient than other positions (reference in original text), although these results were only statistically significant in one of the two trials included"(Berg *et al.*, 2012).*

Generalizability	21	"The results presented here represent some preliminary evidence that the immediate responses to acute pain are increased as a consequence of maternal pregnancy stress. Pain perception is dynamic and labile and is open to manipulation by the social and physical environment throughout life. The extent to which prenatally induced changes might persist throughout an individual's lifetime remains to be determined. However, in rats, PNS increased the severity of the spontaneous behavioural response to formalin through to adulthood" (Rutherford *et al.*, 2009).
Overall evidence	22	"Although the interquartile range was lowest in the SHAM group, the CAST group was also low and did not differ significantly from the SHAM group. SHAM and LIDO piglets had significantly lower entropy than CAST piglets, with the other groups in between. This is quite opposite to the findings of Puppe et al., 2005 (reference in original text), but the direction of differences fits with what we found in the other parameters. Taken together, the vocal parameters are indicative of a pain reducing effect of lidocaine. It is also apparent, however, that not all pain is relieved. This could be due to the fact that intratesticularly applied local anaesthesia probably does not provide full anaesthesia of the cremaster muscle, scrotal ligament and intra-abdominal part of the spermatic cord" (Kluivers-Poodt *et al.*, 2012).

Reporting of Reflect Items That Relate to Objectives and Hypotheses

In the remaining part of the manuscript, we discuss the rationale for a select few REFLECT checklist items so authors are aware of how the information is used by readers; however, a full explanation of the rationale for each REFLECT item is available in Sargeant et al. (2010). Although the objective of the study and sometimes a secondary objective were often provided, very few studies translated the objective into a testable hypothesis that included the metric to be measured (REFLECT checklist item 5).

Table 2: Reporting means and measures of precision, and arm sample size in studies evaluated for complete reporting

Outcome assessed*	Number of relevant study arms	Arms for which data was extracted from figures	Arms with missing summary features	Description of missing summary measures
Intervention: general anesthesia (CO_2/O_2)				
Cortisol 0 to 60 min	8	4	3	2 means, 3 SDs
Cortisol 1 to 24 h	6	2	3	2 means, 3 SDs
β-endorphins 0 to 60 min	9	2	2	2 means, 2 SDs
β-endorphins 1 to 24 h	3	1	2	2 means, 2 SDs, 2 arm sample size
Norepinephrine 0 to 60 min	2	1	1	Arm sample size
Pain-like behaviors 0 to 60 min	8	4	2	1 mean and 2 SDs
Intervention: local anesthesia (lidocaine)				
Cortisol 0 to 60 min	8	7	7	6 SDs and 1 arm sample size
Cortisol 1 to 24 h	6	6	6	6 SDs and 3 arm sample size
Norepinephrine 0 to 60 min	1	0	1	1 mean, 1 SD and 1 arm sample size
Frequency 0 to 60 min²	4	0	3	3 SDs and 1 arm sample size
Energy 0 to 60 min²	4	2	2	1 SD and 2 arm sample size
Rate 0 to 60 min²	8	0	7	7 SDs and 3 arm sample size
Pain-like behaviors 0 to 60 min	3	0	2	1 mean, 2 SDs and 2 arm sample size
Pain-like behaviors 1 to 24 h	1	0	1	Mean, SD, arm sample size
Intervention: NSAID (carprofen, flunixin, ketoprofen, meloxicam)				
Cortisol 0 to 60 min	15	10	2	1 mean, 2 SDs and 1 arm sample size
Cortisol 1 to 24 h	10	4	3	2 means, 3 SDs
Energy 0 to 60 min	5	1	3	1 mean, 1 SD and 3 totals
Pain-like behaviors 0 to 60 min	2	0	2	1 SD and 1 arm sample size
Pain-like behaviors 1 to 24 h	5	0	2	1 SD and 1 arm sample size

NSAID = non-steroidal anti-inflammatory drugs; REFLECT = Reporting guidElines For randomized controLled trials for livEstoCk and food safety.
Examples for the following not included: Introduction (REFLECT item 2), Study flow (REFLECT item 13) Discussion (REFLECT item 20), Generalizability (REFLECT item 21) and Overall evidence (REFLECT item 22).
*For more details of exact outcomes measured refer to Dzikamunhenga *et al.* (2014).

Translating the objective to a hypothesis with a specific metric is important because some metrics may be more valid for specific objectives than others. Therefore, knowing the exact metric that will be tested is important. For example, an objective of a study may be to assess the impact of the intervention on pain mitigation, and this would be assessed using a comparison of the mean frequency (Hertz) of vocalizations in piglets receiving the anesthetic intervention compared with the mean Hertz of vocalizations in piglets without the anesthetic, that is, $H_0 = \text{mean1} - \text{mean2} = 0$. Clarification of the hypothesis ensures that the end-user knows which metric is being used to assess the objective, and should facilitate identification of the primary outcome.

Reporting of Reflect Items That Relate to Outcomes and Sample Size

A clear description of which outcomes were primary or secondary was never explicitly reported by authors who assessed multiple outcomes (REFLECT checklist item 6). The only studies that received a 'yes' for this item reported only one outcome. Another item poorly reported was the primary outcome. Knowledge of the primary outcome is necessary to assess the power of the study. By definition the primary outcome is that used to determine the sample size, so authors need to state the sample size rationale so it is clear what the primary outcome is. Many hypothesis-testing studies, including welfare oriented studies, have multiple outcomes of interest, and in these circumstances the authors should power the study to the outcome with the largest sample size upon which they propose to conduct hypothesis testing and make inference. If the authors instead intend to test the hypotheses about multiple outcomes jointly in a multivariate analysis and make the inference about the joint outcomes, then methods of sample size calculation are available and this can be acknowledged in the report (Huang et al., 2008; Luo, 2014). However, such an approach to analysis is very rare (Kerr, 1998). Unless explicitly declaring that a study is a pilot or making use of animals used for another purpose, assessments of interventions should be hypothesis driven. The hypothesis should be specific enough to enable determination that the number of animals enrolled should be sufficient to enable detection of a clinically meaningful difference in the outcome. Researchers therefore should prospectively design and justify the sample size, which requires knowledge of the primary outcome. Further, if authors do not have an a priori hypothesis about a primary outcome, the potential to 'data mine' for a statistically significant outcome and selective reporting bias is high.

Table 3: Examples of reported 'Methods' items from the trials reported consistent with REFLECT guidelines

Paper section and topic	Item	Example from review studies
Participants	3	'Sows were housed in commercial farrowing crates on a commercial farm in Saxony-Anhalt, Germany' (Marx *et al.*, 2003)
Interventions	4	'Two groups were treated with Flunixin (5 mg); the group termed Flu-30 received an i.m. injection of Flunixin 30 min before castration and of 0.1 ml NaCl (0.9%) immediately before castration, the group termed Flu-0 received 0.1 ml NaCl (0.9%) 30 min before castration and Flunixin immediately before castration' (Reiner *et al.*, 2012)
Objectives	5	'The objective of this study was to evaluate the effect of providing CO_2 anesthesia before castration on the behavior of piglets for up to 8 d after castration in comparison with piglets castrated without anesthesia... The hypothesis of the above study is that piglets will experience less pain and discomfort after castration when anesthetized with CO_2 before castration, thus improving their overall welfare' (Beirendonck *et al.*, 2011)
Outcomes	6	Defining outcomes – 'The primary outcome was Infectious Bovine Keratoconjunctivitis (IBK) cumulative incidence over the study period. The secondary outcome was weaning weight' (Funk *et al.*, 2009)* Methods of measurement and level of assessment' ... and the behavior of each individual pig was recorded using 1 min scan samples (direct observations) for 120 min' (Sutherland *et al.*, 2011)
Sample size	7	'Prior to conducting the study, it was determined that twelve animals per group were required to obtain 80% power to detect a 60% difference in IBK risk between groups based on an expected 10% IBK risk in controls and at least 70% IBK risk in inoculated animals. The test was based on a one sided difference in proportions test for independent binomial data with significance level 0.05. Thus, our aim was to enroll 36 animals. No stopping rules or interim analyses were planned or conducted' (Gould *et al.*, 2013)*
Randomization – sequence generation	8	'Forty steers were randomly assigned to one of five treatment groups as described in Table 1. Calves were ranked in ascending order of bodyweight, blocked into cohorts of five calves, and within each cohort, calves were assigned a random number (Excel, Microsoft Works 2010; Microsoft, Redmond, WA, USA). Random numbers were then assigned (Excel, Microsoft Works 2010, Microsoft) to treatment groups to ensure that bodyweight was equally distributed between treatment groups' (Glynn *et al.*, 2013)*
Randomization – allocation concealment	9	'The individual who generated the allocation sequence was not involved in assessment of eligibility or the outcome' (Gould *et al.*, 2013)*
Randomization – implementation	10	'The allocation status, based on eye and calf, on day 0 was confirmed by two people prior to the allocation. The allocation status of the eye was concealed from the individual responsible for scarification and inoculation process' (Gould *et al.*, 2013)*
Blinding (masking)	11	'Two technicians, who were not blind to the treatments due to practical reasons, performed all measurements. The measurements were split between the two technicians with each technician performing the same measurements in all herds' (Hansson *et al.*, 2011)
Statistical methods	12	'Least square mean estimates for each treatment group and the corresponding estimated SE are reported. Pairwise comparisons were conducted using Bonferroni's method to adjust for multiple comparisons and avoid inflation of Type I error rate. Statistical significance for these multiple comparisons was designated a priori as a P-value $\leqslant 0.05$' (Coetzee *et al.*, 2012)* 'For physiological measures, the main fixed effects were treatment and time. Litter was a random effect. The interactions between treatment by time and treatment by litter were included in the model' (Sutherland *et al.*, 2012)

REFLECT = Reporting guidElines For randomized controLled trials for livEstoCk and food safety.
*Example not selected from the study set.

No studies reported the rationale for the sample size (Checklist item 7). This was surprising, as all studies seemed to purposefully assess the effect of an intervention on an outcome and, therefore, the number of animals needed to detect the magnitude of effect of interest is a prerequisite step in study design. Although reduction of animals included in studies is an important principle of animal research, this concept does not negate the need for sufficient power to detect clinical meaningful changes in the outcome. There are numerous papers devoted to the need for adequately powered animal studies (Cohen, 1997; Chapman and Seidel, 2008). Some would argue that reporting the sample size rationale based on an a priori determined primary outcome is not necessary if the P-values are reported as these indicate the probability of a type 1 and type 2 error. However, knowledge of a priori power is not the only rationale for reporting sample size rationale and the primary outcome, such reporting also guards against authors reporting a different primary outcome based on the results of analyses, a practice colloquially known as HARKing (Hypothesis After the Results are Known) (Kerr, 1998). A practice that might be quite possible, in the area such as animal welfare, where multiple outcomes are of interest.

Table 4: Examples of reported 'Results' items from the trials reported consistent with REFLECT guidelines

Paper section and topic	Item	Example from review studies
Study flow	13	'Nine calves (9/19) in Trial 2 had missing data on d +10 because practical constraints prevented collection of PA-MNT data around scheduled ophthalmic exams and euthanasia. One calf in Trial 1 developed severe respiratory disease and was euthanized on d +7' (Dewell *et al.*, 2014)*
Recruitment	14	'The studies were conducted in two piglet breeding operations (Unit A 550 breeding sows, 2-week production cycle; Unit B 560 sows, 4-week production cycle) from February 2003 to May 2003' (Lahrmann *et al.*, 2006)
Baseline data	15	
Numbers analyzed	16	See Supplementary Figure S1
Outcomes and estimation	17	See Supplementary Figure S1
Ancillary analyses	18	'There was not a significant difference between treatment groups with respect to mortality rate. Piglets receiving meloxicam had a mortality rate of 3.18% and piglets receiving the placebo had a mortality rate of 3.84% ($P = 0.33$). Piglets receiving ketoprofen had a mortality rate of 2.91% whereas piglets receiving the placebo had a mortality rate of 3.94% ($P = 0.27$)' (Tenbergen, 2012)
Adverse events	19	'No adverse effects were noted after IV or oral meloxicam administration' (Kreuder *et al.*, 2012)*

REFLECT = Reporting guidElines For randomized controLled trials for livEstoCk and food safety.
*Example not selected from the study set.

Reporting of Reflect Items that Relate to Confounding: Allocation to Group/Randomization

REFLECT checklist items 8 through 10 (Sequence Generation, Allocation Concealment, and Implementation, respectively) are based on the assumption that the study is randomized. A description of the method of developing the randomization for the sequence generation, allocation concealment and implementation, was not provided in any study. A total of 33 of 52 studies used the term 'randomly' or 'randomized' or 'random' in their description of piglet allocation to treatment group. Occasionally it was unclear if the approach used was truly random, despite a description as such. For example, one study described randomly assigning 245 clinically healthy piglets to one of the 12 experimental groups. However, the sample sizes in each of the seven relevant arms were very different, suggesting a method other than random allocation. Several studies reported using restrictions of randomization. Blocking by continuous covariates or stratification by categorical covariates was reported in 39 studies. Covariates used were weight, litter, weight and litter, sow or weight, or litter and adoption. No study that controlled for weight using blocking explicitly reported the size of the block. Details about the approach to allocation are part of reporting that enables assessment of internal validity as they relate to the exchangeability of groups. If it cannot be determined that groups are exchangeable then it is unclear if the observed differences can be attributed to the intervention. Furthermore, without details of the randomization approach, approaches that are haphazard (lacking any obvious principle of organization) or convenient may be incorrectly reported as random. The importance of random allocation is highlighted by authors of the CONSORT statement which we quote here 'Random assignment is the preferred method; it has been successfully used regularly in trials for more than 50 years. (reference in original text) Randomisation has three major advantages (reference in original text). First, when properly implemented, it eliminates selection bias, balancing both known and unknown prognostic factors, in the assignment of treatments. Without randomisation, treatment comparisons may be prejudiced, whether consciously or not, by selection of participants of a particular kind to receive a particular treatment. Second, random assignment permits the use of probability theory to express the likelihood that any difference in outcome between intervention groups merely reflects chance. (reference in original text) Third, random allocation, in some situations, facilitates blinding the identity of treatments to the investigators, participants, and evaluators, possibly by use of a placebo, which reduces bias after assignment of treatments. (reference in original text) Of these three advantages, reducing selection bias at trial entry is usually the most important. (reference in original text)' (Moher et al., 2010). As many welfare studies are small, it is reasonable that authors would employ restricted randomization tools such as stratification and blocking to increase the power of studies. Regardless of the approach to randomization, it should be described fully so that end-users can assess the potential for bias.

Reporting of Reflect Items that Relate to Performance and Measurement/Information Bias-Blinding

Of the 52 studies, 18 reported blinding as part of their protocol; however, none provided a full description of the approach used to blind the study (REFLECT checklist item 11). Blinding, whether for allocation of treatments or interventions or assessment, was infrequently reported by authors. As blinding is designed to reduce measurement/information bias, it is important to know if outcome assessment is biased. There is some evidence in animal welfare that absence of blinding is associated with more positive outcomes (Tuyttens et al., 2014).

Reporting if Reflect Items Related to Statistical Methods

Statistical methods (REFLECT checklist item 12) were not reported in eight studies. In the remaining 44 studies, statistical methods were considered partially reported because they failed to meet all the criteria described above. Assessment of comprehensive reporting of statistical methods is very difficult; the measure of comprehensiveness is that a reasonably informed individual would be able to assess the validity, although what is 'reasonable' might appear itself subjective. We would encourage authors to consult with documents published previously that describe what should be included in a description of statistical methods (Lang and Altman, 2014).

Reporting of Reflect Items Related to Setting, Study Population Characteristics

Dates relevant to the study recruitment and performance were described in only six studies (REFLECT checklist item 14). Although it is difficult to envision how year or season could affect the response of piglets to pain mitigation, such information is very relevant for other topics, especially those that seek to understand the influence of season or year on an outcome. The same principle can be inferred for study location (i.e., country or region or production system). Baseline demographics and clinical characteristics of each group were generally poorly reported (REFLECT checklist item 15). When weight and age information were presented as summary measures for all enrolled pigs together, we considered this to be partially reported. It is recommended that authors provide demographic information about the groups separately, so that end-users can assess if the groups are comparable, especially given the absence of reporting of allocation methods. Demographic information was frequently reported in the Methods section and not explicitly in the Results section. REFLECT and other statements make the distinction that the methods and materials could, and potentially should, be written before the study is started, therefore the demographic information of the study groups such as the mean age and mean weight (including standard deviations) are a result and should be presented in the Results section.

Reporting of Reflect Items Related to Results of Analyses

The actual number of piglets that contributed to data analyses (REFLECT item 16) was frequently not reported. Presumably authors felt that reporting the number of enrolled animals would suffice because the potential for lossto-follow-up in the subject matter studied was low. For this topic area, this assumption may be valid and failure to report that no loss-to-follow-up occurred may not be a source of bias. Sometimes the unit of analysis was not the same as the number of animals in the study. This was particularly important for the behavior data, which could be reported as number of pigs that demonstrated an activity or the number of time periods when an event was observed. These clearly have different denominators. Similarly, some outcomes appeared to be measured only on a subset of enrolled animals, perhaps because testing all animals was time consuming or expensive. Supplementary Figure S1 is an excerpt of a table (Sutherland et al., 2011) that provides the number of animals included in the analysis. Effect measures regarding outcomes were often poorly reported. Supplementary Table S3 provides examples of the information missing from some studies. Such information would be needed to assess the magnitude of effect so that the balance of benefits and harms could be evaluated (which cannot be evaluated by P-values). If only the P-value is reported, it is not possible to know the magnitude or direction of the effect (i.e., whether the intervention increased or decreased the outcome). Furthermore, measures of variation were often not reported or not reported clearly, especially in figures where it was not always possible to discern if the error bar represented a SEM, a SD or a confidence interval. In studies that used random effects variables to control for clustering, the variance components were never reported, despite their importance for future study design and interpretation. Ancillary analyses (REFLECT checklist item 17, Outcomes and Estimation) were not reported in any study, as no a priori primary and secondary outcomes were reported and no sample size justifications were provided. The rationale for reporting ancillary analyses is to give end-users knowledge of potentially interesting results that arise from data exploration and are therefore hypothesis generating rather than hypothesis testing. Proactive reporting of adverse events was expected in order for end-users to balance the benefits and harms of using pain-mitigating interventions in neonatal piglets. Harms are often rare and therefore often only detected using secondary analyses. Such information would have allowed us to understand whether the reported mortality rate was excessive compared with baseline trends in production. Sometimes adverse events were reported in a way that we could not determine the group to which the animals that experienced the intervention were allocated. Knowledge of the group to which the animal was assigned is vital to interpreting harms. For example, reporting that 10 animals died in the study is not informative, compared with reporting five animals died in each group or one animal died in the control group and nine in the treated group.

	Treatments							
Variable	CON	CUT	LA	SHORT	LONG	CO_2	Pooled s.e.	Treatment P-value
n	10	10	10	10	10	10		
WBC ($10^3/\mu l$)	9.1^{ab}	9.6^{a}	8.7^{ab}	9.5^{a}	8.6^{b}	8.2^{b}	0.33	0.007
Neutrophils ($10^3/\mu l$)	6.4	6.8	6.3	6.9	6.1	6.3	0.24	0.584
Lymphocytes ($10^3/\mu l$)	2.2^{a}	2.0^{ac}	1.8^{bc}	1.9^{ac}	1.8^{b}	1.8^{bc}	0.12	0.001
Neutrophils (%)	68.1^{b}	69.8^{ab}	71.1^{a}	70.7^{c}	71.5^{a}	68.6^{bc}	0.83	0.001
Lymphocytes (%)	24.6^{a}	21.8^{b}	20.5^{b}	21.0^{b}	20.3^{b}	22.5^{ab}	0.85	0.001
N:L	3.0^{b}	3.5^{a}	3.7^{a}	3.9^{a}	3.8^{a}	3.5^{ab}	0.20	0.001

CUT = conventional cutting; LA = CUT with local anesthetic injected immediately before cutting; SHORT = CUT with short-acting anesthetic administered topically onto the tail wound immediately after cutting; LONG = CUT with long-acting anesthetic administered topically onto the tail wound immediately after cutting; CO_2 = CUT while the pig was anesthetized with carbon dioxide gas; CON = sham docked; WBC = total white blood cell count; N:L = neutrophil-to-lymphocyte ratio.
[a,b,c]Least squared means with different superscripts within a row are different ($P < 0.05$).

Supplementary Figure S1: *Leukocyte values of pigs after tail docking using CUT, LA, SHORT, LONG, CO_2 or CON*

DISCUSSION

In the area of animal welfare research we found that, as in other disciplines related to veterinary and animal sciences, reporting of intervention studies was frequently incomplete (Burns and O'Connor, 2008; Sargeant and O'Connor, 2013). Overall, many studies failed to report information that would be needed to assess internal and external validity. There are both ethical and, in some countries, legal reasons for ensuring that scientists using animals must not only adhere to adequately justified methodology but that they should also be able to articulate it according to high reporting standards to their peers and the public. The privilege given to scientists to use research animals entails adherence to rigorous reporting standards that help to ensure compliance with national and international policies that protect the welfare of all research animals. Some of the journals we have assessed might not be considered as truly 'scientific' as these (mostly national/local) journals are periodical magazines intended to inform practitioners on new developments. As an example, the journal 'Der Praktische Tierarzt' has a different audience than Journal of Animal Science. However, if such journals do choose to publish primary research then it seems that the standards of reporting would still apply. Another reason for omissions may be lack of awareness of the need for comprehensive reporting due to the multidisciplinary nature of many projects.

Publication of the results of a scientific study is not the end of the scientific process (Sargeant and O'Connor, 2013). Presumably, researchers publish with the intent that the results of a study will enable generation of new hypotheses, validate current hypotheses, or influence decision making. These secondary uses of primary data rely on the validity of the original study design, analyses and such assessment of validity can only be determined if the report is transparent, accurate and comprehensive (O'Connor et al., 2010; Sargeant and O'Connor, 2013). Further, if incomplete reporting casts doubt over the results of studies then the monetary and ethical value of the animal and financial resources used to generate the data were not fully realized. In addition, animals may have suffered unnecessarily. If the study is considered important enough that the information is needed for decision making, it may even be necessary to repeat the study (Ioannidis et al., 2014; Macleod et al., 2014). In situations where the reporting is so incomplete that useful data cannot be extracted from the original experiment and the study must be repeated, this would be incongruous with the 3Rs (replacement, reduction and refinement) and not be a good use of already scarce research funding. Professionally, the credibility of the authors of the original study could be called into question.

Some might argue that authors, reviewers and editors should be able to 'cherry pick' checklist items from REFLECT or other applicable guidelines that they think are relevant. However, two concepts argue against this idea. The first is that the current body of work was developed under such a system and is lacking in important details of studies. The second reason is that authors, reviewers and editors are not necessarily able to anticipate all the end uses of research. The alternative is to use guidelines that have been developed by a diverse group of developers and end-users, which reduced the need for second guessing what end-users may need. For example, in this data set it might be argued that country is not relevant to pain, however, country may be relevant to a study looking at the use of placebos in animal welfare trials and knowing the country will help the researcher of that question understand differences in placebo use. Clearly authors would not have anticipated such a question, however, consensus groups often have. In human clinical trial reporting, the concept of cherry picking items to report from guidelines is not encouraged. For example, the editorial policy for Nature indicates 'Manuscripts reporting results of a clinical trial must conform to CONSORT 2010 guidelines.' Because numerous guidelines exist and guidelines can change, authors should cite the guideline they used in preparing their report. It is true that precise writing is needed to add the detail requested by guidelines, however, this is likely not a meaningful barrier, as indicated by the widespread adoption of reporting guidelines in human health including high-impact journals such as Nature and Lancet where space is at a premium.

It is unclear why reporting is incomplete. Some might suggest that this is because of lack of awareness of reporting guidelines. However, the concepts of reproducible research and reporting in a manner that reflects the experiment is not new or novel, so lack of awareness cannot explain all of the incomplete reporting (Grindlay et al., 2014). It is imperative that research reporting be complete to enable reproducibility, assessment of the internal and external validity of the study and knowledge translation. Given that animal welfare science is a discipline that often involves interventions that may be perceived as unpleasant to the animal, attention to the quality of reporting is especially critical to advance the field. Comprehensive reporting is an ethical responsibility for researchers undertaking this type of research. For intervention studies, the reader should be able to understand the magnitude and precision of the estimated effect of the intervention, and the probability that the effect is consistent with the null hypothesis. The reader should also be able to assess the potential for bias. We would encourage authors to consider using reporting guidelines to improve reporting. Consistent with the 3Rs, in particular the reduction principle, using reporting guidelines can maximize information from the animals used in the study and minimize the risk of unnecessary studies, therefore reducing further animal use. We are aware that the omission of this information as well as important design characteristics, analyses or results is often unintentional. In addition, we are well aware what constitutes a complete report is not a static list. As knowledge and technology change, the standards for how science is conducted and reported should be expected to change. Given these changing standards, however, the most recent checklists represent minimum current standards. This would not preclude authors from including or editors and peer-reviewers from requesting additional information. Checklists provide guidance for reporting but researchers should adhere to the underlying reporting principles to provide a report that facilitates reuse of the data and enables assessment of bias. With the growing frequency of multiple collaborators involved in manuscript preparation, the final editor may not be aware of all the aspects required for reporting. One reason for an incomplete report might be a lack of knowledge of what and how items should be reported. However, resources are becoming increasingly available to mitigate this problem. Documents specific to animal studies like the REFLECT statement for livestock trials and the ARRIVE guidelines specific to uses of animals undergoing an experimental procedure in a research laboratory or formal test setting (Kilkenny et al., 2010; O'Connor et al., 2010). The ARRIVE guidelines are designed for animals used in experimental settings with a focus on animal populations where the independence assumption is valid. The REFLECT statement is more specific for livestock and provides more focus on non-independent populations such as occur in housed animals. As reporting guidelines are relatively new, the impact on reporting has not been assessed yet. For example, REFLECT was unavailable when many of the papers in this review were published. The standards of reporting observed here are therefore not reflective of the impact of reporting guidelines. The examples provided in Table 3, Table 4 and Supplementary Table S3 can be used as a guideline for how some of the studies we reviewed effectively reported the information requested by the checklist. All three tables should be used together, and are broken into sections here for presentation purposes. Use of a reporting checklist might help reduce the number of items not reported.

CONCLUSION

The overall conclusion after assessing these studies using REFLECT, is that there is (1) an opportunity to improve the reporting and (2) a need to raise awareness of the importance in providing a complete report of how animal welfare studies are conducted. The continued ethical and legal acceptability of using animals is contingent upon accurate and complete reporting. Accurate and complete reporting, in most cases, relates to both high quality research and responsible conduct in animal research.

ACKNOWLEDGMENTS

This project was funded by the National Pork Board (NPB) Grant no. 12-186. The authors have no conflicts of interest to declare.

REFERENCES

1. Beirendonck SV, Driessen B, Verbeke G and Geers R 2011. Behavior of piglets after castration with or without carbon dioxide anesthesia. Journal of Animal Science 89, 3310–3317.
2. Burns MJ and O'Connor AM 2008. Assessment of methodological quality and sources of variation in the magnitude of vaccine efficacy: a systematic review of studies from 1960 to 2005 reporting immunization with Moraxella bovis vaccines in young cattle. Vaccine 26, 144–152.
3. Chapman PL and Seidel GE Jr 2008. Experimental design, power and sample size for animal reproduction experiments. Reproduction, Fertility, and Development 20, 33–44.
4. Coetzee JF, Edwards LN, Mosher RA, Bello NM, O'Connor AM, Wang B, Kukanich B and Blasi DA 2012. Effect of oral meloxicam on health and performance of beef steers relative to bulls castrated on arrival at the feedlot. Journal of Animal Science 90, 1026–1039.
5. Cohen J 1997. Statistical power analysis for the behavioral science. Academic Press, New York, NY, USA.
6. Dawkins MS 2006. A user's guide to animal welfare science. Trends in Ecology & Evolution 21, 77–82.
7. Dewell RD, Millman ST, Gould SA, Tofflemire KL, Whitley RD, Parsons RL, Rowe EW, Liu F, Wang C and O'Connor AM 2014. Evaluating approaches to measuring ocular pain in bovine calves with corneal scarification and infectious bovine keratoconjunctivitis-associated corneal ulcerations. Journal of Animal Science 92, 1161–1172.
8. Dzikamunhenga RS, Anthony R, Coetzee J, Gould S, Johnson A, Karriker L, McKean J, Millman ST, Niekamp SR and O'Connor AM 2014. Pain management in the neonatal piglet during routine management procedures. Part 1: a systematic review of randomized and non-randomized intervention studies. Animal Health Research Reviews 15, 14–38.

9. Funk L, O'Connor AM, Maroney M, Engelken T, Cooper VL, Kinyon J and Plummer P 2009. A randomized and blinded field trial to assess the efficacy of an autogenous vaccine to prevent naturally occurring infectious bovine keratoconjunctivis (IBK) in beef calves. Vaccine 27, 4585–4590.

10. Glynn HD, Coetzee JF, Edwards-Callaway LN, Dockweiler JC, Allen KA, Lubbers B, Jones M, Fraccaro E, Bergamasco LL and KuKanich B 2013. The pharmacokinetics and effects of meloxicam, gabapentin, and flunixin in postweaning dairy calves following dehorning with local anesthesia. Journal of Veterinary Pharmacology and Therapeutics 36, 550–561.

11. Gould S, Dewell R, Tofflemire K, Whitley RD, Millman ST, Opriessnig T, Rosenbusch R, Trujillo J and O'Connor AM 2013. Randomized blinded challenge study to assess association between Moraxella bovoculi and infectious bovine keratoconjunctivitis in dairy calves. Veterinary Microbiology 164, 108–115.

12. Grindlay DJ, Dean RS, Christopher MM and Brennan ML 2014. A survey of the awareness, knowledge, policies and views of veterinary journal Editors-in-Chief on reporting guidelines for publication of research. BMC Veterinary Research 10, 10.

13. Hansson M, Lundeheim N, Nyman G and Johansson G 2011. Effect of local anaesthesia and/or analgesia on pain responses induced by piglet castration. Acta Veterinaria Scandinavica 53, 34.

14. Huang P, Woolson RF and O'Brien PC 2008. A rank-based sample size method for multiple outcomes in clinical trials. Statistics in Medicine 27, 3084–3104.

15. Ioannidis JP, Greenland S, Hlatky MA, Khoury MJ, Macleod MR, Moher D, Schulz KF and Tibshirani R 2014. Increasing value and reducing waste in research design, conduct, and analysis. Lancet 383, 166–175.

16. Kerr NL 1998. HARKing: hypothesizing after the results are known. Personality and Social Psychology Review 2, 196–217.

17. Kilkenny C, Browne WJ, Cuthill IC, Emerson M and Altman DG 2010. Improving bioscience research reporting: the ARRIVE guidelines for reporting animal research. PLoS Biology 8, e1000412.

18. Kreuder AJ, Coetzee JF, Wulf LW, Schleining JA, KuKanich B, Layman LL and Plummer PJ 2012. Bioavailability and pharmacokinetics of oral meloxicam in llamas. BMC Veterinary Research 8, 85.

19. Lahrmann KH, Kmiec M and Stecher R 2006. Die Saugferkelkastration mit der Ketamin/Azaperon-Allgemein-anästhesie: tierschutzkonform, praktikabel, aberwirtschaftlich? (Piglet castration with ketamine/azaperone-anaesthesia: concurring with animal welfare, practical, but economic?) Praktische Tierarzt 87, 802–809.

20. Lang TA and Altman DG 2014. Statistical analyses and methods in the published literature: the SAMPL guidelines*. In Guidelines for reporting health research: a user's manual (ed. David Moher, Douglas Altman, Kenneth Schulz, Iveta Simera and Elizabeth Wager), pp. 264–274. John Wiley & Sons, Ltd., Hoboken, NJ, USA. Luo S 2014. A Bayesian approach to joint analysis of multivariate longitudinal

data and parametric accelerated failure time. Statistics in Medicine 33, 580–594.

21. Macleod MR, Michie S, Roberts I, Dirnagl U, Chalmers I, Ioannidis JP, Al-Shahi Salman R, Chan AW and Glasziou P 2014. Biomedical research: increasing value, reducing waste. Lancet 383, 101–104.
22. Marx G, Horn T, Thielebein J, Knubel B and von Borell E 2003. Analysis of pain-related vocalization in young pigs. Journal of Sound and Vibration 266, 687–698.
23. Moher D, Hopewell S, Schulz KF, Montori V, Gotzsche PC, Devereaux PJ, Elbourne D, Egger M, Altman DG 2010. CONSORT 2010 explanation and elaboration: updated guidelines for reporting parallel group randomised trials. Journal of Clinical Epidemiology 63, 1–37.
24. National Research Council (US) Committee on Recognition and Alleviation of Pain in Laboratory Animals 2009. In Recognition and alleviation of pain in laboratory. Animals National Acameies Press (US), Washington, DC.
25. O'Connor A, Anthony R, Bergamasco L, Coetzee J, Gould S, Johnson AK, Karriker LA, Marchant-Forde JN, Martineau GS, McKean J, Millman ST, Niekamp S, Pajor EA, Rutherford K, Sprague M, Sutherland M, von Borell E and Dzikamunhenga RS 2014. Pain management in the neonatal piglet during routine management procedures. Part 2: grading the quality of evidence and the strength of recommendations. Animal Health Research Reviews 15, 39–62.
26. O'Connor AM, Sargeant JM, Gardner IA, Dickson JS, Torrence ME, Consensus Meeting P, Dewey CE, Dohoo IR, Evans RB, Gray JT, Greiner M, Keefe G, Lefebvre SL, Morley PS, Ramirez A, Sischo W, Smith DR, Snedeker K, Sofos J, Ward MP and Wills R 2010. The REFLECT statement: methods and processes of creating reporting guidelines for randomized controlled trials for livestock and food safety by modifying the CONSORT statement. Zoonoses Public Health 57, 95–104.
27. Reiner G, Schollasch F, Hillen S, Willems H, Piechotta M and Failing K 2012. Effects of meloxicam and flunixin on pain, stress and discomfort in male piglets during and after surgical castration. Berliner und Munchener Tierarztliche Wochenschrift 125, 305–314.
28. Sargeant JM and O'Connor AM 2013. Issues of reporting in observational studies in veterinary medicine. Preventive Veterinary Medicine 113, 323–330.
29. Sargeant JM, O'Connor AM, Gardner IA, Dickson JS and Torrence ME, Consensus Meeting Participants 2010. The REFLECT statement: reporting guidelines for randomized controlled trials in livestock and food safety: explanation and elaboration. Zoonoses Public Health 57, 105–136.
30. Sutherland MA, Davis BL, Brooks TA and Coetzee JF 2012. The physiological and behavioral response of pigs castrated with and without anesthesia or analgesia. Journal of Animal Science 90, 2211–2221.
31. Sutherland MA, Davis BL and McGlone JJ 2011. The effect of local or general anesthesia on the physiology and behavior of tail docked pigs. Animal 5, 1237–1246.

32. Tenbergen R 2012. Investigation of the use of analgesics at the time of castration and tail-docking and following parturition for improving performance and reducing pain in pigs. University of Guelph, Guelph, ON, Canada

33. Tuyttens FAM, de Graaf S, Heerkens JLT, Jacobs L, Nalon E, Ott S, Stadig L, Van Laer E and Ampe B 2014. Observer bias in animal behaviour research: can we believe what we score, if we score what we believe? Animal Behaviour 90, 273–280.

9
CHAPTER

DIFFERENCES IN TOTAL PROTEIN CONCENTRATION BETWEEN FRESH AND FROZEN SERUM AND PLASMA SAMPLES USED TO ASSESS FAILURE OF PASSIVE TRANSFER IN DAIRY CALVES

Aurora Villarroel, Taylor B Miller, Justin K Ward, Elisha D Johnson and Kimberly R Noyes

Keywords: Failure of Passive Transfer (FPT), Calves, Colostrum, Total protein (TP), Refractometer, Serum, Plasma, Fresh frozen

INTRODUCTION

Due to the unique placental interface between cow and calf, there is minimal transfer of immunoglobulins prior to parturition. The majority of the immunoglobulins are absorbed from maternal colostrum through endothelial cells lining the gut of the calf by active endocytosis and transport of macromolecules such as immunoglobulins [1]. Absorption is usually limited to the first 24 hours following birth [2,3]. These immunoglobulins enable the establishment of a targeted immune system within the calf that can protect it from infection until it is capable of producing endogenous immunoglobulins [4,5]. This is why adequate colostrum feeding and management is an important aspect of disease prevention in the neonatal calf.

The term "failure of passive transfer" (FPT) is used to identify calves that do not achieve a specific concentration of Immunoglobulin G (IgG) in serum, which is why it is also termed hypoglubulinemia. This target concentration varies significantly in the literature. The most commonly cited threshold for considering calves as having FPT is established at 1000 mg/dL [6], while others have set the limit at 800 mg/ dL [7], 750 mg/dL [8], or even 350 mg/dL [9]. Whatever the threshold, most studies have shown that calves with hypoglobulinemia are more susceptible to disease than calves with

Citation: Villarroel A, Miller TB, Ward JK, Johnson ED, Noyes KR (2014) Differences in Total Protein Concentration between Fresh and Frozen Serum and Plasma Samples Used to Assess Failure of Passive Transfer in Dairy Calves. Adv Dairy Res 2: 111. doi:10.4172/2329-888X.1000111

adequate passive transfer under the same conditions. Therefore, evaluation of passive transfer as a monitoring system to evaluate adequate colostrum management on dairy farms is an important part of preventive herd health protocols.

Several methods have been reported to evaluate passive transfer status in neonatal calves including zinc sulfate and sodium sulfite turbidity tests [10,11], glutaraldehyde coagulation tests [12], measurement of serum gamma-glutamyl transferase (GGT) activity [13,14], and direct measurement of IgG through single radial immune diffusion (SRID), which is considered the gold standard [11,13,15]. Serum total protein (TP_{serum}) concentration has been shown to correlate with serum IgG concentration [10,16]. This correlation allows fairly rapid assessment of passive transfer in the field with a refractometer using TP_{serum} concentration as a proxy for IgG concentration, avoiding the need for processing samples in a laboratory. However, this correlation is not perfect, which leads to great variability in the threshold of TPserum concentration used to establish FPT throughout the literature. TP_{serum} concentrations between 5.0 g/dL and 5.5 g/dL have been reported corresponding to a serum IgG concentration of 1,000 mg/dL [16-18]. Measurement of TP_{serum} has been reported mostly by use of a manual refractometer [11,19], although the use of a digital refractometer has been recently validated [20].

Most studies reporting TP_{serum} concentration as a measurement to evaluate adequate colostrum intake have used fresh serum samples. However, in some situations, fresh serum samples may not be available. Instead, only frozen serum or plasma samples may be available, usually collected for other reasons, but they could be used to assess passive transfer status if it is known how to interpret the results. Therefore, the correlation between TP_{serum} concentration in fresh serum samples and frozen serum or plasma samples needs to be established, so that appropriate limits can be used for determining adequate colostrum intake in neonatal calves.

The objectives of this study were (1) to determine if TP concentrations differ in fresh and frozen/thawed samples; and (2) to determine the correlation between serum and plasma samples.

MATERIALS AND METHODS

Sample Population

Heifer calves at five commercial dairy farms were enrolled in the study over a period of 18 months, between December 2008 and June 2010. Animals eligible for inclusion in this study were Holstein and Jersey heifer calves between 1 and 3 days of age that appeared physically healthy, and were not visibly dehydrated.

Experimental Design

There were two sampling periods. In 2008 and 2009, only serum samples were obtained as the original question was to measure the difference between fresh and frozen/thawed serum samples. During this time a new question arose as to the correlation between serum and plasma TP concentration. Thus, in 2010 plasma samples were added to the protocol.

Blood samples were obtained from each calf once via jugular vein puncture and collected into a serum-separator tube (BD Vacutainer® SST, BD Diagnostics) to obtain the serum samples and into an EDTA tube (BD Vacutainer® EDTA, BD Diagnostics) to obtain the plasma samples. A unique sampling ID number was assigned to each calf. All samples were processed within four hours of collection. The EDTA and serum separator tubes were centrifuged at 3,000 RPM for 10 minutes to allow removal of plasma and serum, respectively. At this time, fresh sample TP was determined using a digital refractometer (Model 300027, Sper Scientific Ltd.) for both serum and plasma samples [20]. The refractometer was calibrated every 10 samples using distilled water. After measurement of fresh TP concentrations, serum and plasma samples were transferred into 2 mL cryopreservation vials (ML5355, Market lab, Inc.) for storage in a frost-free freezer (-20°C). Frozen/ thawed samples collected between December 2008 and August 2009 were analyzed in September 2009, and those collected during 2010 were analyzed in October 2010. Therefore, samples were stored in the freezer for varying lengths of time before being analyzed, and the effect of length of time stored was evaluated. Measurement of TP in frozen/ thawed samples was performed in the same manner and with the same equipment as prior to freezing, after allowing samples to warm overnight to room temperature, and being vortexed.

As the study evolved and realizing that a difference in serum samples was being observed, a new question arose as to what proteic fraction was responsible for the difference. To assess this, new samples were collected in a serum-separator tube from 20 heifer calves (10 Holstein and 10 Jersey) to perform electrophoresis. The limited number of samples was due to the high expense of the electrophoresis test and budget limitations. Serum samples of all 20 calves were collected on the same day and processed in the same manner and with the same equipment as described above. Then each serum sample was separated into three aliquots. The first aliquot was submitted immediately (overnight shipping) to the Veterinary Diagnostic Laboratory at Cornell University for electrophoresis testing. The remaining two aliquots were frozen in the same freezer as the other samples and submitted for electrophoresis testing two weeks and four months after the initial blood draw, respectively, to evaluate changes at different times of frozen storage.

Statistical Analyses

Correlation between TP concentration in fresh and frozen/thawed samples, as well as between serum and plasma samples was evaluated using Passing and Bablok regression, with frozen/thawed TPserum and TPplasma being the respective dependent variables at a level of significance of 5%. Multivariate linear regression was also performed to determine if breed (dummy variably, Jersey=yes/no) and length of storage of the samples (random variable, months) influenced TP concentration in frozen/thawed samples.

Differences between concentrations of proteic fractions measured by electrophoresis in the tested aliquots (fresh, 2 weeks and 4 months of freezing) were analyzed by paired t-test at a level of significance of 10% due to the small sample size. Specialized statistical software was used for all analyses (MedCalc Software v. 11.2).

RESULTS

A total of 127 heifer calves 1-3 days old were enrolled in the study: 34 Holstein and 93 Jersey.

Fresh vs. Frozen Serum

The simple correlation coefficient between fresh and frozen TP_{serum} was r = 0.896 (P<0.001). The Passing Bablok correlation between fresh TP_{serum} values (x) and frozen/ thawed TP_{serum} values (y) of the same serum sample was represented by the equation (y= 0.30 + 0.97x). The 95% confidence interval (CI) for the intercept was (0.10 to 0.79) and the 95% CI for the slope was (0.889 to 1.000) indicating a systematic difference between fresh and frozen TP_{serum} but no significant deviation from linearity (P<0.680), indicating that the correlation was equal for all values. Additionally, the relationship between fresh and frozen/ thawed TP_{serum} was evaluated using a multivariate linear regression. Both breed (Figure 1) and length of time the samples were frozen (Figure 2) influenced the relationship between fresh and frozen/thawed TP_{serum} concentration. Jersey calves had on average 0.429 ± 0.049 g/ dL higher TP_{serum} values than Holstein calves (P<0.001). For each additional month that the samples were stored in the freezer, TP_{serum} concentration in frozen/thawed samples was 0.116 ± 0.017 g/dL lower than the concentration measured in the fresh sample (P<0.001). The resulting multiple correlation coefficient was 0.895.

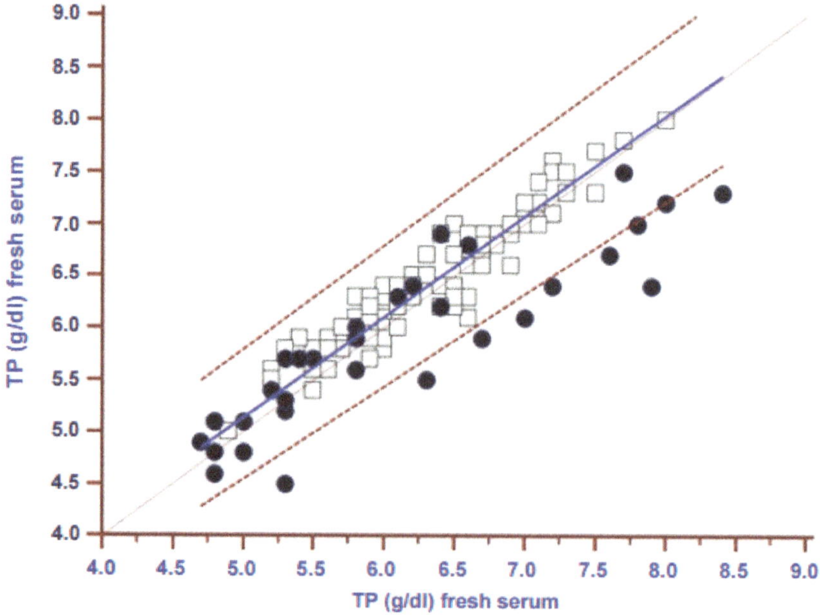

Figure 1: Passing Bablok correlation between total protein (TP) concentration in fresh and frozen/thawed serum samples of 127 calves. Data points are presented for 34 Holstein (•) and 93 Jersey calves (□).

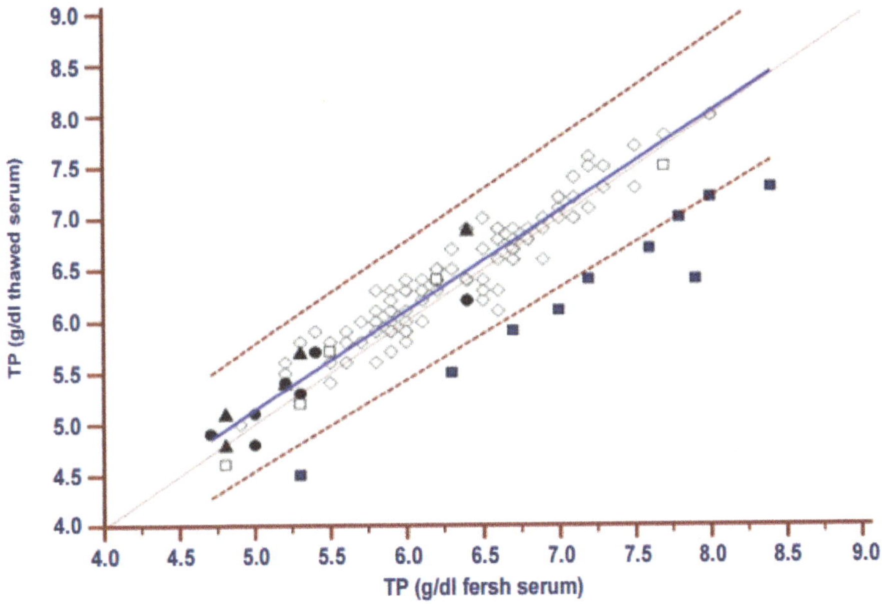

Figure 2: Passing Bablok correlation between total protein (TP) concentrations in fresh and frozen/thawed serum samples of 127 calves. Data points are categorized by length of time that samples were stored in the freezer: ≤1 month (●), 2 months (□), 3 months (▲), 5 months (◊) and 8 months (■).

Electrophoresis performed on a sample of 10 Holstein and 10 Jersey calves to determine whether specific protein fractions differed between fresh and frozen/thawed serum samples showed a larger decrease in albumin concentration as length of storage increased (Table 1). However, there was large variability between calves in all proteic fractions limiting the reliability of the results.

Serum vs. Plasma

The simple correlation coefficient between TP_{serum} and TP_{plasma} was r=0.912 (P<0.001). Plasma values were only available for 58 calves sampled in 2010 (6 Holstein and 93 Jersey). The Passing Bablok regression model obtained for the correlation between fresh TPserum values and fresh TP_{plasma} values of the same calf was explained by the equation(y= 0.60 + 1.00x). The 95% CI for the intercept (0.059 to 1.543) was different than zero, indicating that TP_{plasma} concentration was consistently higher than TP_{serum} concentration. The 95% CI for the slope (0.857 to 1.091) did include one, indicating that there was no proportional difference as concentration of TP_{serum} increased (Figure 3). The cusum test for linearity showed no significant deviation from linearity (P=0.79). Breed did not influence this relationship.

Fresh vs. Frozen Plasma

The simple correlation coefficient between fresh and frozen TP_{plasma} was r=0.959 (P<0.001). The Passing Bablok regression model obtained for the correlation between fresh TP_{plasma} values (x) and frozen TP_{plasma} values (y) was explained by the equation (y=0.10 + 1.00x). The 95% CI for the intercept (0.100 to 0.600) did not include zero, indicating a systematic difference between fresh and frozen/thawed samples. The slope (0.923 to 1.000) included one, indicating that there was no proportional difference between fresh and frozen/thawed plasma samples. However, there was a significant deviation from linearity (P=0.01), indicating that the differences were not consistent throughout the range in TP concentration (Figure 4). Breed had no influence in this relationship, and the length of time the samples were frozen was the same, and thus did not allow evaluating any differences.

Table 1: Electrophoresis results of serum samples from 10 Holstein and 10 Jersey calves at different times of storage under freezing conditions

	Fresh serum g/dL	Frozen (2 weeks) g/dL	Frozen (4 months) g/dL
Total protein	5.92 ± 0.82^{a}	5.89 ± 0.84^{b}	5.77 ± 0.78^{c}
Albumin	2.45 ± 0.17^{a}	2.40 ± 0.19^{b}	2.31 ± 0.19^{c}
Alpha	1.16 ± 0.12^{a}	1.19 ± 0.13^{b}	1.17 ± 0.12^{b}
Beta 1	0.99 ± 0.16^{a}	1.01 ± 0.15^{b}	0.98 ± 0.15^{a}
Beta 2	0.44 ± 0.15	0.46 ± 0.16	0.46 ± 0.14
Gamma	0.88 ± 0.29^{a}	0.83 ± 0.29^{b}	$0.85 \pm 0.26^{a,b}$

[a, b, c] Values within the same row with different superscripts differ at a level of significance of 10%.

DISCUSSION

To our knowledge, this is the first study that compares TP concentrations in samples before and after freezing. Our data showed a lack of a direct linear relationship between fresh and frozen/thawed TP_{serum} concentration in the same samples, meaning the relationship was inconsistent throughout the range of TP concentration, mostly due to the presence of some outliers that had a significantly reduced TP concentration after freezing. Multiple samples, dropped below 4 g/dL after frozen storage, which is the minimum recorded in fresh serum samples from calves that have not received colostrum [2,21]. Additionally, we saw a correlation between how long the sample had been frozen and the degree of variation between TP_{serum} concentrations before and after freezing. Samples frozen for a shorter period of time appeared to have less difference than those samples frozen for longer periods. These results show that freezing can detrimentally affect TP concentrations, especially during longer freezing periods, as other studies have shown in different species [22] or with other blood analytes [23].

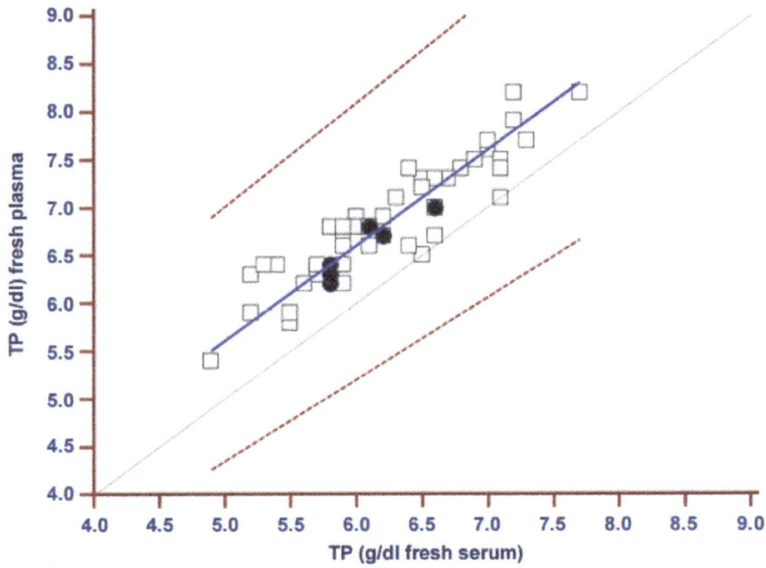

Figure 3: Correlation between total protein (TP) concentrations in fresh serum and plasma samples of 58 calves. Data points are presented for 6 Holstein (•) and 52 Jersey calves (□).

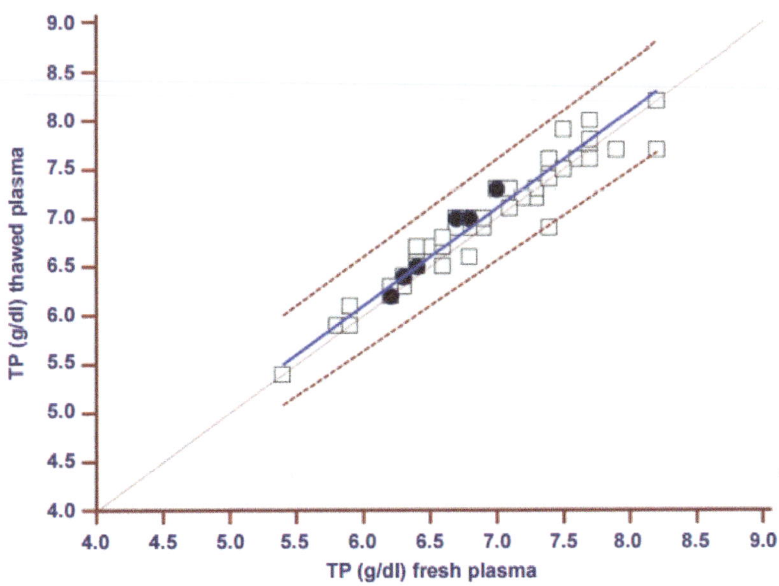

Figure 4: Passing Bablok correlation between total protein (TP) concentration in fresh and frozen/thawed plasma samples of 58 calves. Data points are presented for 6 Holstein (•) and 52 Jersey calves (□).

To determine which protein fractions were affected during the freezing/thawing process that could explain our results, electrophoresis was performed in another 20 samples. Data showed that all protein fractions changed after freezing, but these changes were not consistent. Some fractions increased, at one point and decreased at another. The only consistent change was a decrease in albumin concentration, which represents the largest portion of TP in serum. Most of the variability was due to changes in the concentration of albumin and IgG. The concentration of the different protein fractions in this study were similar to those found by others [21], showing external validity.

The effect of storage time could not be tested for plasma samples because all plasma samples had the same length of freezing time (5 months), which was similar to the second longest time used for serum samples. Data in this study showed a similar lack of a direct linear relationship between fresh and frozen/thawed TP_{plasma} concentration as with serum samples, but the correlation was higher than for serum samples. In contrast to abnormally low TP_{serum} samples, there were no samples with TP_{plasma} concentrations below minimum recorded in neonates before receiving colostrum (4.0 g/dL) [21]. All results together may suggest that plasma samples are more stable under freezing conditions than serum samples. This is an important finding that can help in the monitoring of appropriate colostrum management in dairy calves, as plasma samples could be stored for longer periods of time without the apparent risk of altering TP concentrations. Therefore, the next question would be how TP_{plasma} concentrations correlate to TP_{serum} concentrations, given that most studies on passive transfer evaluation in dairy calves have been performed using serum samples.

Our data indicate that TP_{plasma} concentrations were consistently higher than TP_{serum} concentrations from the same calf. This is likely because plasma still contains fibrinogen and other coagulation proteins which are lost when processing serum. The significance of this finding is that when measuring TP_{plasma} to determine the effectiveness of colostrum management, the commonly used cut-off values for determination of FPT should be increased (by ~0.60 g/dL) compared to what would be expected in a serum sample. Further study is needed to show the correlation between TP_{plasma} and IgG concentration. Otherwise, most calves will be categorized as having adequate passive transfer of immunoglobulins, when in fact they may be marginal or even deficient.

CONCLUSION

Plasma sample sallow for easier monitoring of colostrum management when testing is performed in the field compared to serum samples, because they can be centrifuged immediately as opposed to serum samples, which have to wait for initial clotting before being centrifuged. Additionally, clotting time for serum samples depends on ambient temperature and can be significantly delayed in cold weather, making field management more difficult. The use of plasma will save time when testing TP in the field, although we realize this may not be possible in all situations. The apparent higher stability of plasma samples compared to serum samples during long periods of frozen storage provide a new option for veterinarians and producers, along with commonly used fresh serum samples, to determine the efficacy of colostrum management protocols established on farms.

ACKNOWLEDGEMENTS

This project was possible thanks to a generous grant from the AJCC Foundation from the American Jersey Association. We would like to thank Dr. Gabriel Calderon (3 Mile Canyon Dairy) and the staff at the Veterinary Diagnostic Laboratory at Oregon State University for the help provided during this study.

REFERENCES

1. Deutsch HF, Smith VR (1957) Intestinal permeability to proteins in the newborn herbivore. Am J Physiol 191: 271-276.
2. McCoy GC, Reneau JK, Hunter AG, Williams JB (1970) Effects of diet and time on blood serum proteins in the newborn calf. J Dairy Sci 53: 358-362.
3. Stott GH, Marx DB, Menefee BE, Nightengale GT (1979) Colostral immunoglobulin transfer in calves I. Period of absorption. J Dairy Sci 62: 1632-1638.
4. Sangild PT (2003) Uptake of colostral immunoglobulins by the compromised newborn farm animal. Acta Vet Scand Suppl 98: 105-122.
5. Stott GH, Marx DB, Menefee BE, Nightengale GT (1979) Colostral immunoglobulin transfer in calves. III. Amount of absorption. J Dairy Sci 62: 1902-1907.
6. Saskatoon, Saskatchewan (1983) Failure of passive transfer of colostral immunoglobulins and neonatal disease in calves: a review, Veterinary Infectious Disease Organization, Canada.
7. Wittum TE, Perino LJ (1995) Passive immune status at postpartum hour 24 and long-term health and performance of calves. Am J Vet Res 56: 1149-1154.
8. Penhale WJ, Christie G, McEwan AD, Fisher EW, Selman IE (1970) Quantitative studies on bovine immunoglobulins. II. Plasma immunoglobulin levels in market calves and their relationship to neonatal infection. Br Vet J 126: 30-37.
9. Berge AC, Besser TE, Moore DA, Sischo WM (2009) Evaluation of the effects of oral colostrum supplementation during the first fourteen days on the health and performance of preweaned calves. J Dairy Sci 92: 286-295.
10. Naylor JM, Kronfeld DS (1977) Refractometry as a measure of the immunoglobulin status of the newborn dairy calf: comparison with the zinc sulfate turbidity test and single radial immunodiffusion. Am J Vet Res 38: 1331-1334.
11. Pfeiffer NE, McGuire TC, Bendel RB, Weikel JM (1977) Quantitation of bovine immunoglobulins: comparison of single radial immunodiffusion, zinc sulfate turbidity, serum electrophoresis, and refractometer methods. Am J Vet Res 38: 693-698.
12. Tyler JW, Besser TE, Wilson L, Hancock DD, Sanders S, et al. (1996) Evaluation of a whole blood glutaraldehyde coagulation test for the detection of failure of passive transfer in calves. J Vet Intern Med 10: 82-84.

13. Ameri M, Wilkerson MJ (2008) Comparison of two commercial radial immunodiffusion assays for detection of bovine immunoglobulin G in newborn calves. J Vet Diagn Invest 20: 333-336.
14. Parish SM, Tyler JW, Besser TE, Gay CC, Krytenberg D (1997) Prediction of serum IgG1 concentration in Holstein calves using serum gamma glutamyltransferase activity. J Vet Intern Med 11: 344-347.
15. Dawes ME, Tyler JW, Hostetler D, Lakritz J, Tessman R (2002) Evaluation of a commercially available immunoassay for assessing adequacy of passive transfer in calves. J Am Vet Med Assoc 220: 791-793.
16. Quigley JD 3rd, Kost CJ, Wolfe TM (2002) Absorption of protein and IgG in calves fed a colostrum supplement or replacer. J Dairy Sci 85: 1243-1248.
17. Donovan GA, Dohoo IR, Montgomery DM, Bennett FL (1998) Associations between passive immunity and morbidity and mortality in dairy heifers in Florida, USA. Prev Vet Med 34: 31-46.
18. Tyler JW1, Hancock DD, Wiksie SE, Holler SL, Gay JM, et al. (1998) Use of serum protein concentration to predict mortality in mixed-source dairy replacement heifers. J Vet Intern Med 12: 79-83.
19. Calloway CD, Tyler JW, Tessman RK, Hostetler D, Holle J (2002) Comparison of refractometers and test endpoints in the measurement of serum protein concentration to assess passive transfer status in calves. J Am Vet Med Assoc 221: 1605-1608.
20. Wallace MM, Jarvie BD, Perkins NR, Leslie KE (2006) A comparison of serum harvesting methods and type of refractometer for determining total solids to estimate failure of passive transfer in calves. Can Vet J 47: 573-575.
21. Tennant B, Harrold D, Reina-Guerra M, Laben RC (1969) Neonatal alterations in serum gamma globulin levels of Jersey and Holstein-Friesian calves. Am J Vet Res 30: 345-354.
22. Thoresen SI, Tverdal A, Havre G, Morberg H (1995) Effects of storage time and freezing temperature on clinical chemical parameters from canine serum and heparinized plasma. Vet Clin Pathol 24: 129-133.
23. Tothova C, Nagy O, Seidel H, Kovac G (2012) The effect of storage temperature and time on the concentrations of bovine serum amyloid A and its mammary associated isoform. Vet Med Int 1-6: 861458.

10
CHAPTER

FACTORS AFFECTING SERUM TOTAL PROTEIN AND IMMUNOGLOBULIN G CONCENTRATION IN REPLACEMENT DAIRY CALVES

Aurora Villarroel, Taylor B Miller, Elisha D Johnson, Kimberly R Noyes and Justin K Ward

Keywords: Failure of passive transfer (FPT), Calve, Colostrums, Total protein (TP), Serum, Jersey, Breed, Age, Weight, Hydration

INTRODUCTION

Evaluation of dairy replacement calf health is generally divided into the pre-weaning and post-weaning periods; with an average weaning age of 8.6 weeks [1]. In 2007, the NAHMS Dairy study found that the pre-weaning period accounts for about 81% of total dairy replacement calf mortality [1]. Accordingly, management strategies to optimize health during this period will have the greatest impact on overall dairy calf mortality. Cows do not confer immunity transplacentally, but rather calves acquire protection from pathogens through ingestion of colostrum in a process called passive transfer [2]. While morbidity and mortality risks are not exclusively correlated with passive transfer status, multiple studies have reported an increased risk for morbidity and mortality in calves with low levels of circulating immunoglobulins [3-5]. Colostrum has great amounts of immunoglobulins, with the majority being Immunoglobulin G (IgG), although IgA and IgM are also present and may be important [2,6-8]. Immunoglobulins are absorbed by enterocytes in the gut and aid in disease prevention until the calf's own immune systems develops [2]. In addition to immunoglobulins, colostrum contains other diseasefighting constituents, such as leukocytes, lactoferrin, lysozyme [2], and cytokines [2,9]. Colostrum also serves as an important source of nutrition for the neonate, offering richer fat and vitamin contents than whole milk [10]. Higher concentration

Citation: Villarroel A, Miller TB, Johnson ED, Noyes KR, Ward JK (2013) Factors Affecting Serum Total Protein and Immunoglobulin G Concentration in Replacement Dairy Calves. Adv Dairy Res 1:106. doi:10.4172/2329-888X.1000106.

of immunoglobulins have been associated with improved production within the herd, specifically improved average daily gain [11] and increased milk yield [12].

There are three primary immunoglobulins found in colostrum: IgA, IgG and IgM. IgG is found in the highest concentration, and represents up to 90% of the immunoglobulins found in colostrum [13], though both IgA and IgM remain valuable contributors to passive immunity in the calf [7,8]. When adequate amounts of immunoglobulins are not absorbed to transfer immunity, the calf is determined to have failure of passive transfer (FPT), which has been convened to be defined in terms of serum IgG concentration <1000 mg/dL [14]. However, some researchers have used different cut off limits to define FPT such as 800 mg/dL [15], 750 mg/dL [16], and even as low as 350 mg/dL [17]. A strong correlation has been observed between IgG concentrations and TPin serum [18,19]. This relationship allows measuring serum TP concentration via refractometer under field conditions as a proxy for IgG concentration. Using the commonly accepted cutoff limit of 1,000 mg/dL of IgG to define FPT has resulted in a broad variation in cutoff values for serum TP in Holstein calves between 5.0 and 5.5 g/dL [5,20-24].

FPT has been associated through univariate analysis studies with up to a 90% increased risk rate in morbidity and up to 39% of calf mortality [5,25], and up to 180 days of age [11,20]. However, multivariate analysis studies differ in their results; some studies showed no association between IgG or TP in serum and morbidity or mortality [26,27], while others did find an association [17,20] although not for all diseases [20].

This variability in results, in combination with the absence of data for Jersey calves under current feeding conditions, calls for research to determine if factors such as calf breed, age, weight and temperature affect IgG and TP concentration in serum. Some differences known between Holstein and Jersey cattle include significant discrepancy in immunoglobulin content in colostrum [28] and decreased total volume of distribution due to the smaller size of Jersey calves [29]. Immunoglobulin concentration in Holstein cow colostrumhas been reported to be on average 55.9 g/L compared to the much higher concentration of 90.4 g/L in Jersey cow colostrum [28,30]. The differences in colostrum properties could potentially alter immunoglobulin absorption and create significant differences in serum TP and IgG levels that have been previously reported between the two breeds [31,32]. A study from 1969 showed that Jersey calves that were allowed to suckle from their dams up to 3 days of life had significantly higher serum IgG concentration during the first 16 weeks of life compared to Holstein calves with the same management [31]. However, to our knowledge there are no studies published using current standards of calf rising away from the dam.

It is reasonable to assume that when two calves of different body weights are fed the same amount of colostrum, the final concentration of IgG and TP in the smaller calf would be higher than in the larger calf due to the smaller volume of distribution in the smaller calf. However, previous studies found no association between body weight and IgG concentration among Holstein calves [11,20]. In review, the current body of research supports evaluation of the passive transfer status of calves younger than one week of age, although a specific timeframe has not been scientifically proven.

The objectives of this study were (1) to determine if calf breed, age, weight, temperature and hydration status influence serum TP concentration and IgG in dairy calves, and (2) whether commonly accepted cutoff values for FPT in Holstein calves are applicable to Jersey calves.

MATERIALS AND METHODS

Study Design

This was a cross-sectional study in which each calf was sampled only once.

Sampling Population

Heifer calves from three commercial dairy farms in Oregon were enrolled in the study between December 2008 and June 2010. Animals chosen for inclusion in the study were purebred Jersey or Holstein heifer calves between 1 and 30 days of age. Colostrum management protocols on all three farms specified delivery of at least 4 quarts of colostrum by esophageal tube prior to 12 hours of life. The exact number, timing, and volume of individual feedings varied between animals and farms.

Biological Tests

Blood samples were obtained via jugular venipuncture and collected into serum separator tubes (BD VACUTAINER® SST, Becton Dickinson Diagnostics, Franklin Lakes, NJ) and EDTA tubes (BD VACUTAINER® EDTA, Becton Dickinson Diagnostics, Franklin Lakes, NJ). A unique sampling identification number was assigned to each calf to track individual animals throughout the study. Parameters recorded included animal ID, birth date, date and time of blood draw, rectal temperature (digital thermometer), weight via girth tape (Weight-By-Breed Dairy Management Tape, C06070N, Nasco, Fort Atkinson, WI)). All samples were processed within four hours of collection. During the second year only, blood from each EDTA tube was drawn into a capillary tube to determine PCV. The EDTA and serum separator tubes were centrifuged at 3,000 RPM for 10 minutes to allow removal of serum, respectively. Serum samples were transferred into 2 mL cryopreservation vials (ML5355, Market Lab, Inc., Detroit, MI) for storage. At this time, fresh sample TP was determined using a digital refractometer (Model 300027, Sper Scientific Ltd., Scottsdale, AZ). Serum samples were stored frozen until laboratory analyses were performed.

For laboratory analyses, samples were allowed to warm to room temperature via overnight exposure to ambient air, and then TP and IgG were measured in thawed samples in the same manner and with the same equipment as before freezing. Serum samples were tested for IgG concentration using a commercial kit for single radial immunodiffusion assay (SRID, Single Radial Immunodifusion Kit – Bovine IgG, Catalog No. 240-60, VMRD, Inc. Pullman, WA). The SRIDkit included 4 standards with IgG concentrations of 400, 800, 1,600 and 3,200 mg/dL. Each well was inoculated with 3 μL of sample. After incubating at room temperature for approximately 24 hours, the precipitation ring of each well was read using an immunoviewer (TRANSIDYNE

General Corporation Calibrating Viewer, Ann Arbor, MI). IgG concentration in the test samples was determined according to manufacturer instructions using the curve generated by the standards in each kit.

Statistical Analyses

Stepwise multivariate linear regression analysis was used to determine the effect of breed, body weight and age on serum TP and IgG concentrations in dairy calves. Alpha to enter and leave the model was established at 15%. All analyses were performed using commercial statistical software (Minitab 15, Minitab Inc. State College, PA).

RESULTS

In total, 673 calves were included in the study; 411 Jersey and 262 Holstein. Serum TP and IgG were moderately correlated (r=0.670, P<0.001, Figure 1).The resulting TP cutoff value corresponding to an IgG concentration of 1,000 mg/dL was 4.7 g/dL. In fact, it varied between Holstein calves (5.06 g/dL) and Jersey calves (4.3 g/dL).

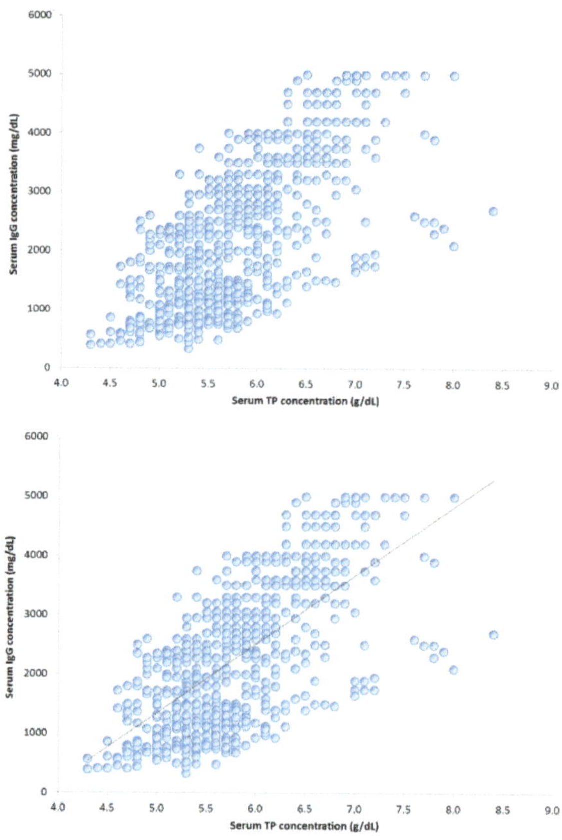

Figure 1: Correlation between serum IgG and Total Protein (TP) concentrations in 673 dairy replacement heifer calves 1 to 30 days of age.

Table 1: Regression analysis results for factors associated with IgG concentration (mg/dL) in dairy replacements calves at different ages (alpha to enter was 15%).

	1-30 days of age N=673	1-14 days of age N=483	1-7 days of age N=257
Constant	1278	978.6	1861
Jersey breed	1759	2080	1986
Age (days)	-66.2	-104.8	-74.0
Weight (lbs)	8.4	12.6	-

Only calves sampled during the second year of the study (N=434) had hematocrit information, and their data was used for stepwise regression to analyze the effect of dehydration on TP and IgG concentration in serum. Univariate analysis on these animals showed that hematocrit was not a significantly associated with serum TP(P=0.951) or serum IgG concentration (P=0.779). Regression analysis of the other factors on all 673 calves showed that age, breed and weight were associated with IgG concentration (Table 1), while age, breed and temperature were associated with serum TP concentration (Table 2). None of the variables were highly correlated to each other. The highest correlation was between age and weight (r=0.468).

The effect of age on IgG (Figure 2) and TPconcentration (Figure 3) was consistent in both breeds. However, this relationship was not linear, as can be observed in the graphs. The highest values for TP and IgG concentration were obtained at days 2 and 3 of age (Figures 2 and 3). The decrease was significant up to approximately 2 weeks of age (P<0.001), at which point TP remained similar in both breeds at approximately at 5.5 g/dL, while IgG concentration remained constantly higher in Jersey calves. To eliminate a potential effect that older age may have had on the other factors in the regression analyses, they were repeated including only calves up to 7 or 14 days of age. Temperature was no longer associated with TP concentration (Table 2). Factors associated with IgG concentration remained the same, although the parameters changed (Table 2). These results emphasize the need for sampling at a young age, and to be consistent to be able to compare results over time.

All Jersey calves had IgG and TP concentrations above commonly accepted standard of 1000 mg/dL to determine presence of FPT (minimum was 1150 mg/dL in a 10 day old calf). However, 20/411 (4.9%) of the Jersey calves had TP values <5.0 g/dL; all but one were over 10 days old.

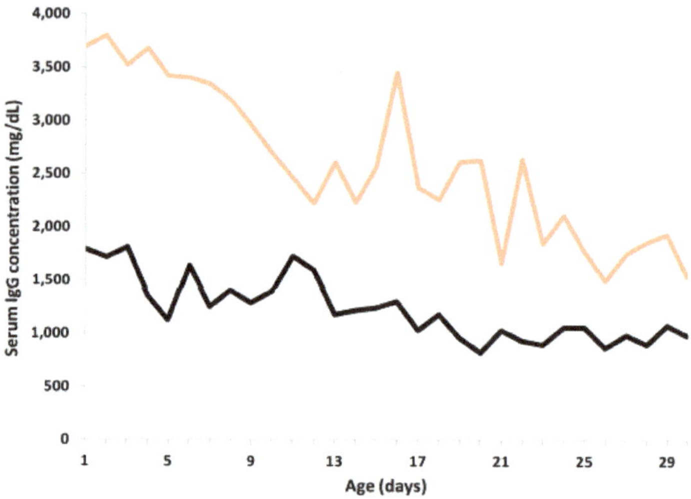

Figure 2: Distribution of serum IgG concentration by agein411Jersey (grey line) and 262 Holstein (black line) calves between 1 and 30 days of age.

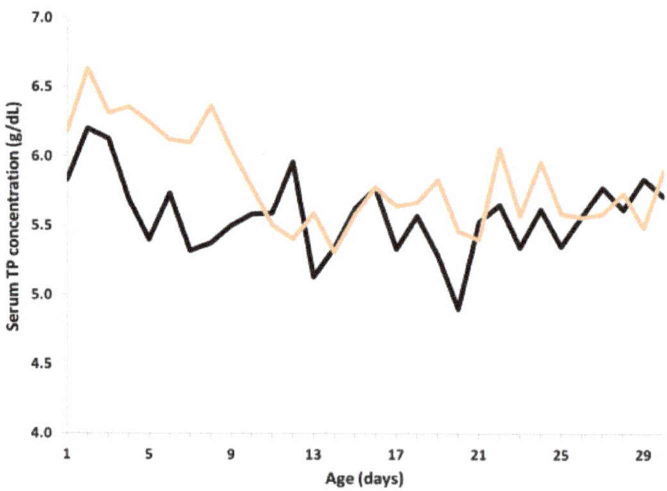

Figure 3: Distribution of serum Total Protein (TP) concentration by agein411 Jersey (grey line) and262Holstein (black line) calves between 1 and 30 days of age.

Morbidity and mortality information could only be obtained for 91 Jersey calves that were sampled at 2-3 days of age in this study. Of these, 33 developed scours, 15 developed pneumonia, and 20 were afflicted with both, leaving 23 healthy calves for comparison. Only one calf died during the study, making statistical analyses for mortality risk impossible. Calves were diagnosed with pre-weaning scours at an average of 6.83 ± 1.81 days, and with pneumonia at an average of 36.71 ± 9.05 days (Figure 4). Mean ±

SD for IgG and serum TP concentrations calves are presented in Table 3. There was no significant difference in serum TP (P=0.924) or IgG (P=0.348) between healthy calves and those that developed scours, pneumonia, or both. Using the commonly accepted definition of FPT of 1,000 mg/dL of IgG, there was no difference in risk of scours, risk for pneumonia or risk for developing both diseases compared to healthy calves. There was also no difference in risk of morbidity using standard TP cutoff values of 5.0 g/dLor 5.5 g/ dL. Survival analyses showed no difference in median time to develop morbidity between calves that were above or below 5.5 g/dL (Figure 5) or 6.0 g/dL (data not shown). Because there was only one calf that died, FPT could not be evaluated as a risk factor for mortality.

DISCUSSION AND CONCLUSIONS

Traditional research in Holstein calves has established cutoff values for FPT at IgG concentration at 1,000 mg/dL in serum [14], which has corresponding values of serum TP concentration between 5.0 and 5.5 g/dL [5,20-24]. As in previous studies [19,21,22], we found that serum IgG and TP concentrations were correlated, but less so than in other studies. The correlation was similar to other studies for Holstein calves, but differed in Jersey calves. Cutoff values for TP concentration corresponding to 1,000 mg/dL of IgG varied greatly between breeds. In fact, this cutoff value had to be estimated in Jersey calves, as none had IgG concentration below 1,150 mg/dL. Another likely reason for this discrepancy was the inclusion of older calves in this study. The less than perfect correlation between IgG and TP concentration would have resulted in 4.9% of Jersey calves to have been classified as having FPT using the proxy of serum TP<5.0 g/dL as the cutoff limit, but 0% if IgG<1,000 mg/dL would have been used. This poses a practical problem for veterinarians working with Jersey calves.

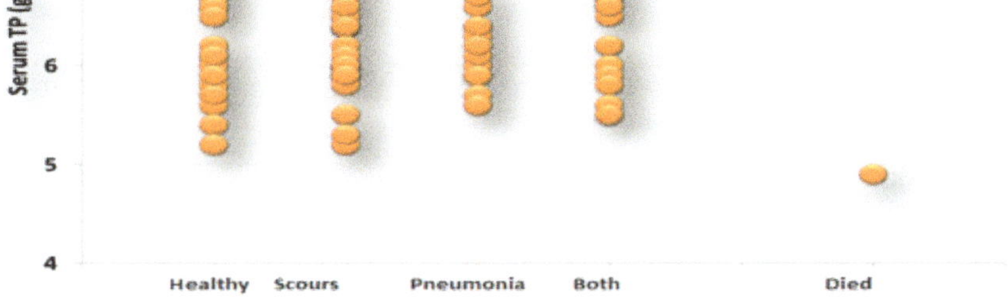

Figure 4: Comparison of serum TP concentration among 91 Jersey calves (1-3 days old at sampling) according to morbidity group.

Table 2: Regression analysis results for factors associated with total protein concentration (g/dL) in dairy replacements calves at different ages (alpha to enter was 15%).

	1-30 days of age N=673	1-14 days of age N=483	1-7 days of age N=257
Constant	10.67	6.19	6.06
Jersey breed	0.28	0.36	0.49
Age (days)	-0.024	-0.074	-0.067
Temperature (°F)	-0.047	-	-

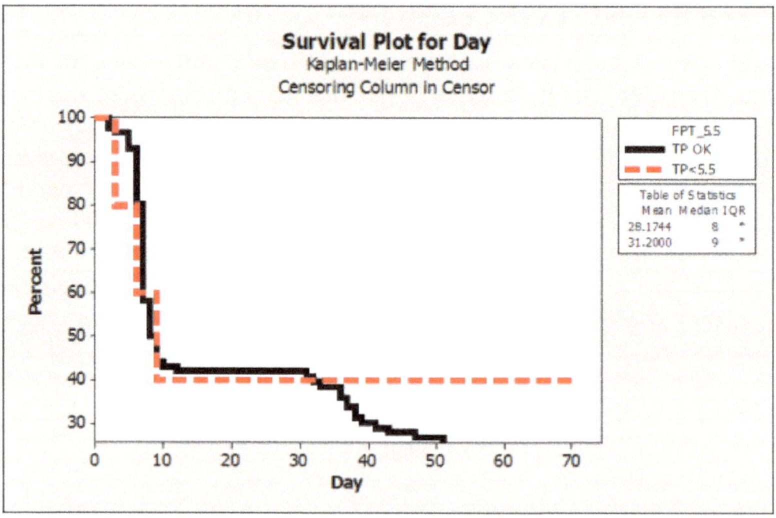

Figure 5: Kaplan-Meier survival curve for morbidity among 91 Jersey calves using 5.5 g/dL of TP as the cutoff limit for determining FPT.

Our study found that Jersey calf serum IgG and TP concentrations were significantly and consistently higher than those in Holstein calves. In fact, IgG concentrations were more than twice that of Holstein calves at most ages. TP concentrations were on average approximately 0.5 g/dL higher than in Holstein calves throughout the first week of age (Table 2). These findings indicate that applying Holstein cutoff values to manage Jersey calves could result in inadequate monitoring programs that can have detrimental consequences such as increased morbidity and mortality in these calves. Actual biological implications of these differences were not apparent as there was no difference in IgG or TP concentrations between healthy calves and those that developed diarrhea, pneumonia or both. Only one calf died, making evaluation of mortality impossible.

As seen in Figures 2 and 3, the differences in IgG and TP concentrations between the two breeds decreased over time. These results are similar to those obtained by Tennant et al. [31], which supports an inherent breed difference that may not be addressed by using

Holstein cutoff values in Jersey calves. This is most obvious when comparing TP and IgG differences between breeds. TP concentrations seemed to equalize approximately at 2 weeks of age, while IgG concentrations remained constantly different.

The effect of age on both TP and IgG concentration was important (Figures 2 and 3). These results emphasize the need for sampling not only at a young age, but also to be consistent in the selection of calves within that age range to be able to compare results over time. TP and IgG concentration were highest at 2 and 3 days of age. This peak represents the systemic absorption and circulation of maternal antibodies, which occurs primarily in the first 24 to 48 hours, needing some lag time to show in serum. Sampling calves at 2-3 days of age will guarantee estimation of the highest TP and IgG concentrations achieved on a farm. Our findings indicate that calves up toone week old could be sampled to evaluate FPT, but cutoff values to establish inadequate TP concentrations should be adjusted as they decrease approximately by 0.07 g/dL each day within the first week of age. IgG concentration decreased by 74 mg/dL each day.

Our data also supports the hypothesis that weight influences IgG concentration, but not TP concentration. This result is further proof of the limited use of TP as a proxy for IgG concentration, and that interpretation of low TP concentration as an indicator of FPT in calves (especially Jersey calves) has to be done with caution. We agree that these measurements can be used to indicate whether there is a problem with the colostrum management program by determining the proportion of calves with low values. However, absolute values cannot be used to predict morbidity or mortality risk.

Limited amount of research in Jersey calf management has been published in the last 25 years. This has resulted in the necessity to assume that standards of care acceptable for Holstein calves are applicable to Jersey calves. While the farms in this study did not experience significant differences in morbidity and mortality between calves categorized as having FPT or being healthy using Holstein cutoff values, it is a limited cohort and may not represent all types of dairy management practices across the US.

Table 3: Comparison of serum total protein and IgG between calves with different morbidity.

Parameter	N	Total Protein	Range	Ig G	Range
Healthy calves	23	6.42 ± 0.62	5.2-7.3	3,904 ± 616	2,950 –5,000
Scours	33	6.33 ± 0.69	4.9 –8.0	3,561 ± 826	2,000 –5,000
Pneumonia	15	6.40 ± 0.63	5.6 –8.0	3,547 ± 802	2,100 –5,000
Both	20	6.45 ± 0.59	5.5 –7.5	3,770 ± 738	2,600 –5,000
All calves	91	6.39 ± 0.63	4.9 –8.0	3,688 ± 753	2,000 –5,000

However, our findings indicate that there are significant breed differences, demonstrating the need for future research to focus specifically on Jersey calves.

REFERENCES

1. USDA-APHIS-VS (2007) Dairy 2007. Part II: Changes in the U.S. Dairy Cattle Industry, 1991-2007.
2. Godden S (2008) Colostrum management for dairy calves. Veterinary Clinics of North America Food Animal Practice 24: 19-39.
3. McGuire TC, Pfeiffer NE, Weikel JM, Bartsch RC (1976) Failure of colostral immunoglobulin transfer in calves dying from infectious disease. J Am Vet Med Assoc 169: 713-718.
4. Naylor JM, Kronfeld DS, Bech-Nielsen S, Bartholomew RC (1977) Plasma total protein measurement for prediction of disease and mortality in calves. J Am Vet Med Assoc 171: 635-638.
5. Tyler JW, Hancock DD, Wiksie SE, Holler SL, Gay JM, et al. (1998) Use of serum protein concentration to predict mortality in mixed-source dairy replacement heifers. J Vet Intern Med 12: 79-83
6. Larson BL, Heary HL Jr, Devery JE (1980) Immunoglobulin production and transport by the mammary gland. J Dairy Sci 63: 665-671.
7. Bush LJ, Staley TE (1980) Absorption of colostral immunoglobulins in newborn calves. J Dairy Sci 63: 672-680.
8. Stott GH, Marx DB, Menefee BE, Nightengale GT (1979) Colostral immunoglobulin transfer in calves. III. Amount of absorption. J Dairy Sci 62: 1902-1907.
9. Hagiwara K, Kataoka S, Yamanaka H, Kirisawa R, Iwai H (2000) Detection of cytokines in bovine colostrum. Vet Immunol Immunopathol 76: 183-190.
10. Foley JA, Hunter AG, Otterby DE (1978) Absorption of colostral proteins by newborn calves fed unfermented, fermented, or buffered colostrum. J Dairy Sci 61: 1450-1456.
11. Robison JD, Stott GH, DeNise SK (1988) Effects of passive immunity on growth and survival in the dairy heifer. J Dairy Sci 71: 1283-1287.
12. DeNise SK, Robison JD, Stott GH, Armstrong DV (1989) Effects of passive immunity on subsequent production in dairy heifers. J Dairy Sci 72: 552-554.
13. Elfstrand L, Lindmark-Månsson H, Paulsson M, Nyberg L, Åkesson B (2002) Immunoglobulins, growth factors and growth hormone in bovine colostrum and the effects of processing. Int Dairy J 12: 879-887.
14. Veterinary Infectious Disease Organization (1983) Failure of passive transfer of colostral immunoglobulins and neonatal disease in calves: a review. Saskatoon, Saskatchewan, Canada.
15. Wittum TE, Perino LJ (1995) Passive immune status at postpartum hour 24 and long-term health and performance of calves. Am J Vet Res 56: 1149-1154.
16. Penhale WJ, Christie G, McEwan AD, Fisher EW, Selman IE (1970) Quantitative studies on bovine immunoglobulins. II. Plasma immunoglobulin levels in market calves and their relationship to neonatal infection. Br Vet J 126: 30-37.

17. Berge AC, Besser TE, Moore DA, Sischo WM (2009) Evaluation of the effects of oral colostrum supplementation during the first fourteen days on the health and performance of preweaned calves. J Dairy Sci 92: 286-295.

18. Naylor JM, Kronfeld DS (1977) Refractometry as a measure of the immunoglobulin status of the newborn dairy calf: comparison with the zinc sulfate turbidity test and single radial immunodiffusion. Am J Vet Res 38: 1331-1334.

19. Tyler JW, Hancock DD, Parish SM, Rea DE, Besser TE, et al. (1996) Evaluation of 3 assays for failure of passive transfer in calves. J Vet Intern Med 10: 304-307.

20. Donovan GA, Dohoo IR, Montgomery DM, Bennett FL (1998) Associations between passive immunity and morbidity and mortality in dairy heifers in Florida, USA. Prev Vet Med 34: 31-46.

21. Quigley JD, Kost CJ, Wolfe TM (2002) Absorption of protein and IgG in calves fed a colostrum supplement or replacer. J Dairy Sci 85: 1243-1248.

22. Tyler JW, Hancock DD, Thorne JG, Gay CC, Gay JM (1999) Partitioning the mortality risk associated with inadequate passive transfer of colostral immunoglobulins in dairy calves. J Vet Intern Med 13: 335-337.

23. Tyler JW, Hancock DD, Wilson L, Muller F, Krytenberg D, et al. (1999) Effect of passive transfer status and vaccination with Escherichia coli (J5) on mortality in comingled dairy calves. J Vet Intern Med 13: 36-39.

24. Weaver DM, Tyler JW, VanMetre DC, Hostetler DE, Barrington GM (2000) Passive transfer of colostral immunoglobulins in calves. J Vet Intern Med 14: 569-577.

25. Ibrahim A, Lemma A (2009) Relationships between serum protein concentration and passive transfer of immunity, morbidity and mortality of dairy calves in market oriented urban dairy farms. Revue de Méd Vét 180: 394-399.

26. Pare J, Thurmond MC, Gardner IA, Picanso JP (1993) Effect of birthweight, total protein, serum IgG and packed cell volume on risk of neonatal diarrhea in calves on two California dairies. Can J Vet Res57: 241-246.

27. Virtala AM, Grohn YT, Mechor GD, Erb HN (1999) The effect of maternally derived immunoglobulin G on the risk of respiratory disease in heifers during the first 3 months of life. Prev Vet Med 39: 25-37.

28. Muller LD, Ellinger DK (1981) Colostral immunoglobulin concentrations among breeds of dairy cattle. J Dairy Sci 64: 1727-1730.

29. Quigley JD, Drewry JJ, Martin KR (1998) Estimation of plasma volume in Holstein and Jersey calves. J Dairy Sci 81: 1308-1312.

30. Quigley JD 3rd, Martin KR, Dowlen HH, Wallis LB, Lamar K (1994) Immunoglobulin concentration, specific gravity, and nitrogen fractions of colostrum from Jersey cattle. J Dairy Sci 77: 264-269.

31. Tennant B, Harrold D, Reina-Guerra M, Laben RC (1969) Neonatal alterations in serum gamma globulin levels of Jersey and Holstein-Friesian calves. Am J Vet Res 30: 345-354.

32. Jaster EH (2005) Evaluation of quality, quantity, and timing of colostrum feeding on immunoglobulin G1 absorption in Jersey calves. J Dairy Sci 88: 296-302.

11
CHAPTER

TRANSCRIPTOMICS COMPARISONS OF MAC-T CELLS VERSUS MAMMARY TISSUE DURING LATE PREGNANCY AND PEAK LACTATION

Afshin Hosseini, Rekha Sharma, Massimo Bionaz and Juan J Loor

Keywords: Mac-T cells, Bovine Mammary Tissue, Transcriptomics

INTRODUCTION

Cell culture systems have been useful to help answer basic questions of mammary gland biology in normal and diseased states [1]. Several immortalized cell lines have been established from primary mammary epithelial cells. The primary cells or tissue explants are likely more representative of the in vivo system due to the temporal proximity to the original tissues/cells having, likely, more similar molecular milieu (e.g. more similar epigenome and transcriptome); however, it is cumbersome to consistently obtain tissue from non-laboratory species under controlled conditions (e.g. age of animal, stage of glandular development and parity). Therefore, despites the limitation in modeling the in vivo system the use of established cell lines can be helpful in the study of epithelial mammary gland biology.

Several ruminant mammary epithelial cell lines (five bovine, two ovine and one caprine) have been established and well-characterized [2-9]. Bovine cell lines BMEC+H (Bovine Mammary Epithelial Cells of the Hormone-adapted) [4], HH2A (spontaneously immortalized bovine mammary epithelial cell line) [10], and ET-C (epithelial and myoepithelial-like characteristics) [11] do not express lactationspecific proteins. Other bovine mammary epithelial cell lines such as BME-UV [5] express

Citation: Hosseini A, Sharma R, Bionaz M, Loor JJ (2013) Transcriptomics Comparisons of Mac-T cells Versus Mammary Tissue during Late Pregnancy and Peak Lactation. Adv Dairy Res 1:103. doi:10.4172/2329-888X.1000103.

a low level of α-lactalbumin and αs1-casein, whereas Mac-T (Mammary Alveolar Cells) have a variable expression level of milk proteins [2,12,13]. Compared with other immortalized bovine mammary epithelial cells, the Mac-T cells can express and secrete α- and β-caseins [2]. These cells can be passed to >350 without sign of senescence [2]. The Mac-T cells, as ET-C and BME-UV cells, were established by stable integration of the Simian Virus large T– antigen (SV-40 large T-antigen) gene to induce immortalization. The SV40LTA is known to bind retinoblastoma protein and p53, thereby potentially modulating both the cell cycle and apoptotic pathways leading to an immortal phenotype. While this is useful in establishing and maintaining the cell line it is not clear how this modification affects the mammary epithelial-specific phenotype.

To date the Mac-T cells have been widely used in order to provide preliminary data for the effect of growth factors or hormones [12,14-16], fatty acids [17], amino acids [18,19], and bacterial cell wall components such as LTA and LPS [20]. However, the development of Mac-T cells was performed using a mixture of primary cells from slaughtered cows [2] without considering or reporting the stage of lactation of the cows. Therefore, it remains to be determined how reliable the Mac-T cells are for providing preliminary in vitro data relevant to mammary epithelial cells in general; more specifically, we aimed to determine how reliable the Mac-T cells cultured in lactogenic media on 2 dimensional plastic plates would be as a model to study in vivo mammary tissue biology. To that aim, we performed direct transcriptomics comparison between Mac-T cells and non-lactating (4 wk prior to parturition; -30 day in milk (DIM)) or lactating mammary tissue (+60 DIM) followed by a thorough bioinformatics analysis.

MATERIALS AND METHODS

Samples

The total RNA from Mac-T cells was from 4 replicates of the control group maintained in lactogenic medium from a previous experiment [17]. The total RNA from mammary tissue of 3 multiparous Holstein cows at 1 month prior parturition (-30 DIM) and at peak lactation (+60 DIM) was from a previous experiment [21].

Microarray Analysis

A specific bovine 13K oligo array previously described [22] was used. This platform is publicly accessible at the National Center for Biotechnology Information (NCBI) Gene Expression Omnibus (GEO) database (GSE4304). The synthesis of cDNA, the protocol for microarray analysis, and data quality assessment were as previously described [23]. Microarrays were run in a dye-swap design as previously described [22] with the cDNA obtained from the Mac-T cells as reference. Few additional modifications compared with previous methods were introduced: prior to hybridization labeled cDNA of Mac-T cells was pooled for each dye and the same amount of labeled cDNA from mammary and Mac-T was added to two separate tubes prior to vacuum-drying in the dark in order to prepare the specimens for co-hybridization. The microarray data presented in this manuscript have been deposited at NCBI's Gene Expression Omnibus [24] and are accessible through GEO Series accession number GSE46476.

Verification of microarray data using quantitative real-time PCR (qPCR)

The results form microarray were validated performing qPCR for a selected panel of 32 genes considered important for the mammary gland functioning, e.g. growth and proliferation, transport, metabolic path ways, and regulatory elements (see primer pairs and amplicon sequencing results in Tables S1 and S2 in File S1 in supplementary online material). Methods for primer pair design and validation and qPCR were as previously described [21].

Essential for reliability of qPCR data is normalization using appropriate internal control genes (ICGs). Microarray data were mined to identify suitable ICGs as previously described [25]. A panel of more than 32 genes with high stability in expression between mammary and Mac-T (ratio ≈ 1) plus previously tested ICGs [26] were evaluated for co-regulation using Ingenuity Pathway Analysis (IPA, Ingenuity® Systems, www.ingenuity.com). Among those VPS4A, SLC26A11, ABHD11, GAPDHS, ZBTB2, SLC41A3, RECQL5, COL23A1, SH3BGR, MRPL39, MTG1, RPS15 and UXT did not have known co-regulation. A qPCR analysis using all samples was performed for these 13 genes. The raw qPCR results from tested genes were used to select the most stable ICGs as well as the number of ICGs which should be used for accurate normalization using the geNorm software [27]. The results from geNorm suggested that MRPL39, UXT, and MTG1 were the most reliable ICGs and the geometric mean of the three was used for qPCR normalization.

Statistical Analysis

The GSDAT files were uploaded into Gene Spring GX7 (Agilent, USA), normalized using Lowess, and the statistical package included in the software was used to provide a list of similar (SEG) and differentially expressed genes (DEG) using ANOVA. Due to the microarray experimental design a direct comparison between the microarray signal from mammary tissue and Mac-T cells was performed using a Benjamini-Hochberg false discovery rate (FDR). The SEG and DEG between mammary tissue and Mac-T were determined by using an FDR cut-off ≥ 0.20 and 0.05, respectively.

The qPCR data were log2 transformed prior statistical analysis. The data were analyzed using a Generalized Linear Model (GLM) using SAS (v. 9.3) with cell/tissue type (Mac-T cells, mammary tissue at -30 DIM, and mammary tissue at +60 DIM) as main effect. Significance was declared at $P<0.05$.

Data Mining

Data were mined by means of IPA (analysis performed the 12/15/2010) and Database for Annotation and Visualization and Integrated Discovery (DAVID; performed the 12/10/2010) v6.7 [28]. All the data from oligos on the array were uploaded into IPA with associated annotations plus expression ratios and FDR. Each annotated gene was mapped to its corresponding gene object in the IPA Knowledge Base and the whole annotated bovine microarray was used as background. Genes from the dataset that met the FDR $\leq$ 0.05 or FDR>0.20 and were associated with biological functions in the IPA Knowledge Base were used for the analysis. The functions and pathways in IPA were considered significantly enriched and discussed at a Benjamini-Hochberg FDR $\leq$ 0.05. The interpretation of IPA analysis was performed as previously described [23]. For DAVID analysis the lists of DEG and SEG with associated bovine Entrez gene ID were up-loaded into the system and the whole annotated microarray was used as background. The DAVID default annotation databases plus chromosome and UP-tissue were downloaded as Functional Annotation Chart and Functional Annotation Cluster with an Easy score $\leq$ 0.01.

For analysis of the KEGG pathways of DEG between mammary tissue and Mac-T cells we also used the novel Dynamic Impact Approach (DIA) [29]. The analysis of DEG was as previously described [29]. In order to increase reliability of the DIA results only KEGG pathways that had: 1) $\geq$ 4 bovine genes mapped to the pathway and 2) annotated genes in microarray that covered $\geq$ 30% of all bovine genes mapped in the pathway by the bovine genome were used for analysis.

RESULTS

The complete dataset with expression ratio, p-value adjusted by FDR between each comparison, and a complete list of genes verified by qPCR is available in File S2.

Expression of several selected genes by qPCR

In Figure 1 are reported the results of genes verified with qPCR (plus CCND2 and PPARG, not present on the microarray) clustered based on main functions related to mammary biology. Overall >70% of genes measured with qPCR had a result deemed similar (considering also the statistical analysis) to microarray data (File S2 and Figure 1). Among the verified genes it was clear that Mac-T cells had a more pronounced expression of components of cell cycle, carbohydrate and lipid catabolism, and energy production compared with mammary gland (Figure 1). The mammary gland compared with Mac-T cells had a greater expression of genes involved in transport, lipid anabolism (particularly milk fat synthesis), caseins, and lactose synthesis. Among genes involved in regulatory functions mammary tissue vs. Mac-T cells had greater expression of PPARG, PRLR, and STAT5B, while expression of AKT1 and SREBF1 was similar or slightly greater in Mac-T cells compared with mammary tissue (Figure 1).

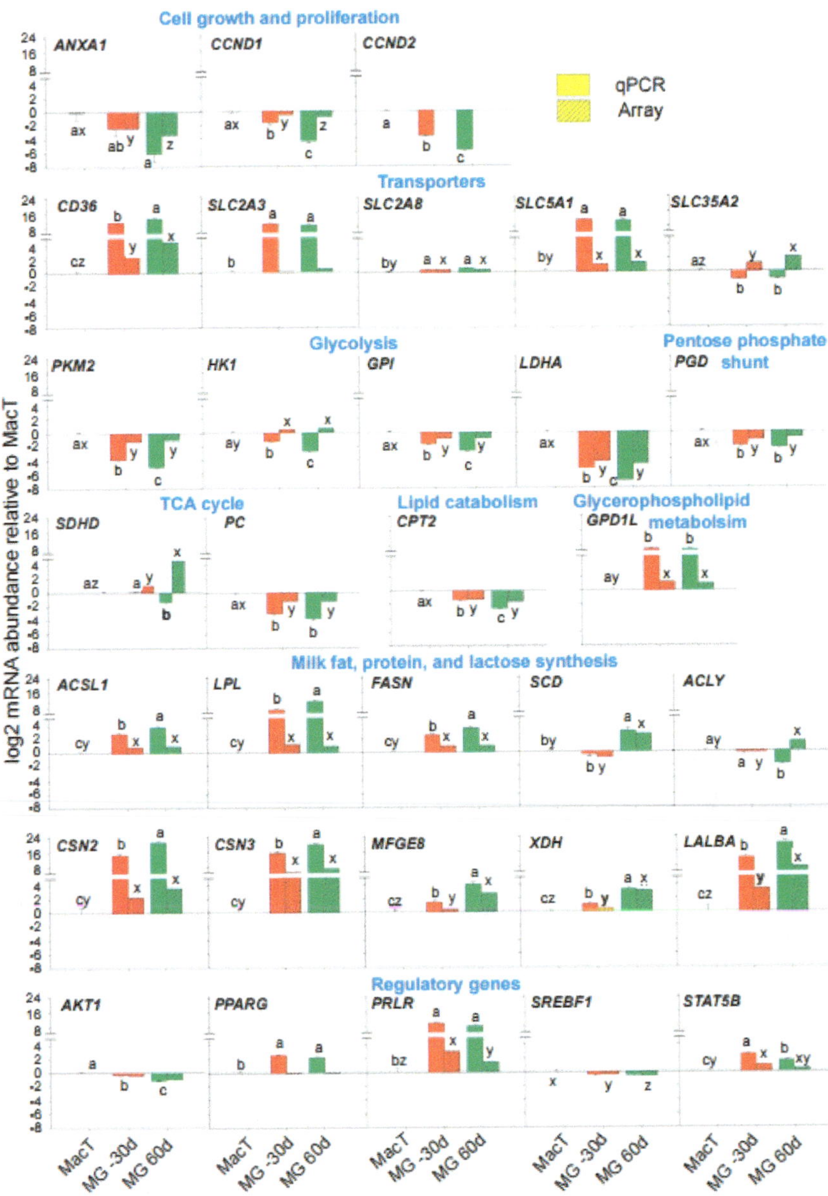

a,b,c denote differences with $p<0.05$ in qPCR data and x,y,z denote differences with $p<0.05$ in microarray data.

Figure 1: qPCR analysis of selected transcripts and comparison with microarray data. Transcripts are grouped based on main functions. Several of selected transcripts were present and differentially expressed between Mac-T and mammary tissue in the microarray data (File S2), others such as PPARG and CCND2 were run with qPCR but were no present in the microarray data.

Number of similarly expressed genes (SEG) and differentially expressed genes (DEG) in the Microarray Data

Overall there were ≥37% of SEG between mammary tissue and Mac-T cells at an FDR ≥ 0.20, and >53% at an FDR ≥ 0.05 (Table 1). We observed >46% overall DEG in mammary tissue vs. Mac-T cells (FDR<0.05). When comparing mammary tissue at -30 DIM and +60 DIM vs. Mac-T cells separately, the highest number of DEG was observed at -30 (ca. 44%) compared with +60 DIM (ca. 37%). The transcriptome of the Mac-T cells was more similar (FDR>0.20) to the transcriptome of the lactating (i.e., +60 DIM; >45%) compared to the non-lactating (i.e., -30 DIM) mammary tissue.

FUNCTIONAL ANALYSIS OF SEG AND DEG

Enrichment Analysis

SEG between Mac-T cells and Mammary Tissue

Enriched biological functions in IPA: No functions were significantly enriched with a Benjamini-Hochberg (B-H) FDR<0.05. Among the many functions enriched with a more liberal Exact Fisher test P<0.05 and using the "effect on function" analysis in IPA (Table S3 in File S1), we deemed interesting for our comparison those associated with lipid metabolism, carbohydrate metabolism, energy metabolism, and cell morphology. Within lipid metabolism were vitamin A and beta-estradiol. Apparently, vitamin A was more enriched in SEG at -30 DIM (Table S4) and beta-estradiol in SEG at +60 DIM (Table S5). In addition, the SEG between Mac-T cells and mammary tissue at -30 DIM were significantly enriched by functions related to induction of progesterone response (Table S4 in File S1).

The genes coding for proteins involved in increasing D-glucose import and ATP production were highly enriched in overall SEG (Table S3 in File S1), whereas genes involved in ATP production and AMP biosynthesis were significantly enriched among SEG at -30 DIM (Table S4). Other enriched lipid-related functions within SEG between Mac-T cells and mammary tissue at -30 DIM were modification and esterification of fatty acids (Tables S4 in File S1). Genes involved in cell morphology and cell-to-cell signaling, including cytoskeleton modification, vesicle formation, and tight junctions were enriched in SEG between Mac-T cells and mammary tissue at +60 DIM (Tables 3S and 5S in File S1).

Enriched pathways in IPA: Among overall SEG the most enriched pathways uncovered by IPA was 'Fatty Acid Elongation in Mitochondria' (File 4S in File S1). The 'Synaptic Long Term Depression', 'Keratan Sulfate Biosynthesis', and 'CD27 Signaling in Lymphocytes' were the most enriched pathways in SEG between Mac-T cells and mammary tissue at +60 DIM (File 4S in File S1). 'Taurine and Hypotaurine Metabolism' was the most enriched in genes with similar expression between Mac-T cells and mammary tissue at -30 DIM (File 4S in File S1).

Enriched biological functions and pathways in DAVID: Except 'SH3 domain' (SRC Homology 3 Domain) in SEG between Mac-T cells and mammary tissue at +60 DIM, no biological terms were significantly enriched with FDR $\leq$ 0.05 among SEG between Mac-T cells and mammary (Files S5 and S6).

DEG more expressed in Mac-T cells vs. Mammary Tissue

Enriched biological functions in IPA: Functions related to cellular migration and movement, cell death, cell growth and proliferation, cell morphology and development (including epithelial cells), cell-to-cell signaling and interaction (particularly cell adhesion and cytoskeleton organization), protein synthesis and modification, and amino acid transport were the most enriched (B-H FDR<0.05) in IPA for DEG with greater expression in Mac-T cells vs. mammary tissue (Tables S6 and S7 in File S1). All these functions, with exception of cell death, were more induced in Mac-T cells vs. mammary tissue. Interestingly, protein synthesis was the most enriched function in DEG with greater expression in Mac-T cells vs. mammary tissue at +60 DIM and apparently more induced in Mac-T cells.

Enriched pathways in IPA: The DEG with greater expression in Mac-T cells vs. mammary tissue (both overall, at -30, and + 60 DIM) were significantly enriched by pathways related to energy production in mitochondria (e.g. 'Oxidative Phosphorylation', 'Ubiquinone Biosynthesis'), cell-to-cell interaction, amino acid and glucose metabolism (e.g. 'Glycine, Serine and Threonine Metabolism', 'Glycolysis/ Gluconeogenesis'), cell proliferation (e.g. 'p53 signaling'), and protein synthesis (e.g. 'Aminoacyl-tRNA Biosynthesis') (File S3 and Table 2).

Enriched biological functions and pathways in DAVID: Similarly to IPA, the most enriched functions and pathways in DAVID among DEG with greater expression in Mac-T cells vs. mammary tissue were related to protein synthesis, energy production, and cell-tocell interaction (particularly cell adhesion, extra-cellular matrix, and cytoskeleton organization; Table 3 and details in Files S5 and S6). The analysis also uncovered that DEG with greater expression in Mac-T cells vs. mammary tissue were highly enriched by genes located on the chromosome 29 (Table 3 and File S5).

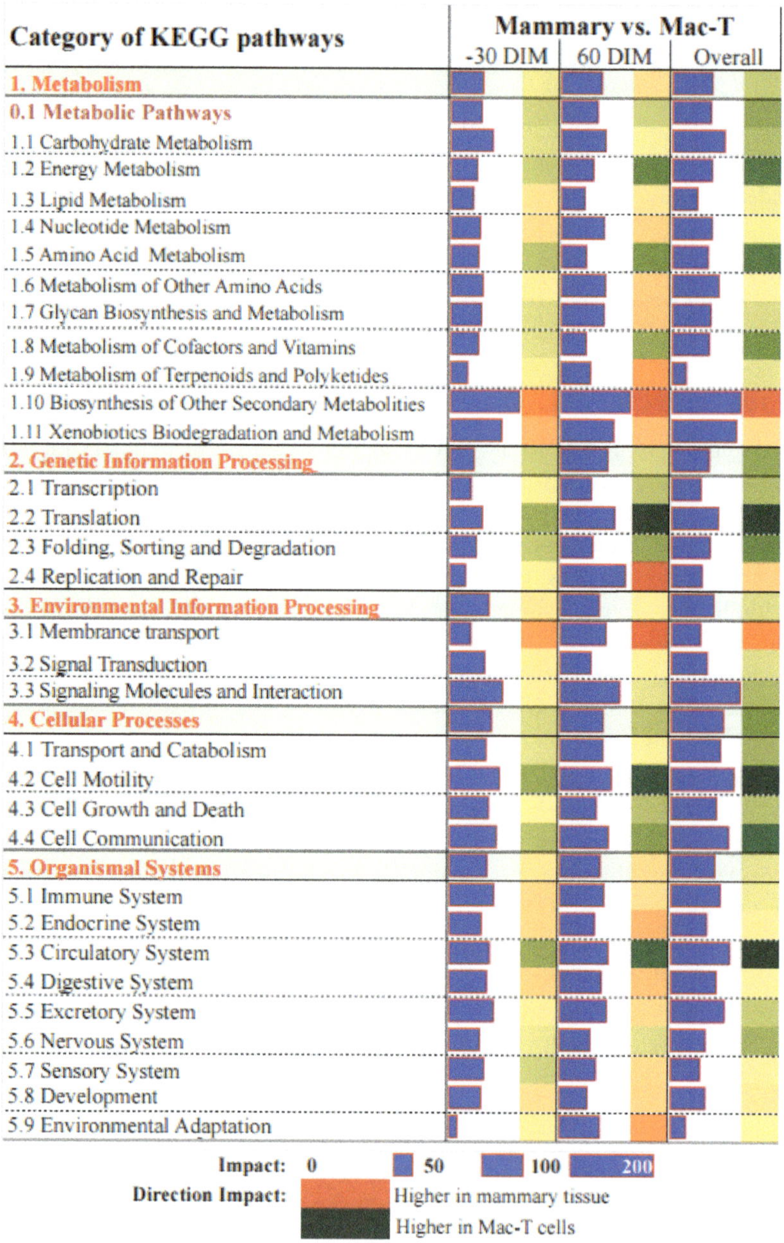

Figure 2: Overall calculated flux and impact of KEGG pathways by DIA of DEG between mammary tissue and Mac-T cells. Shown are the main categories of KEGG pathways (e.g., 'Metabolism', 'Genetic Information Processing') with related sub-categories. For each category and sub-category are reported the impact (horizontal bar; larger the bar larger the impact) and the direction of the impact (red more induced in mammary tissue vs. Mac-T cells and green more induced in Mac-T cells vs. mammary tissue).

Table 1: Number of transcripts similarly (FDR ≥ 0.05 and FDR ≥ 0.20) and differentially (FDR<0.05) expressed in mammary tissue at -30 and +60 DIM (or analysis when those were combined) vs. Mac-T cells. The number of genes was determined using the 10,308annotated transcripts on the microarray that passed all the criteria to select reliable data (see Materials and Methods).

FDR cut-off	Higher expression	Day in milk					
		-30	% array	+60	% array	Combined	% array
≥ 0.05		5,771	56.0	6,495	63.0	5,525	53.6
≥ 0.20		3,811	37.0	4,698	45.6	3,839	37.2
<0.05		4,537	44.0	3,813	37.0	4,783	46.4
	Mammary	2,184	21.2	1,564	15.2	2,106	20.4
	Mac-T	2,353	22.8	2,249	21.8	2,677	26.0

Table 2: Most enriched canonical pathways (-log10Benjamini-Hochberg FDR corrected P-value >2) by DEG in various comparisons using Ingenuity Pathways Analysis. Reported are the name of the pathway, the main general category of the pathway, the -log10 Fisher's- exact test P-value of the enrichment (p-value), the -log10 of the Benjamini-Hochberg FDR corrected P-value (FDR), the ratio of DEG/total number of genes in the pathway, and number of DEG in the pathway (Genes).

Pathway	Main category	p-val*	FDR*	Ratio	Genes
DEG with greater expression in Mac-T cells vs. mammary tissue overall					
Integrin Signaling	Cell-to-cell interaction	8.7	6.5	0.17	34
Actin Cytoskeleton Signaling		5.5	3.8	0.12	27
Ephrin Receptor Signaling		3.9	2.3	0.11	22
Regul. of Actin-based Motility by Rho		3.4	2.1	0.14	13
14-3-3-mediated Signaling	Signal transduction	3.6	2.1	0.14	16
Leukocyte Extravasation Signaling	Immune-related	3.5	2.1	0.11	21
Clathrin-mediated Endocytosis	Endocytosis	3.3	2.0	0.11	19
Oxidative Phosphorylation	Energy metabolism	8.2	6.2	0.17	26
Mitochondrial Dysfunction		6.5	4.7	0.13	23
Ubiquinone Biosynthesis		3.5	2.1	0.10	10
Gly, Ser and Thr Metabolism	Amino acid metabolism	3.2	2.0	0.06	9
Glycolysis/Gluconeogenesis	Glucose metabolism	3.2	2.0	0.08	12
DEG with greater expression in Mac-T cells vs. mammary tissue at -30 DIM					
Integrin Signaling	Cell-to-cell interaction	8.4	6.1	0.22	43
Ephrin Receptor Signaling		5.0	3.4	0.17	32
Actin Cytoskeleton Signaling		3.7	2.3	0.14	31
Tight Junction Signaling		3.6	2.3	0.16	26
Axonal Guidance Signaling		3.3	2.0	0.11	44
14-3-3-mediated Signaling	Signal transduction	4.6	3.1	0.20	23
fMLP Signaling in Neutrophils	Immune-related	3.6	2.3	0.17	21
Leukocyte Extravasation Signaling		3.2	2.0	0.14	27
PI3K/AKT Signaling	Cell cycle/protein synthesis	5.3	3.7	0.20	27
Oxidative Phosphorylation	Energy metabolism	7.7	5.7	0.20	32
Mitochondrial Dysfunction		6.3	4.5	0.17	29
Ubiquinone Biosynthesis		3.7	2.3	0.12	13
Methane Metabolism		3.2	2.0	0.08	5
Phenylalanine Metabolism	Amino acid metabolism	3.3	2.0	0.08	9
DEG with greater expression in Mac-T cells vs. mammary tissue at +60 DIM					
Integrin Signaling	Cell-to-cell interaction	6.7	4.9	0.21	41
Actin Cytoskeleton Signaling		5.2	3.5	0.16	36
Ephrin Receptor Signaling		3.8	2.4	0.15	30
Reg. of Actin-based Motility by Rho		3.6	2.3	0.20	18
Cell Cycle: G1/S Checkpoint Reg.	Proliferation	3.3	2.1	0.22	13
Wnt/β-catenin Signaling	Development/ differentiation	4.6	3.1	0.18	30
TGF-β Signaling		3.4	2.1	0.20	17
ERK/MAPK Signaling		3.3	2.1	0.16	30
Oxidative Phosphorylation	Energy metabolism	12.8	10.5	0.26	41
Mitochondrial Dysfunction		7.4	5.4	0.19	32
Ubiquinone Biosynthesis		4.8	3.2	0.14	15
Hyp. Signal. Cardiovascular System	Other	3.9	2.5	0.24	17

Enrichment analysis of DEG more expressed in Mammary Tissue vs. Mac-T cells

Enriched biological functions in IPA: There were no functions significantly enriched with a B-H FDR<0.05 in DEG with greater expression in mammary tissue compared with Mac-T cells (Table S8 in File S1). The use of a more liberal Exact Fisher's test P<0.05 indicated that functions related to carbohydrate and lipid metabolism were the most enriched and more induced in mammary vs. Mac-T (Table S8 in File S1). The analysis indicated a greater triacylglycerol (TAG), phospholipid, and ganglioside synthesis in mammary tissue vs. Mac-T cells (Table S8 in File S1). Also transport of constituents such as TAG, fatty acids, amino acids, and calcium, and cell death were apparently more induced in mammary tissue vs. Mac-T cells (Table S8 in File S1).

Among DEG with greater expression in mammary tissue at -30 DIM vs. Mac-T cells we observed enrichment (Exact Fisher's test P<0.05) of genes coding for proteins involved in lipid metabolism, particularly TAG synthesis, immune system activity and inflammation, cell death, transport of folic acid and mobilization of calcium (Table S10 in File S1). All those functions were more induced in mammary tissue vs. Mac-T cells.

In DEG with greater expression in mammary tissue at +60 DIM vs. Mac-T cells we observed enrichment (Exact Fisher's test P<0.05) of genes coding for proteins involved in tissue development, cellular growth and proliferation, immune cell trafficking, and inflammatory response (Table S11 in File S1). The data also indicated a significant enrichment of protein trafficking and lipid, carbohydrate, amino acid, nucleic acid, and calcium metabolism but the significance of the enrichment was lower compared with the above functions. In all cases the terms were overall more induced in mammary tissue compared with Mac-T cells.

Enriched pathways in IPA: As for the biological functions, no pathways were significantly enriched with a B-H FDR<0.05 by DEG with greater expression in mammary tissue vs. Mac-T cells. Considering a more liberal Exact Fisher's test P<0.05, the most enriched pathways were related to protein metabolism (e.g. 'Glutathione metabolism'), immune system (e.g. 'Antigen presentation pathway'), and regulation of lipid metabolism (e.g. 'LXR/RXR signaling'), particularly for the DEG with greater expression in mammary tissue at +60 DIMvs. Mac-T cells (File S3).

Enriched biological functions and pathways in DAVID: Similarly to IPA also in DAVID very few terms were significantly enriched in DEG with greater expression in mammary tissue vs. Mac-T cells (Files S5 and S6). Among the few significant enriched terms were the gene ontology terms related to extracellular matrix and mammary gland as UP-tissue (Files S5 and S6 and Table 3).

Table 3: Results from DAVID analysis for terms enriched with a Benjamini-Hochberg FDR<0.05. Additional results are available in Files S5 and S6. Category denotes the main group of terms in DAVID (BP=biological process, CC=cellular component, KEGG and UP Tissue=Uniprot Tissue). The count denotes the number of genes among the DEG in the specific term. No significant terms were observed for the DEG with greater expression in mammary tissue at -30 DIM vs. Mac-T cells and genes similarly expressed between mammary tissue and Mac-T with FDR>0.20.

Category	Term	Count	B-H*
DEG with greater expression in Mac-T cells vs. mammary tissue at -30 DIM			
Chromosome	29	66	<0.001
GOterm_BP_FAT	GO:0006412~translation	56	0.002
GOterm_CC_FAT	GO:0042470~melanosome	21	0.002
GOterm_CC_FAT	GO:0070469~respiratory chain	21	0.009
GOterm_CC_FAT	GO:0005739~mitochondrion	111	0.015
KEGG_pathway	bta00190:Oxidative phosphorylation	33	0.002
KEGG_pathway	bta04512:ECM-receptor interaction	20	0.016
KEGG_pathway	bta04510:Focal adhesion	38	0.016
KEGG_pathway	bta03010:Ribosome	22	0.022
KEGG_pathway	bta04810:Regulation of actin cytoskeleton	35	0.031
UP_tissue	Ileum	187	<0.001
UP_tissue	Heart	28	0.034
DEG with greater expression in Mac-T cells vs. mammary tissue at +60 DIM			
Chromosome	29	60	0.002
Chromosome	3	111	0.002
Chromosome	18	91	0.015
Chromosome	23	57	0.020
Chromosome	5	90	0.021
Chromosome	2	76	0.041
GOterm_BP_FAT	GO:0006412~translation	95	<0.001
GOterm_BP_FAT	GO:0006091~gener. precur. metab. and energy	50	<0.001
GOterm_BP_FAT	GO:0022900~electron transport chain	27	0.010
GOterm_BP_FAT	GO:0006414~translational elongation	9	0.017
GOterm_CC_FAT	GO:0005840~ribosome	73	<0.001
GOterm_CC_FAT	GO:0030529~ribonucleoprotein complex	95	<0.001
GOterm_CC_FAT	GO:0022627~cytosolic small ribosomal subunit	10	<0.001
GOterm_CC_FAT	GO:0070469~respiratory chain	24	<0.001
GOterm_CC_FAT	GO:0005743~mitochondrial inner membrane	52	0.003
GOterm_CC_FAT	GO:0042470~melanosome	19	0.013
KEGG_pathway	bta03010:Ribosome	60	<0.001
KEGG_pathway	bta00190:Oxidative phosphorylation	38	<0.001
UP_tissue	Ileum	229	<0.001
UP_tissue	Heart	37	<0.001
UP_tissue	Lymphoid epithelium	20	<0.001
DEG with greater expression in mammary tissue at +60 DIMvs. Mac-T cells			
GOterm_CC_FAT	GO:0005576~extracellular region	96	0.01
UP_tissue	Mammary gland	16	0.02

*Benjamini-Hochberg false discovery rate corrected P-value.

Dynamic Impact Approach Analysis of KEGG Pathways and Chromosome

Impact on main KEGG pathway categories: The overall view of the KEGG categories of pathways (Figure 2) evidenced an overall greater induction of pathways in Mac-T cells compared with mammary tissue with few exceptions such as pathways related to the biosynthesis of secondary metabolites, xenobiotics degradation, and membrane transport that were more induced in mammary tissue vs. Mac-T cells. The metabolic pathways were overall more induced in Mac-T cells, particularly in comparison with mammary tissue at -30 DIM. The 'Genetic Information Processing' was evidently more induce in Mac-T cells vs. mammary tissue, with a very strong impact and larger induction of the 'Translation' while the 'Replication and Repair' was more induced in mammary tissue (Figure 2). The Mac-T cells also had an apparent greater activation compared with mammary tissue of pathways related to 'Cellular Processes', particularly evident for the 'Cell Motility' and 'Cell Communication' (Figure 2). Several pathways related to 'Organismal Systems' were more induced in mammary tissue vs. Mac-T cells such as the ones related to the 'Endocrine System', 'Digestive System', and "Environmental Adaptation', particularly when considering mammary tissue at +60 DIM (Figure 2).

Integrated view of KEGG pathways using DIA: The DIA analysis of pathways indicated that the overall metabolism was greater in Mac-T cells compared with mammary tissue (Files S7 and S8). Most of the pathways related to carbohydrate metabolism were more induced in Mac-T vs. mammary tissue, particularly at +60 DIM. The Mac-T cells had a greater induction of pathways related to the use of glucose for energy purpose (e.g. 'Glycolysis/Gluconeogenesis', 'Pyruvate metabolism', 'Pentose and glucuronate interconversions'; File S8) while the mammary tissue had a more induced "Galactose metabolism' and 'TCA cycle' compared with Mac-T cells.

The lipid catabolism was overall more induced in Mac-T cells compared with mammary (e.g. 'Fatty acid metabolism' and 'Synthesis and degradation of ketone bodies') while mammary tissue (especially at +60 DIM) vs. Mac-T cells had a greater induction of lipid anabolism, particularly for synthesis of fatty acids and TAG (e.g. 'Glycerolipid metabolism', and 'Biosynthesis of unsaturated fatty acids') but not for sterols (Files S7 and S8).

With exception of 'Taurine and hypotaurine' and 'Glutathione metabolism' that were more induced in mammary tissue, the metabolism of amino acids was more pronounced in Mac-T cells compared with mammary tissue, chiefly the catabolism of Cys and Met and the biosynthesis of Phe, Tyr, and Trp (Files S7 and S8).

The synthesis of several glycans was more induced in mammary tissue vs. Mac-T cells, especially the synthesis of O-mannosyl, keratan sulfate glycosaminoglycan, glycosylphosphatidylinositol-anchor, and lacto and neolacto series of glycosphingolipids (Files S7 and S8). As for the above pathways this difference with Mac-T cells was more pronounced in lactating (i.e., at +60 DIM) compared with nonlactating mammary tissue.

Metabolic pathways related to cofactors and vitamins were overall more induced in Mac-T cells compared with mammary tissue, with a very large induction of 'One carbon pool of folate' in Mac-T cells (File S7). The pathways related to the biosynthesis of secondary metabolism and the degradation of xenobiotics were more induced in mammary vs. Mac-T cells, particularly concerning the 'Caffeine metabolism' and the 'Drug metabolism - cytochrome P450' (Files S7 and S8). The large induction of 'Caffeine metabolism' in mammary tissue vs. Mac-T was due to the larger expression of xanthine dehydrogenase (XDH, see File S2).

Among non-metabolic related pathways the DIA analysis uncovered a greater induction in Mac-T cells vs. mammary tissue of pathways related to transcription and, with an even greater induction, translation which was more pronounced in Mac-T cells vs. mammary tissue at +60 DIM (Files S7 and S8). All pathways related to DNA replication and repairs were more induced in mammary tissue vs. Mac-T cells (File S8). In accord with this observation was the greater induction of cell cycle in mammary tissue vs. Mac-T cells, particularly considering the non-lactating mammary tissue confirmed also by a greater induction in Mac-T of the 'p53 signaling pathway'(File S7).

The membrane transport involving ABC transporters was greater in mammary tissue vs. Mac-T cells (File S7) while cell-to-cell interaction and signaling was overall more pronounced in Mac-T cells vs. mammary tissue with few exceptions (File S7). Among these pathways 'Calcium signaling' and 'Jak-STAT signaling' were more induced in mammary tissue vs. Mac-T cells. The KEGG pathway analysis by the DIA also indicated that Mac-T cells vs. mammary tissue were characterized by having larger cell adhesion involving chiefly extracellular matrix interaction, focal adhesion, adherens junctions, and cytoskeleton regulation (File S7).

The overall greater induction in mammary tissue vs. Mac-T cells of the 'Organismal Systems' categories of pathways was mostly due to several pathways related to the innate immune system (File S7 and S8). The mammary tissue at +60 DIMvs. Mac-T cells had a larger induction of 'Complement and coagulation cascades', pathways related to the differentiation of immune-cells (i.e., 'Hematopoietic cell lineage'), sensitivity to bacteria and virus DNA (i.e., 'Cytosolic DNA-sensing pathway') and, particularly, response to antigens (i.e., 'Intestinal immune network for IgA production'), but a less induced 'Antigen processing and presentation', pathways related to the migration of leukocytes (i.e., 'Leukocyte trans endothelial migration'), and 'NODlike receptor signaling pathway' (Files S7 and S8). The non-lactating mammary tissue had instead a more induced 'Antigen processing and presentation' compared with Mac-T cells (File S7 and S8)

Among endocrine-related pathways the 'PPAR signaling', the 'Renin-angiotensin system', and, with a minor impact, the 'GnRH signaling', were more induced in mammary tissue (chiefly at +60 DIM) vs. Mac-T cells (File S7). The pathways related to excretory system appeared to be more induced in mammary tissue vs. Mac-T cells, particularly for calcium and water reabsorption (Files S7 and S8)

For other pathways with likely less biological relevance in the present experiment (e.g. 'Human Disease' category of pathways), it is interesting to point out that the pathway 'Bacterial invasion of epithelial cells' was more induced in Mac-T cells compared with mammary tissue (Files S7 and S8).

In summary the overall analysis of KEGG pathways using the DIA indicated that:

- Mac-T cells compared with mammary tissue had a greater catabolism in order to produce energy, a greater protein synthesis and degradation capacity, a greater induction of cell-to-cell interaction heavily involving the cytoskeleton, and greater response to bacterial invasion.
- Mammary tissue compared with Mac-T cells was characterized by having a greater anabolism particularly involving glucose and lipid, production of secondary metabolites, degradation of xenobiotic, DNA replication and cell cycle, innate immune capacity, and signaling pathways known to be important for milk fat and protein synthesis.

Most impacted chromosomes uncovered by DIA: Overall, the BTA6 was the chromosome with the greatest impact due to DEG between mammary tissue and Mac-T cells with an obvious greater induction of transcription in mammary tissue (File S7).

This was mostly due to the greater expression of casein genes in mammary tissue vs. Mac-T cells, but not exclusively; few genes, several involved in transport, such as SLC39A8 (Zn transporter) and SLC34A2 (phosphate transporter), are also located in the BTA6 and had greater expression in lactating mammary tissue vs. Mac-T cells (File S2 and Figure S1). Among the most impacted chromosomes was BTA14 in which transcription overall was more induced in mammary tissue vs. Mac-T cells (File S7). Several other chromosomes, including BTA29 and BTA23, were among the most impacted by the DEG between mammary tissue and Mac-T cells and transcription was overall more induced in the latter (File S7 and Figure S2). Several of the chromosomes with greater impact in the DIA analysis were also significantly enriched in the DAVID analysis (Table 3).

DISCUSSION

Immortalized mammary epithelial cell lines can be useful to conduct in vitro studies to obtain data pertinent to the bovine mammary gland. The immortalized Mac-T cells have been extensively used to investIgAte the effect of fatty acids [12,14-16], growth factors or hormones [17], and amino acids [18,19], to study the milk fat [17] and protein synthesis [30], and inflammatory response [20,31]. The usefulness of the data obtained using immortalized cell lines is related to the similarity with mammary tissue. In order to evaluate this similarity, in the present study we performed a comparison at the transcriptomics level between Mac-T cells cultivated in a 2 dimensional (2D) in vitro system and mammary tissue during the end of pregnancy and peak lactation.

The use of a 2D culture system in the present study is an important point considering that the 3 dimensional (3D) structure has a strong effect on determining the identity, therefore function, of cells [32]. This appears to be even more important for mammary epithelial cells that are strongly polarized. It has been clearly shown that mammary epithelial cells, including the ones from bovine, cultured on 2D systems have an impaired expression of the main milk protein genes compared with cells cultivated in collagen, where they can form the 3D structure [33]. The importance of a 3D culture for a full differentiation of Mac-T cells was already demonstrated in the first paper describing the Mac-T cells [2].

A direct comparison between mammary explants and primary cells from the same animals indicated a greater quantity and secretion of β-casein in explants vs. expanding primary cells in response to somatotropin and insulin like growth factor I (IGF-I); however, the difference decreased when cells reached confluence [34]. The authors concluded that "primary cell cultures are comparable to explant cultures when used to study mechanisms of DNA and milk protein synthesis and secretion". The mammary explants stem compared with isolated cells preserves mammary tissue composition, the 3D structure, and the extracellular matrix; all factors essential to obtain a fully functional lactating mammary epithelial [35]. In another study Mac-T cells had a very similar response compared with primary mammary epithelial cells isolated from prepubertal heifers when treated with IGF-I or inflammation-related proteins such as IL-6, IL-1b, MCP-1, and PAI- 1 [36]. Similarly, the transcriptomics response to growth hormone in Mac-T cells, bovine mammary tissue explants, and primary mammary epithelial cells isolated from milk was similar [37]. In a recent study, a similar effect of amino acids and a greater effect of insulin on mTOR signaling was observed in Mac-T cells compared with mammary tissue explants [19]. All the above studies provided evidence that Mac-T cells are an adequate system to study bovine mammary in vitro; however, a large transcriptomics analysis might provide a more comprehensive comparison.

Similarities between Mac-T and Mammary Tissue

Overall our data indicated a more similar transcriptome in Mac-T cells induced to lactation and lactating mammary tissue. The DIA analysis of SEG also supported a closer functional similarity between Mac-T cells and lactating than non-lactating mammary tissue. Due to the very large number of SEG compared to the analysis of DEG, it is not surprising that no biological terms were significantly enriched at a B-H<0.05 [29,32]. However, this might also indicate that Mac-T cells are not "specialized" compared with the mammary tissue either at -30 or +60 DIM.

Differences between Mac-T cells and mammary tissue

The functional analysis of DEG compared with the analysis of SEG has to be considered more important for our objectives. Overall, the large number of DEG between Mac-T cells and mammary tissue (Table 1) raises doubts about the reliability of the former to study the latter.

Are Mac-T cells a good model for studying milk synthesis?

The qPCR analysis of selected genes (Figure 1) clearly indicated that the Mac-T cells cultivated in a 2D system do not closely resemble the lactating mammary tissue, particularly for the tasks that the mammary tissue does best: synthetizing milk. The expression of all measured transporters of fatty acids, amino acids, and glucose, the main milk proteins, and genes involved in milk fat synthesis had greater expression in mammary tissue compared with Mac-T cells (Figure 1). For most of those genes the greater expression in mammary tissue vs. Mac-T cells was observed even at -30 DIM. This is not novel, particularly for expression of caseins [33]. Even though it has been reported that Mac-T cells can express and synthetize caseins [13,38] the amount produced is often below the limit of detection [2]; certainly not at the level of mammary tissue. Therefore, contrary to previous reports [19,34], the qPCR data in the present study indicated that the usefulness of Mac-T cells for studying milk protein synthesis, particularly if expression of milk proteins is considered, is limited.

The Mac-T cells also cannot be considered a good model for studying milk fat and lactose synthesis. This was evident both by using qPCR (Figure 1) and by the functional analysis of microarray data, particularly with the DIA analysis (File S7). These differences appear to be not only related to production of milk protein, fat, and lactose, but also other milk synthesis-related functions. This is supported by the large differences between Mac-T cells and mammary tissue observed in the present study for pathways previously shown to be highly activated during lactation such as glutathione metabolism, GPI-anchor biosynthesis, TCA cycle, and caffeine metabolism (Files S3 and S7) [39]

Interesting also is the marked translational capacity in Mac-T cells vs. mammary tissue. It was previously observed the paradoxical phenomenon of a strong decrease in expression of ribosomes from pregnancy to lactation both in bovine and mouse mammary tissue [33,39], despite an overall increase in protein synthesis. We have interpreted such findings as a way for the mammary gland to prioritize the translation of transcripts coding for proteins for milk production, i.e., a way for the mammary tissue to specialize [33,39]. This observation seems to be not only related to mammary tissue but also other tissues such as adipose [40]. Therefore, if we accept this hypothesis, the data in the present experiment further indicate that the Mac-T cells were less "specialized" than mammary tissue, particularly for milk synthesis.

Are Mac-T cells a good model for studying milk fat and protein regulation?

Despite the large differences observed for the genes involved in milk synthesis, the Mac-T cells have been often used successfully to study regulation of milk protein [12,14,18] and milk fat [17,41,42] synthesis.

Milk protein synthesis regulation: All the studies involving regulation of milk protein synthesis where Mac-T cells were compared with mammary explants suggested a very similar response between both systems [18,19,37]. In mammals the main pathway involved in the regulation of translation is the insulin-mTOR pathway. This also seems to be the case for the bovine mammary tissue [33]. Our data indicated that both mTOR and insulin signaling were slightly more induced in Mac-T cells vs. mammary at +60 DIM, indicating that Mac-T cells might be more sensitive to regulation by insulin. This also has been shown by recent data [19]. It has to be considered that the Mac-T cells in the present study were cultured in a supra physiological dose of insulin [17].

The regulation of casein expression is known to be under control of the Jak-STAT signaling. This has been clearly established in rodents and bovine [43,44]; however, for the latter this remains somewhat controversial [33]. Recently from transcriptomics data it was observed a large induction of the Jak-STAT signalling from pregnancy to lactation [39]. In the present study the DIA analysis suggested the Jak-STAT signalling was overall more induced in mammary tissue compared with Mac-T cells (File S7). This might partly explain the greater expression of milk protein genes in the former (Figure 1). In summary, it appears that Mac-T cells cultivated in 2D might be an adequate model to study regulation of protein synthesis, particularly for the insulin-mTOR signaling.

Milk fat synthesis regulation: Three main transcription factors have been reported to control milk fat synthesis: SREBP1 [45], LXRα [46,47], and PPARγ [17,48]. Our data, both from qPCR and microarray, indicated a marked degree of similarity between Mac-T cells and mammary tissue for the SREBF1 expression (Figure 1). Unfortunately our microarray did not contain probes for co-factors of SREBP1 such as SCAP and INSIG1 (however, expression of INSIG2 was not different between Mac-T cells and mammary at +60 DIM, see File S2). Similarly, we only had data for the LXRβ (or NR1H2) for which the expression was greater in Mac-T cells vs. mammary tissue (File S2). Even though only with an Exact Fisher's test <0.05, functional analysis in IPA indicated that LXR/RXR activation was the most enriched pathway in genes with greater expression in mammary at +60 DIM vs. Mac-T cells (File S3). This might indicate an overall greater sensitivity of the mammary tissue compared to Mac-T cells to activation of LXR.

The PPARG expression was ca. 4-fold greater in mammary compared with Mac-T cells (Figure 1). In addition, the 'PPAR signaling' was among the most induced pathways in lactating mammary tissue compared with Mac-T cells (File S7). If the observations for LXR and, particularly, PPARγ, hold true, the findings in Mac-T cells about activation of LXR and PPARγ might be underestimating the response of the mammary tissue in vivo. In summary, it appears that Mac-T cells are a good model to study transcription regulation of milk fat synthesis, particularly for SREBP1.

Metabolic differences between Mac-T cells and Mammary Tissue

The results of the enrichment and DIA analyses of pathways indicated that the overall metabolism was greater in Mac-T cells compared with mammary tissue. The functional analyses indicated an overall greater utilization of glucose, amino acids, and fatty acids for energy production (particularly in mitochondria) in Mac-T cells compared with mammary tissue. The marked utilization of those molecules for energy production might be partly due to a reduction in anabolism associated with low expression of milk-related genes (Figure 1) and/or the higher availability of glucose from the medium (>4 g/L in the medium vs. <1 g/L in blood), despite the lower expression of glucose transporters (Figure 1, File S2). However, the apparent greater glucose utilization for energy production in Mac-T cells in 2D, therefore deprived of ECM, compared with mammary tissue, that contains the ECM, seems to contradict with results of a previous study [49]. In that study it was observed detachment of human mammary epithelial cells from ECM negatively affected ATP production mostly due to a decrease in glucose import into cells; therefore, glucose availability and ATP production [49]. Therefore, the suggested greater use of glucose as energy source by Mac-T cells as a result of the greater expression of the genes involved in production of energy might be a reaction of the cells to an inner lower capacity, due to absence of ECM, for the utilization of energy compared with mammary tissue.

Differences in Cell-to-Cell Communication

The Mac-T cells also had a more pronounced cell-to-cell interactions and cell movement, particularly involving cytoskeleton and ECM compared with mammary tissue. This might be a consequence of the effect of the in vitro system on the epithelial cells. The ECM, together with the 3D formation of the alveolus and the establishment of the polarity, is essential for the formation of a functional mammary epithelium, primarily via integrins and cytoskeleton [50-52]. The ECM, through integrins, might affect the expression of milk-related genes via epigenetic mechanisms [53]. Therefore, the large differences observed between Mac-T cells and mammary tissue in the present experiment might be partly explained by the lack of ECM and 3D structure [54].

It is interesting in this regard the differences in the type of glycosaminoglycan biosynthesis observed (File S7). Intriguingly, the data indicated a more induced formation of heparin sulfate in Mac-T cells, considered to be absent in undifferentiated mammary epithelial cells [54], and a more induced formation of chondroitin sulfate in mammary tissue, which is highly abundant in undifferentiated mammary cells (File S7) [54]. This observation, together with the greater induction in Mac-T cells vs. mammary tissue of the cell-tocell interaction, might indicate a reaction of the cultured cells to a lack of functional structure. This reaction might be aimed to recover the original alveolus structure. This idea is supported by the observed high capacity of the mammary epithelial cells, even the immortalized, to easily form lobule-like structure when cultured in laminin and/ or collagen [2,55]. It will be interesting, from this point of view, to compare the transcriptome between Mac-T cells cultured in 3D and bovine mammary tissue.

Membrane Transport and Signaling

The membrane transporters, particularly the ABC transporters, and pathways of excretory systems appeared to be more induced in mammary tissue vs. Mac-T cells, particularly for calcium and water reabsorption (Files S7 and S8). The expression of the ABC transporters is mainly controlled by LXR/RXR signaling to regulate cholesterol homeostasis [56]. Among the ABC transporters the expression of ABCG2 dramatically increased from pregnancy to lactation in bovine mammary [21] and a single-point mutation strongly affected milk yield [57]. This gene was >4-fold greater in expression in mammary tissue at +60 DIM vs. Mac-T cells (File S2).

Extensive calcium loads occur in lactating mammary gland to support the requirements of milk calcium. Besides the provision for producing milk, the calcium in mammary gland also plays pivotal roles as a signaling molecule for regulation of proliferation, differentiation, and apoptosis [58].

Do Mac-T cells respond to inflammation as mammary epithelial cells in vivo?

Mac-T cells have been used to study the in vitro response of mammary epithelial cells to bacteria, their cell wall components [20], and activated neutrophils [59,60]. Our analysis indicated that potential inflammatory response was not similar between mammary tissue and Mac-T cells. Most of the functions and pathways uncovered to be more induced in mammary tissue (especially the lactating one) vs. Mac-T cells were immune cells-rather than epithelial cells-related (e.g. 'Intestinal immune network for IgA production'). This might be partly due to the presence of a relatively important number of immune cells in mammary tissue that tends to increase during lactation, particularly for macrophages [39,61].

Not considering the presence of immune cells, our data indicate a potentially greater response of Mac-T cells to bacteria and viruses. This was suggested by the superior induction in Mac-T cells of 'NOD-like receptor signaling' pathway (Files S7 and S8), one of the most activated pathways during mastitis induction through intramammary infusion of Escherichia coli or Staphylococcus uberis [61], and 'Leukocyte trans endothelial migration' pathway (Table 2 and Files S7 and S8).

It is interesting in this regard the observation made from transcriptomics analysis of the bovine mammary tissue from pregnancy to end of subsequent lactation [39]. Overall functional analysis of the data clearly indicated that the lactating vs. non – lactating mammary tissue had an overall decrease of sensitivity to bacteria invasion due to evident inhibition of the antigen presentation pathway but had put a large effort to keep a prepared immune system during lactation. This strong reduction of the antigen presentation was likely pertinent to epithelial cells [61]. Therefore, this indicates that in vivo epithelial cells are likely having an overall low responsiveness to bacteria invasion.

Generally, our data indicate that the Mac-T cells have a potentially stronger response to inflammatory challenges, particularly using bacteria, compared with the lactating epithelia cells in vivo. In addition, the data indicated a stronger capacity of the Mac-T cells compared with mammary tissue to recruit and to allow invasion of immune cells; however, this might be also a consequence of an inexistent 3D structure and a strong cell-to-cell interactions and cytoskeleton activity for the Mac-T cells that can be considered independent from any inflammatory response.

CONCLUSIONS

Results suggest that the transcriptome between Mac-T cells and mammary tissue differs substantially, but with a slightly greater degree of similarity of the former with lactating compared to non-lactating mammary tissue as also suggested by the analysis of SEG. But this overall similarity was not associated with the lactation phenotype. In fact, when functional analysis of DEG was performed the Mac-T cells appeared to be more different than lactating compared with nonlactating mammary tissue, particularly for lactation-specific functions/ pathways. This was concluded by the greater catabolism of glucose and lipid for production of energy in Mac-T cells, rather than anabolism for production of major milk constituents.

Concerning milk constituents, the data clearly indicated that Mac-T cells cultured in a 2D plastic system differ fundamentally from lactating mammary tissue because of an impaired ability to synthesize lactose, not adequate milk fat synthesis and secretion (including the control of it through PPAR signaling), a compromised expression of major milk proteins and control of milk protein synthesis (e.g. lower Jak- STAT signaling, very large translational machinery), and mechanisms concerning uptake and metabolism of calcium. Furthermore, several pathways and chromosomes previously found to be important during lactation (e.g. GPI- anchor biosynthesis, BTA6, BTA14) were also the ones with greater differences between lactating mammary tissue and Mac-T cells.

Our data also indicated that the Mac-T cells do not substantially differ from mammary tissue with regards to the regulation of milk protein synthesis by the insulin-mTOR pathway. With the exception of SREBP1, our analysis indicated that results from studies of milk fat synthesis regulation using Mac-T cells are likely underestimating the in vivo mammary response. Lastly, the Mac-T cells appeared to be more immune responsive than lactating mammary tissue; therefore, likely overestimating the in vivo response.

The present study suffers from the limitation of having Mac-T cells cultivated in a 2D plastic system. As previously demonstrated, the Mac-T cells can be a better model for in vivo mammary tissue if cultivated in a 3D system. In this regard, it would be interesting to perform the same study presented here on Mac-T cells cultivated in a 3D system.

ACKNOWLEDGEMENTS

We extend our gratitude to Dr. Harris Lewin (University of California, Davis) and Dr. Robin Everts (presently at Sequenom Inc.) for providing access to the microarray platform and annotation. Partial support for the cell culture and gene expression work was provided by the National Institute of Food and Agriculture (NIFA) under Hatch project ILLU-538-392. Afshin Hosseini (HO 4596/1-1) received fellowship support from the German Research Foundation (DFG).

REFERENCES

1. Ip MM, Darcy KM (1996) Three-dimensional mammary primary culture model systems. J Mammary Gland Biol Neoplasia 1: 91-110.
2. Huynh HT, Robitaille G, Turner JD (1991) Establishment of bovine mammary epithelial cells (MAC-T): an in vitro model for bovine lactation. Exp Cell Res 197: 191-199.
3. Gibson CA, Vega JR, Baumrucker CR, Oakley CS, Welsch CW (1991) Establishment and characterization of bovine mammary epithelial cell lines. In Vitro Cell Dev Biol 27A: 585-594.
4. Schmid E, Franke WW, Grund C, Schiller DL, Kolb H, et al. (1983) An epithelial cell line with elongated myoid morphology derived from bovine mammary gland. Expression of cytokeratins and desmosomal plaque proteins in unusual arrays. Exp Cell Res 146: 309-328.
5. Zavizion B, van Duffelen M, Schaeffer W, Politis I (1996) Establishment and characterization of a bovine mammary myoepithelial cell line. In Vitro Cell Dev Biol Anim 32: 149-158.
6. German T, Barash I (2002) Characterization of an epithelial cell line from bovine mammary gland. In Vitro Cell Dev Biol Anim 38: 282-292.
7. Pantschenko AG, Woodcock-Mitchell J, Bushmich SL, Yang TJ (2000) Establishment and characterization of a caprine mammary epithelial cell line (CMEC). In Vitro Cell Dev Biol Anim 36: 26-37.
8. Düchler M, Schmoll F, Pfneisl F, Brem G, Schellander K (1998) OMEC II: a new ovine mammary epithelial cell line. Biol Cell 90: 199-205.
9. Ilan N, Barash I, Gootwine E, Shani M (1998) Establishment and initial characterization of the ovine mammary epithelial cell line NISH. In Vitro Cell Dev Biol Anim 34: 326-332.
10. Huynh H, Pollak M (1995) HH2A, an immortalized bovine mammary epithelial cell line, expresses the gene encoding mammary derived growth inhibitor

(MDGI). In Vitro Cell Dev Biol Anim 31: 25-29.

11. Zavizion B, Duffelen M, Schaeffer W, Politis I (1995) Use of microinjection to generate an immortalized bovine mammary cell line with both epithelial and myoepithelial characteristics. Methods Cell Sci 17: 271-282.

12. Johnson TL, Fujimoto BA, Jiménez-Flores R, Peterson DG (2010) Growth hormone alters lipid composition and increases the abundance of casein and lactalbumin mRNA in the MAC-T cell line. J Dairy Res 77: 199-204.

13. Zavizion B, Gorewit RC, Politis I (1995) Subcloning the MAC-T bovine mammary epithelial cell line: morphology, growth properties, and cytogenetic analysis of clonal cells. J Dairy Sci 78: 515-527.

14. Burgos SA, Cant JP (2010) IGF-1 stimulates protein synthesis by enhanced signaling through mTORC1 in bovine mammary epithelial cells. Domest Anim Endocrinol 38: 211-221.

15. Romagnolo D, Akers RM, Wong EA, Boyle PL, McFadden TB, et al. (1993) Lactogenic hormones and extracellular matrix regulate expression of IGF-1 linked to MMTV-LTR in mammary epithelial cells. Mol Cell Endocrinol 96: 147-157.

16. ZarzyÅ„ska J, Gajewska M, Motyl T (2005) Effects of hormones and growth factors on TGF-beta1 expression in bovine mammary epithelial cells. J Dairy Res 72: 39-48.

17. Kadegowda AK, Bionaz M, Piperova LS, Erdman RA, Loor JJ (2009) Peroxisome proliferator-activated receptor-gamma activation and long-chain fatty acids alter lipogenic gene networks in bovine mammary epithelial cells to various extents. J Dairy Sci 92: 4276-4289.

18. Appuhamy JA, Knoebel NA, Nayananjalie WA, Escobar J, Hanigan MD (2012) Isoleucine and leucine independently regulate mTOR signaling and protein synthesis in MAC-T cells and bovine mammary tissue slices. J Nutr 142: 484-491.

19. Appuhamy JA, Bell AL, Nayananjalie WA, Escobar J, Hanigan MD (2011) Essential amino acids regulate both initiation and elongation of mRNA translation independent of insulin in MAC-T cells and bovine mammary tissue slices. J Nutr 141: 1209-1215.

20. Yu C, Shi ZR, Chu CY, Lee KH, Zhao X, et al. (2010) Expression of bovine granulocyte chemotactic protein-2 (GCP-2) in neutrophils and a mammary epithelial cell line (MAC-T) in response to various bacterial cell wall components. Vet J 186: 89-95.

21. Bionaz M, Loor JJ (2008) Gene networks driving bovine milk fat synthesis during the lactation cycle. BMC Genomics 9: 366.

22.

23. Loor JJ, Everts RE, Bionaz M, Dann HM, Morin DE, et al. (2007) Nutrition-induced ketosis alters metabolic and signaling gene networks in liver of periparturient dairy cows. Physiol Genomics 32: 105-116.

24. Piantoni P, Bionaz M, Graugnard DE, Daniels KM, Everts RE, et al. (2010) Functional and gene network analyses of transcriptional signatures characterizing pre-weaned bovine mammary parenchyma or fat pad uncovered novel inter-tissue signaling networks during development. BMC Genomics 11: 331.
25. Edgar R, Domrachev M, Lash AE (2002) Gene Expression Omnibus: NCBI gene expression and hybridization array data repository. Nucleic Acids Res 30: 207-210.
26. Tramontana S, Bionaz M, Sharma A, Graugnard DE, Cutler EA, et al. (2008) Internal controls for quantitative polymerase chain reaction of swine mammary glands during pregnancy and lactation. J Dairy Sci 91: 3057-3066.
27. Bionaz M, Loor JJ (2007) Identification of reference genes for quantitative real-time PCR in the bovine mammary gland during the lactation cycle. Physiol Genomics 29: 312-319.
28. Vandesompele J, De Preter K, Pattyn F, Poppe B, Van Roy N, et al. (2002) Accurate normalization of real-time quantitative RT-PCR data by geometric averaging of multiple internal control genes. Genome Biol 3: RESEARCH0034.
29. Huang da W, Sherman BT, Lempicki RA (2009) Systematic and integrative analysis of large gene lists using DAVID bioinformatics resources. Nat Protoc 4: 44-57.
30. Bionaz M, Periasamy K, Rodriguez-Zas SL, Hurley WL, Loor JJ (2012) A novel dynamic impact approach (DIA) for functional analysis of time-course omics studies: validation using the bovine mammary transcriptome. PLoS One 7: e32455.
31. Prizant RL, Barash I (2008) Negative effects of the amino acids Lys, His, and Thr on S6K1 phosphorylation in mammary epithelial cells. J Cell Biochem 105: 1038-1047.
32. Boutet P, Sulon J, Closset R, Detilleux J, Beckers JF, et al. (2007) Prolactin-induced activation of nuclear factor kappaB in bovine mammary epithelial cells: role in chronic mastitis. J Dairy Sci 90: 155-164.
33. Huang da W, Sherman BT, Lempicki RA (2009) Bioinformatics enrichment tools: paths toward the comprehensive functional analysis of large gene lists. Nucleic Acids Res 37: 1-13.
34. Bionaz M, Hurley WL, Loor JJ (2012) Milk Protein Synthesis in the Lactating Mammary Gland: Insights from Transcriptomics Analyses. In: Hurley WL, editor. Milk Protein: InTech. pp. 285-324.
35. Keys JE, Cifrian E, Guidry AJ, Farrell HM (1997) Bovine mammary explant versus primary cell cultures: effect of bovine somatotropin and insulin-like growth factor-I on DNA content and protein synthesis. In Vitro Cell Dev Biol Anim 33: 206-211.
36. Muschler J, Streuli CH (2010) Cell-matrix interactions in mammary gland development and breast cancer. Cold Spring Harb Perspect Biol 2: a003202.
37. Thorn SR, Purup S, Vestergaard M, Sejrsen K, Meyer MJ, et al. (2008) Regulation

of mammary parenchymal growth by the fat pad in prepubertal dairy heifers: role of inflammation-related proteins. J Endocrinol 196: 539-546.

38. Lisa McDonnell, Tasha Johnson, Daniel G. Peterson (2010) Comparison of the Effects of Growth Hormone on MAC-T Cells and Primary Mammary Cells. FASEB Journal 24: 629.9.
39. Zhou Y, Akers RM, Jiang H (2008) Growth hormone can induce expression of four major milk protein genes in transfected MAC-T cells. J Dairy Sci 91: 100-108.
40. Bionaz M, Periasamy K, Rodriguez-Zas SL, Everts RE, Lewin HA, et al. (2012) Old and new stories: revelations from functional analysis of the bovine mammary transcriptome during the lactation cycle. PLoS One 7: e33268.
41. Bionaz M, Loor JJ (2012) Ruminant metabolic systems biology: reconstruction and integration of transcriptome dynamics underlying functional responses of tissues to nutrition and physiological state. Gene Regul Syst Bio 6: 109-125.
42. McFadden JW, Corl BA (2009) Activation of AMP-activated protein kinase (AMPK) inhibits fatty acid synthesis in bovine mammary epithelial cells. Biochem Biophys Res Commun 390: 388-393.
43. Peterson DG, Matitashvili EA, Bauman DE (2004) The inhibitory effect of trans-10, cis-12 CLA on lipid synthesis in bovine mammary epithelial cells involves reduced proteolytic activation of the transcription factor SREBP-1. J Nutr 134: 2523-2527.
44. Yang J, Kennelly JJ, Baracos VE (2000) The activity of transcription factor Stat5 responds to prolactin, growth hormone, and IGF-I in rat and bovine mammary explant culture. J Anim Sci 78: 3114-3125.
45. Yang J, Kennelly JJ, Baracos VE (2000) Physiological levels of Stat5 DNA binding activity and protein in bovine mammary gland. J Anim Sci 78: 3126-3134.
46. Bauman DE, Harvatine KJ, Lock AL (2011) Nutrigenomics, rumen-derived bioactive fatty acids, and the regulation of milk fat synthesis. Annu Rev Nutr 31: 299-319.
47. Oppi-Williams C, Suagee JK, Corl BA (2013) Regulation of lipid synthesis by liver X receptor Î± and sterol regulatory element-binding protein 1 in mammary epithelial cells. J Dairy Sci 96: 112-121.
48. McFadden JW, Corl BA (2010) Activation of liver X receptor (LXR) enhances de novo fatty acid synthesis in bovine mammary epithelial cells. J Dairy Sci 93: 4651-4658.
49. Shi H, Luo J, Zhu J, Li J, Sun Y, et al. (2013) PPAR Î³ Regulates Genes Involved in Triacylglycerol Synthesis and Secretion in Mammary Gland Epithelial Cells of Dairy Goats. PPAR Res 2013: 310948.
50. Grassian AR, Metallo CM, Coloff JL, Stephanopoulos G, Brugge JS (2011) Erk regulation of pyruvate dehydrogenase flux through PDK4 modulates cell proliferation. Genes Dev 25: 1716-1733.

51. Katz E, Streuli CH (2007) The extracellular matrix as an adhesion checkpoint for mammary epithelial function. Int J Biochem Cell Biol 39: 715-726.

52. Zoubiane GS, Valentijn A, Lowe ET, Akhtar N, Bagley S, et al. (2004) A role for the cytoskeleton in prolactin-dependent mammary epithelial cell differentiation. J Cell Sci 117: 271-280.

53. Streuli CH (2003) Cell adhesion in mammary gland biology and neoplasia. J Mammary Gland Biol Neoplasia 8: 375-381.

54. Lelièvre SA (2009) Contributions of extracellular matrix signaling and tissue architecture to nuclear mechanisms and spatial organization of gene expression control. Biochim Biophys Acta 1790: 925-935.

55. Bissell MJ, Ram TG (1989) Regulation of functional cytodifferentiation and histogenesis in mammary epithelial cells: role of the extracellular matrix. Environ Health Perspect 80: 61-70.

56. Streuli CH, Bailey N, Bissell MJ (1991) Control of mammary epithelial differentiation: basement membrane induces tissue-specific gene expression in the absence of cell-cell interaction and morphological polarity. J Cell Biol 115: 1383-1395.

57. Schmitz G, Langmann T (2005) Transcriptional regulatory networks in lipid metabolism control ABCA1 expression. Biochim Biophys Acta 1735: 1-19.

58. Cohen-Zinder M, Seroussi E, Larkin DM, Loor JJ, Everts-van der Wind A, et al. (2005) Identification of a missense mutation in the bovine ABCG2 gene with a major effect on the QTL on chromosome 6 affecting milk yield and composition in Holstein cattle. Genome Res 15: 936-944.

59. Lee WJ, Monteith GR, Roberts-Thomson SJ (2006) Calcium transport and signaling in the mammary gland: targets for breast cancer. Biochim Biophys Acta 1765: 235-255.

60. Lauzon K, Zhao X, Bouetard A, Delbecchi L, Paquette B, et al. (2005) Antioxidants to prevent bovine neutrophil-induced mammary epithelial cell damage. J Dairy Sci 88: 4295-4303.

61. Boulanger V, Zhao X, Lacasse P (2002) Protective effect of melatonin and catalase in bovine neutrophil-induced model of mammary cell damage. J Dairy Sci 85: 562-569.

62. Loor JJ, Moyes KM, Bionaz M (2011) Functional adaptations of the transcriptome to mastitis-causing pathogens: the mammary gland and beyond. J Mammary Gland Biol Neoplasia 16: 305-322.

12
CHAPTER

RECENT ADVANCES IN REPRODUCTIVE BIOTECHNOLOGIES IN SHEEP AND GOAT

Basil Alexander, Gabriela Mastromonaco and W. Allan King

Keywords: Reproductive biotechnologies, In vitro/in vivo fertilization, Artificial insemination, Somatic cell nuclear transfer, Transgenesis

CONTENTS

This review focuses on the platform of recent advances in reproductive biotechnologies in sheep and goat. Here, we discuss some aspects of veterinary and biomedical importance in artificial insemination using sexed semen, in vitro/in vivo fertilization, interspecies in vitro fertilization, multiple ovulation and embryo transfer (MOET), somatic cell nuclear transfer and transgenic animal production in sheep and goat. Some of these technologies are already widely used in the cattle industry, and some are new to the goat and sheep industry.

INTRODUCTION

The world sheep and goat populations in 2006 were recorded as 416 and 489 million, respectively [1]. Reproductive biotechnologies are playing an increasingly important role in the production and management of these species. Animal biotechnology encompasses the application of science and engineering principles to the processing or production of materials to provide services or goods for human use [2]. Examples of current animal biotechnologies include artificial insemination using fresh, frozen or sexed semen, embryo transfer using fresh or frozen in vivo or in vitro produced embryos, cloning using somatic cell nuclear transfer procedures and production of transgenic animals. Recent advances in reproductive biotechnologies have shown important developments in the technologies and their delivery that promise to yield significant benefits for their application within the livestock industry as well as in biomedical sciences. This review focuses on the recent advances of biotechnologies in sheep and goats.

Citation: Alexander B, Mastromonaco G, King WA (2010) Recent Advances in Reproductive Biotechnologies in Sheep and Goat. J Veterinar Sci Technol 1:101. doi:10.4172/2157-7579.1000101.

METHODS

Artificial Insemination (AI)

Artificial insemination is the procedure involving the mechanical deposition of pre-collected semen into the uterus of an estrous female by a technician. The widespread commercial application of AI, a robust tool in livestock breeding, began in earnest in the 1950's after researchers developed methods to successfully freeze cattle semen [3]. This technology, which is the most extensively applied reproductive biotechnology in the world, has many valuable advantages such as reducing venereal diseases, rapidly increasing genetic merits of the production animals through selective breeding and eliminating lethal alleles. The potential to pass high genetic merits of a selected male to thousands of females makes this a far more efficient technology for producing large numbers (1.5 x 108) of offspring per year compared to female based technologies, such as embryo transfer (ET), which can only produce a few progeny from a selected female. More recently, the advent of separation of sperm into X- and Y-chromosome fractions using flow cytometry has added a new dimension to livestock production [4].

Since the first report of live rabbit pups of predetermined sex using X and Y-enriched sorted semen over 20 years ago [5], there have been more than 40,000 births of live animals resulting from the so- called Beltsville sperm sexing technology [4]. This sperm sexing technology has proven to be a reliable and more efficient method of obtaining predetermined sex in animals compared to other relatively wasteful methods of embryo sexing and pre-implantation genetic diagnosis [6 - 8]. Live offspring born from fresh sex- sorted semen have been reported in many species including cattle [9], pigs [10,11], horses [12,13] and sheep [14,15]. As the sorting procedure can separate sperm only at a speed of approximately 10 million per hour, inseminations have to be carried out using limited numbers of sperm in each insemination dose compared to the conventional standards for AI. To obtain high success rates with sorted semen AI, most trials have been carried out using fresh sorted semen, nevertheless, use of frozen sorted semen is ultimately more beneficial to the livestock industry as it can be stored for long duration and easy to transport anywhere in the world.

Semen Sexing Technology

Semen sexing is the process of separating spermatozoa into two subpopulations containing X-chromosome and Y-chromosome bearing spermatozoa. In theory, use of the X-bearing spermatozoa fraction for insemination will result in female progeny, while the Y-bearing spermatozoa fraction will give male progeny. Considerable efforts have been made in this area over the past 30-40 years, but only recently has there been considerable progress, particularly in cattle [16]. The most commonly applied technique for the separation of a fresh semen sample into X and Y-chromosome bearing fractions is fluorescence activated cell sorting (FACS). The concept behind this technique is described below:

Fluorescence Activated Cell Sorting (FACS)

This technique involves the separation of X- and Y- bearing spermatozoa in small quantities based on the DNA content of the spermatozoa so that they can then be used in in vitro fertilization and AI programs [17]. The procedure is relatively slow and requires expensive infrastructure at least for mass scale commercial application. The technique is based on the biological principle that the DNA content of the X and Y chromosomes as X-chromosome has 3.8% more DNA than Y-chromosome [9]. This difference can, nonetheless, be detected with FACS technique. Before the spermatozoa are passed through the cell sorter, the sample is exposed to a nontoxic DNA dye that specifically binds to DNA to allow measurement of the DNA content of each spermatozoa. The cell suspension is then passed through an extremely fine nozzle that vibrates at high frequency causing the fluid stream to disperse into micro-drops which, optimally holds only one spermatozoa after the conditions are calibrated. When the stream of micro-droplets is passing through a UV laser beam, the already stained DNA will start to glow or emit fluorescence. The strength (intensity) of the florescence is measured by a detector, and if the florescence is in the range designated for the Y-bearing spermatozoa, that micro-drop is given an electric charge. Furthermore, if the intensity is within the range for X-bearing spermatozoa, the micro-drop is given an opposite electric charge. The charged micro-drops fall between two charged plates, resulting in two groups of droplets to be separated into different pools (Figure 1). Droplets that fall outside of the fluorescence range for X or Y are collected into a third group where the sex cannot be determined. The disadvantages of the technique are its relatively slow process and increased wastage of unsorted spermatozoa. The advantage is it can give up to 95% purity sexed semen samples.

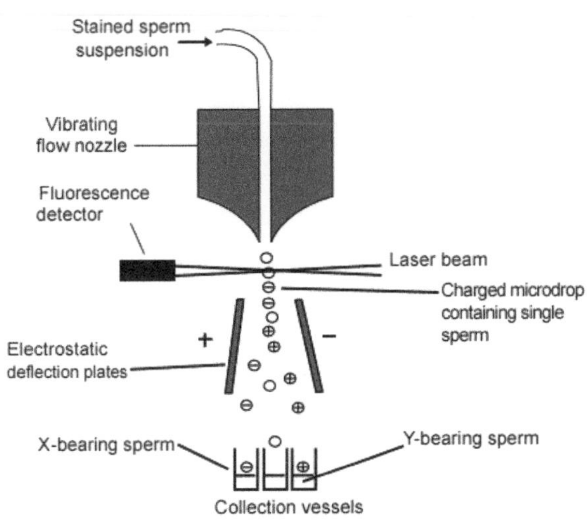

Figure 1: Schematic representation of the concept behind the Florescent Activated Cell Sorting technique used in separation of X- and Y-bearing chromosomes in a fresh semen sample.

Application of Sexed Semen in Goat and Sheep Production

Although the sexed semen technology is gaining in popularity in commercial cattle industries, its use has yet to be developed in the sheep and goat industries. Nevertheless recent attempts to separate the X- and Y-chromosomal fractions using flow cytometry on ram semen has been attempted and yielded fraction purity greater than 90% [18]. In flow cytometry, the sperm extending medium plays a critical role in the efficiency of separation and the viability of the sperm after sorting [17,19,20]. Ram sperm have to be diluted 20- fold before sorting and a further 200-fold dilution takes place when it passes through the cell sorter resulting in capacitation-like changes that lead to reduced fertility of sexed sheep semen [17,20]. It has also been shown that suspension of ram semen in Tris-based medium without egg- yolk, improves the staining of the DNA and efficiency of separation of sperm without reducing its viability [8,18,21].

When working with frozen sex- sorted semen in sheep, it is believed that the semen should be deposited in the oviduct or deep within the uterus to obtain high success rates, although it appears to be not as critical in cattle [16]. In sheep, frozen semen is usually deposited laparoscopically into the uterus and these same methods have been used effectively to produce offspring from frozen sexsorted semen [15]. Oviductal deposition of frozen semen in sheep is not practical if large numbers of sheep are to be inseminated. To our knowledge, there have been no reports of attempts to inseminate sheep or goat using sex- sorted semen in the cervix or transcervically.

The minimum number of sperm in an intrauterine AI dose in sheep varies between 5 x 106 and 25 x 10^6, but in Australia it has been suggested as 25 x 10^6 motile spermatozoa. Although it is not known exactly the minimum number of frozen sex-sorted sperm for intrauterine insemination of sheep, it is expected that it should be in the range of standard commercial intrauterine AI semen dose [20]. However, it is acceptable that the number of sperm in frozen sex sorted semen should be increased in line with the negative effects of the sorting process to obtain high efficiency.

The timing of AI also plays a significant role in AI success in sheep but recent findings have suggested that the best results can be achieved with sex–sorted frozen sperm when inseminated just before the expected ovulation time in ovulation-synchronized progesterone treated ewes. Using sex sorted frozen semen (40 million sperm per AI dose), optimal results can be expected when ewes are inseminated at 58 h after sponge removal or 36 h after GnRH injection [20].

Embryo Production and Transfer

Multiple ovulation and embryo transfer (MOET): This breeding model has been extensively used to multiply genetically superior goats and sheep and many protocols have been developed to optimize its application in research as well as in commercial practice [22-25]. In most domestic farm animals, especially in cattle, the genetic gain is limited on the female side due to low reproductive rate with only one calf being produced per year and also relatively long intervals between generations. To overcome these limitations MOET has been developed. Using this method, a female is hormonally stimulated to have multiple ovulations (superovulation). Most MOET programs in sheep and goat consist of a 14 to 17 day progesterone exposure, with FSH injections beginning 2 days before the removal of the progesterone releasing device. Following six or eight FSH injections given twice daily, the animals are allowed to breed naturally or by artificial insemination. The ovulated ova are allowed to fertilize in vivo following artificial insemination or natural mating. After fertilization and initiation of embryo development, 6-7 day old embryos (morula/ blastocyst stage) are recovered from within the uterus using embryo flushing techniques. In sheep, this is done surgically [26]. These embryos are then transferred to non-pregnant estrus-synchronized recipients 6-7 days post estrus or frozen in liquid nitrogen (-196° C) for future use.

Table 1: World embryo transfer activity in small ruminants in 2008.

Species	Transferable embryos	Transferred embryos		
		Fresh	Frozen	Total
Sheep	18828	4793	433	5226
Goat	3141	824	278	1102
Cervids	980	840	0	840

It is well-known that the superovulation protocols in small ruminants do not guarantee the recovery of quality transferable embryos due to various factors that affect the success of the program [25]. These factors can be listed as extrinsic and intrinsic factors. The extrinsic factors include: a) source and purity of the gonadotrophins, b) protocol of gonadotrophin injection, and c) the type of progesterone treatment protocol [25]. Intrinsic factors include: age, breed, nutrition and reproductive status of the animals [25]. However, despite all the achievements in the MOET programs in small ruminants so far, the success of obtaining high quality transferable embryos still remains low and more studies are needed to investigate the optimization of these programs.

Embryo Transfer (ET)

According to the available data from the International Embryo Transfer Society (IETS), the number of embryo transfers carried out in small ruminants is shown in Table 1. In sheep and goat, ET is usually carried out by laparoscopic surgery 6-7 days post estrus [27], (Figure 2). Estrus is generally synchronized in sheep as follows. On day-1 of the program, progesterone releasing intravaginal sponges (Veramix®, 60mg medroxyprogesterone acetate, Pharmacia & Upjohn, Orangeville, Canada) are inserted in to the vagina. On day- 14, the sponges are removed and 300 IU PMSG (Folligon®, Intervet) are injected intramuscularly to each ewe at the time of sponge withdrawal. Vasectomised rams are introduced on day-15 and ewes in estrus are identified between 12-36 h post PMSG injections. Surgical embryo transfer is then performed on day-22 of the program [27].

The embryos are usually evaluated for morphology according to the IETS criteria [28] and only the excellent quality blastocysts/ morulae are transferred into recipient ewes. Ewes are anesthetized using xylazine (0.2mg/kg) injection intravenously, and are secured on a laparoscopic cradle. An incision of approximately 1 cm is made in the skin approximately 3 cm to the left of the mid line and approximately 5 cm from the udder, using a scalpel blade (Figure 2B-D). A trochar and cannula are inserted into the abdominal cavity through the incision made in the skin (Figure 2E-F). The uterine horns and the ovaries with corpora lutea are visualized using a laparoscope (Figure 2G). The uterine horn is punctured close to the utero-tubular junction using a blunt 18-gauge needle and the embryos are loaded into an embryo transfer catheter and transferred into the lumen of the uterine horn (Figure 2H). The ewes are then monitored postoperatively for a few hours and released to the paddocks.

Recently, a goat ET group in Sri Lanka reported that they have produced ET born goat kids (Figure 2B) from fresh embryos in a field embryo transfer program without using sophisticated equipment [22], thus validating the field application of the technique in veterinary practice to multiply genetically valuable goats to establish an elite goat herd under local production conditions. Moreover, ET in small ruminants is still carried out using the conventional ET techniques employing laparoscopic techniques with an average pregnancy rate of 60% with fresh embryo transfers. The pregnancy rates can vary due to many factors such as donor, recipient, embryo and operator. The transcervical ET procedure seems to be not practical in these species due to the problems associated with handling of the cervix and its anatomy. Recently, a research group [29] reported that anestrus sheep cervix can be dilated using prostaglandin 12 h prior to the insertion of AI gun into the uterus suggesting possibility of transcervical ET in sheep. However, more studies are needed to optimize this technique in small ruminants to obtain high success rates.

In vitro Embryo Production in Sheep and Goat

In vitro embryo production in sheep and goat provides an alternate source of low cost embryos for transfer in breeding programs such as MOET. The process consists of 3 main steps: 1) oocyte collection and In vitro maturation (IVM) of cumulus-oocyte complexes (COCs), 2) In vitro fertilization (IVF) and 3) in vitro culture of embryos. These steps have been reviewed in detail by others [30 - 33]. Here we will highlight the important differences between species.

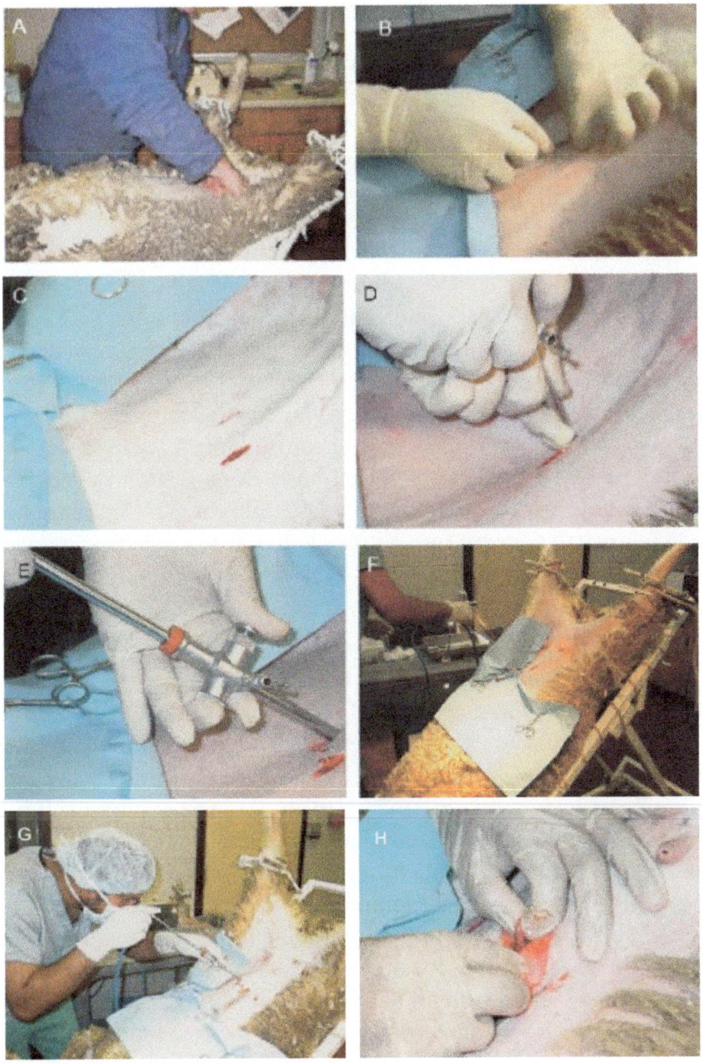

Figure 2: Embryo Transfer in sheep and goat. Embryo Transfer in sheep and goats (A–H): (A), preparation of anesthetized ewe on a cradle; (B) and (C), two skin incisions (each 1cm) are made on either side of the mid ventral line, 5 cm away from the udder; (D, E and F), insertion of trocar and cannula, cannula is left on the abdomen through which endoscope is inserted into the abdomen; (G), insertion of the endoscope through the cannula to visualize the ovaries, corpora lutea and reproductive tract; (H), a uterine horn is punctured close to the utero-tubular junction using a blunt-end 18 Gauge needle, embryos are loaded into an ET catheter and transferred into the lumen.

Oocyte Collection and in Vitro Maturation

Prior to in vitro fertilization, COCs are collected either by minimally invasive laparoscopic ovum pick up (LOPC) or postmortem from slaughtered females [32,34]. Cumulus-oocyte complexes are collected from live sheep by a laparotomy followed by aspiration of oocytes from ovaries using an aspiration pump. This method is often called as laparoscopic ovum pick-up (LOPU) and seems to be very efficient for the recovery and quality of the COCs obtained in sheep and goats. Usually, in sheep, a short repeated gonadotrophin treatment is followed by LOPU [35-38]. In contrast, goats can be stimulated with a 'one shot' regime of gonadotrophins (one shot of follicular stimulating hormone (FSH) plus equine chorionic gonadotrophin (eCG), compared to the traditional multi-dose gonadotrophin regimes. The results revealed no differences in the efficiency of COCs recovered and their quality [36,39]. It has also been shown that 'one shot' regime can be performed as a 4-day treatment regime followed by LOPU in sheep and goats without reducing the efficiency in oocyte recovery and quality [40]. The most common method of oocyte collection is from abattoir derived ovaries from sheep or goat. Ovaries are transported to the laboratory at 37°C in phosphate buffered saline (PBS) after which they are finely sliced with a scalpel blade and rinsed in oocyte collection medium consisting of 2% steer serum in Ham's F10 medium [34].

In vitro maturation of abattoir derived oocytes from sheep and goats (for comprehensive reviews see 32, 31) rely on methods based on bi-carbonate buffered TCM 199 medium supplemented with L-glutamine, ovine FSH (10 μg/ml), ovine LH (10 μg/ml, estrogen (1 μg/ml) and 10% fetal bovine serum [34]. Some researchers have used mature follicular fluid supplementation in the IVM medium to enhance the maturation of sheep oocytes [31]. Under these conditions, goat and sheep oocytes tend to extrude first polar body between 16-24 h of in vitro maturation [41]. In contrast, it is accepted that supplementation of the glutathione precursor cysteamine (100 μM) in IVM medium enhances the in vitro maturation process of goat oocytes [41]. In addition, a recent study [42] showed that cysteamine and BSA added to IVM medium for sheep, resulted in high blastocyst rates (42% blastocyst rate from total cultured oocytes) and high survival rate of blastocysts after embryo transfer. It is obvious that artificial oocyte maturation and culture systems have detrimental effects on embryo development compared to their in vivo counterparts. Recent fluorescent in situ hybridization cytogenetic studies from our laboratory using whole chromosome painting probes have revealed that high incidence of chromosomal abnormalities in sheep blastocysts derived from IVP systems, compared to in vivo derived blastoctsts [43] emphasizing the inherent differences in the two production methods.

In Vitro Fertilization

In vitro fertilization of the oocytes is usually carried out following 24 h maturation in the IVM medium. Surrounding cumulus cells are removed by gentle pippetting and washing in fertilization medium and groups of 40-50 oocytes are placed in 4-well dishes in 500 µl of synthetic oviductal fluid (SOF) covered by 200 µl of mineral oil [31,43]. Motile sperm are obtained by centrifugation of frozen-thawed semen in Percoll gradient (45%/90%) at 500 g for 10 min in room temperature. Percoll gradient separation of spermatozoa seems to be an effective means of yielding motile sperm from frozen-thawed semen [44]. Motile sperm are collected from the bottom of the 90% fraction and diluted to a concentration of 107 spermatozoa/ml in media. Recent studies have shown that incubation of spermatozoa for a further 1 h period in media supplemented with heat-inactivated estrus serum enhances the capacitation process of spermatozoa (sheep and goat : [41]; deer : [45,46]). In our laboratory we routinely add heat – inactivated sheep serum (2%) to the IVF medium and incubate oocytes with motile spermatozoa for 17 hours resulting acceptable levels of blastocyst development [43]. Although 17 h co-incubation is practiced in IVF, it has been shown that even 4 h co-incubation was enough to obtain acceptable cleavage and blastocyst rates in sheep [41]. Further application of in vitro fertilization in obtaining sex pre-determined offspring will be an advantage for the livestock industry [20,47]. In vitro fertilization of prepubertal oocytes and oocytes recovered from LOPU [37] are also important areas to be developed further.

In Vitro Culture

Following in vitro fertilization, embryos are washed several times and then placed in in vitro culture (IVC) media for further development. There are 3 methods available for embryo culture after fertilization, such as co-culture with somatic cells, culture in semi-defined media and culture of embryos in in vivo systems (oviducts) of the same or different species [41]. Co-culture systems in sheep and goat are usually based on including oviductal cells in TCM-199 medium. Semi-defined medium is also used often in small ruminant IVC based on the components of the sheep oviductal fluid (SOF), a medium containing essential and non-essential amino acids supplemented with BSA [43,48]. In our laboratory, sheep embryos cultured in SOF medium supplemented with amino acids and BSA at 38°C in 5% O_2, 5% CO_2, 90% N2 humidified atmosphere resulted in 19.2% blastocyst rates from total oocytes cultured [43]. This rate is confirmed by [38], who showed that SOF culture systems supplemented with BSA and amino acids produce high quality blastocysts in sheep. Addition of fetal bovine serum (FBS) on day 2 or 3 of culture does result in high embryo survival post-embryo transfer and normal viable offspring [41]. However, some studies have reported that sheep embryos reduce their viability by 15%–25% when produced in SOF systems supplemented with serum compared to co-culture [48] or in in vivo systems of culturing the embryos in oviducts [49]. It has been noted that adding steer serum or human serum to sheep and goat IVC media may attribute to many abnormalities in the offspring [50,51].

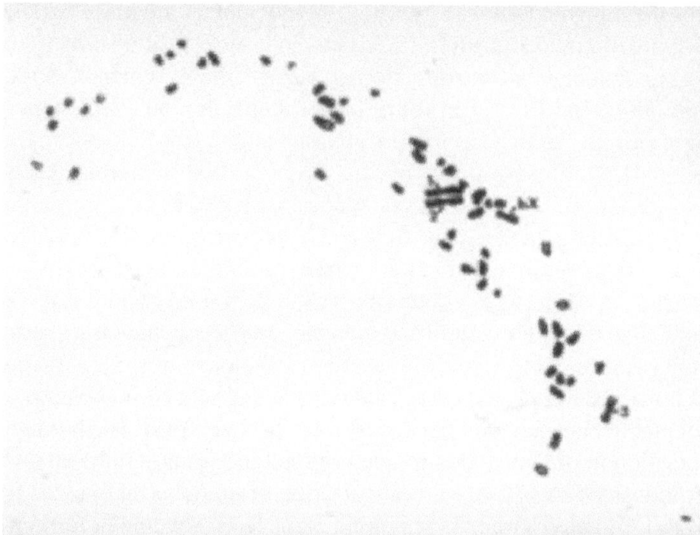

Figure 3: Metaphase spread from a 4-cell sheep-cattle hybrid embryo showing 57 chromosomes (30 cattle and 27 sheep). Note the 3 metacentric sheep chromosomes labeled 1,2 and 3 and the cattle X chromosome labeled bX. (King and Kelk, unpublished data).

Interspecies in Vitro Fertilization

Investigating interspecies fertilization provides valuable insights into the role of cytoplasmic and nuclear events that regulate full term development. Due to species specific molecular, morphological and chronological markers these embryo models are unique for examining the paternal and maternal interactions and their vital role in development. Interesting findings from our laboratory revealed that ram spermatozoa is not able to fertilize bovine oocytes in vivo, however, in vitro fertilization occurs and the resulting hybrid embryos, confirmed by cytogenetic analysis (Figure 3), arrested at 4-8 cell stage [52] even when zygotes were transferred to sheep oviducts for early development. Similar developmental arrest occurs with the reverse cross. Hybridization between sheep and goats through cross or chance breeding like many species, appears to be polar with the success or efficiency dependent of the parent of origin [53]. Goats bred to rams conceive and pregnancies will progress to second trimester while in the case of the reverse cross, sheep oocytes exposed to buck sperm in vivo do not fertilize [52]. However, a few cases of sheep X goat hybrids born to goats have been documented [54]. In vitro sheep and goat oocytes, could be fertilized by buck or ram sperm and embryos developed to the blastocyst stage, hatched and, following transfer into the uteri of sheep or goats, they established pregnancies. However, retarded fetal growth, defects in placentome formation and pregnancy failure during the second month of gestation were observed [52]. The in vitro production of hybrid embryos is particularly amenable to facilitate further study of the reproductive barriers and species differences in imprinted and epigenetic, regulation of early embryo development that are key to embryo survival and development.

Somatic Cell Nuclear Transfer (Cloning)

A new era in reproductive biotechnology began when Wilmut and colleagues produced the first somatic cell nuclear transfer (SCNT) derived viable sheep, "Dolly", from transplanting an adult somatic cell into an enucleated sheep oocyte [55]. Following this breakthrough, theoretically, it seems to be a promising technology to produce unlimited numbers of genetic copies from an adult animal or a fetus; and so far many species including farm, pet and endangered animals have been cloned using this technology [56]. Obtaining a viable offspring through this technology still imposes a great challenge for researchers due to its relatively low success rate and in some instances compromised health status of offspring [57,58]. At present, cloning is carried out for biomedical and research purposes, and agriculture production. Although extensive reviews are available on animal cloning, here we emphasize some recent findings that are important in small ruminant cloning.

The standard SCNT technique in sheep/goat in most of the laboratories follows some common steps such as oocyte maturation, removal of cumulus cells from oocytes, enucleation, transplantation of the somatic cell into the perivitelline space of the oocyte, electrofusion of the cell couplet, activation of the embryo and in vitro embryo culture. When the embryos develop to blastocyst, they are transplanted into the uteri of estrus-synchronized surrogate animals 6-7 day post estrus [27]. In most cases, the oocytes are collected from abattoir ovaries using the slicing method and undergo IVM for 17h (goat: [59-61], Sheep: [27,34]). Some laboratories that conduct goat SCNT programs preferred in vivo derived oocytes either following ovulation [62,63] or LOPU [64,65]. This may be due to the decreased availability of goat abattoirs in different countries or the higher success rates from in vivo matured oocytes. Following maturation, the cumulus cells are removed by vortexing in 3% Sodium citrate medium [27] for 1-2 minute or in 0.01% hyaluronidase. Cumulus –free oocytes with first polar body extrusion are selected and after a short exposure to cytochalasin B and Hoechst stain, enucleation is carried out in a micromanipulator equipped with UV light [27,61,64,65]. Following oocyte enucleation, cell cycle-synchronized donor cells are transplanted into the perivitelline space of the oocytes in a manner that both cell membranes attach firmly to each other. The cellcouplets are then electrofused in a 0.3 M mannitol fusion medium. One hour following fusion, cell-couplets are activated by culturing for 5 min in ionomycin containing medium followed by culturing them in 6-DMAP or cycloheximide containing medium [27,34,61]. Recent studies in our laboratory revealed that a greater incidence of chromosomal abnormalities occur when SCNT-sheep embryos are cultured in 6-DMAP containing medium compared to cycloheximide [27,34]. After activation, embryos are cultured in IVC medium and are surgically transferred into uteri of estrus –synchronized ewes [27] or into the oviducts of the does [61]. Pregnancy confirmation is done by ultrasound scanning beginning on day 35 of pregnancy. Pregnancy loss in sheep following SCNT-derived embryo transfer can be monitored by ultrasound imaging combined with serum progesterone levels [27].

Recently, there has been some controversy on the biological age of the SCNT-derived offspring, and whether they may susceptible for pre-mature death compared to their age-matched control counterparts [56]. A good example of this was the premature death of "Dolly". The hypothesis behind this is that donor somatic cells from adults may correlate with offspring carrying shortened telomere lengths compared to normal controls. Telomeres are located at the end of the chromosomes and become shorter in each cell cycle when somatic cells replicate. When the telomere length reaches a critical length the cell either enters senescence or enters a 'crisis' state associated with the development of cancer cells. This phenomenon does not happen in the germ line and in embryonic cells as these cells posses telomerase, an enzyme that adds telomere sequences to chromosomes during cell replication thus preventing telomere shortening from occurring. In most somatic cells, the telomerase enzyme is not present, leading to continuous telomere shortening when cells replicate [66].

Recent findings in telomere lengths in animal clones revealed cases of shorter [67, 68] not different [69-71] and surprisingly even longer lengths compared to those measured in their age matched control counterparts [72]. These findings have triggered much speculation on the fate of clones. However, this makes animal clones derived from adult somatic cells appropriate models for studying cellular rejuvenation and aging in vivo since they are generated from aged and culture propagated somatic cells known to contain shorten telomeres. Following SCNT, the differentiated donor nucleus undergoes extensive remodeling and reprogramming by the recipient oocyte cytoplasm [73], which has been shown to cause altered gene expression profiles [74] and chromatin modifications [75] that likely contribute to the high incidence of abnormalities, prenatal and post natal losses found in clones[73]. Altered expression of telomerase activity [71] and donor cell type-dependent telomere length differences [76] further support inappropriate genome reprogramming, however, the precise etiology of altered telomere lengths in vivo are not clear for clones. Recent studies from our laboratory on SCNT-derived small ruminants showed that shortened telomere lengths in sheep cloned using adult donor cells compared to controls [67] and a similar phenomenon was observed in SCNT derived goats [77]. Interestingly, the offspring derived from natural breeding between cloned parents, did not show any discrepancies in telomere length [67] suggesting that telomere length was reset in the germ line of the cloned offspring. To date we do not possess enough data on telomere length of clones and "Dolly" cannot be considered the only incidence in this nature, thus more research is needed to investigate the lifespan of cloned offspring; at this stage the data available is inconclusive.

Production of Transgenic Sheep and Goats

Transgenic animals are produced either to carry an exogenous gene or to have a "knock out" gene from their own genome. Many transgenic animals have been produced for biomedical research interests as well as to test their application in livestock production [78]. There are many techniques available for the production of transgenic animals, such as pronuclear microinjection [37,79,80], sperm-mediated transgenesis [81] and somatic cell nuclear transfer[65,82,83]. During the last decade, research has focused on producing transgenic animals which secrete recombinant proteins in their milk.

These animals produce relatively high quantities of recombinant proteins compared to other conventional methods due to the physiological nature of the mammary gland for its unique ability to synthesize proteins [78,84,85,86,87,88]. A few examples of transgenic small ruminants producing recombinant proteins in milk for pharmaceutical and biomedical interest are sheep producing human clotting factor VIII [89] and factor IX [82] and goats synthesizing human antithrombin III [90], human alpha-fetoprotein [91], human monoclonal antibodies [92, 93], malaria antigens [94] spider silks [95], human granulocyte colony stimulating factor (96) and human butyryl-cholinesterase [97,98]. Although these animals have shown high expression of the transgene product in the milk, a recent report suggests that the transgene may have some negative effects on the physiology of mammary gland function, leading to a compromised situation [87]. However, the first protein produced by transgenic animals, ATryn, a recombinant human anti-thrombinproduced in transgenic goats' milk, received approval for use in humans in 2009.

CONCLUSIONS

In this review, we have focused on some aspects of reproductive biotechnology that are of veterinary and biomedical importance in sheep and goat that have been reported in the recent past. It is well known that most biotechnologies take many years to come into full practice after enormous amounts of laboratory and field research. Therefore, more focus should be carried out in emerging areas like artificial insemination of sheep and goat using sexed semen, in vitro and in vivo embryo production systems, transgenic animal production and cloning using stem cells.

ACKNOWLEDGEMENTS

The authors gratefully acknowledge the assistance provided by Ed R Reyes, Elizabeth St. John, Pamela Hasson in the University of Guelph, the MITACS Accelerate internship program, Canada, the Canadian Commonwealth Scholarship Plan and the University of Peradeniya in preparation of this manuscript.

REFERENCES

1. Zygoyiannis D (2006) Sheep production in the world and in Greece. Small Rumin Res 62: 143-147.
2. Ghadar F, Spindler H (2005) The growth of biotechnology. Industrial Management 47: 19-26.
3. Polge C (1968) Frozen semen and the A.I. Programme in Proc. 2nd Tech. Conf. Artif. Insem. Reprod. NAAB, Columbia, MO: 46-51.
4. Johnson LA, Rath D, Vazquez JM, Maxwell WMC, Dobrinsky JR (2005) Preselection of sex of offspring in swine for production: current status of the process and its application. Theriogenology 63: 615-624.
5. Johnson LA, Flook JP, Hawk HW (1989) Sex preselection in rabbits: live births from X and Y sperm separated by DNA and cell sorting. Biol Reprod 41: 199-203.

6. Garner DL (2001) Sex-sorting mammalian sperm: concept to application in mammals. J Androl 22: 519-526.
7. Seidel GE Jr, Garner DL (2002) Current status of sexing mammalian spermatozoa. Reproduction 124: 733-743.
8. Evans G, Hollinshead FK, Maxwell WMC (2004) Preservation and artificial insemination of sexed semen in sheep. Reprod Fertil Dev 16: 455-464.
9. Seidel GE, Schenk JL, Herickhoff LA, Doyle SP, Brink Z, et al. (1999) Insemination of heifers with sexed sperm. Theriogenology 52: 1407-1420.
10. Vazquez JM, Martinez EA, Parrilla I, Roca J, Gill MA, et al. (2003) Birth of piglets after deep intrauterine insemination with flow cytometrically sorted boar spermatozoa. Theriogenology 59: 1605-1614.
11. Rath D, Ruiz S, Sieg B (2003) Birth of female piglets following intrauterine insemination of a sow using flow cytometrically sexed boar semen. Vet Rec 152: 400-401.
12. Buchanan BR, Seidel GE, McCue PM, Schenk JL, Heickhoff LA, et al. (2000) Insemination of mares with low numbers of either unsexed or sexed spermatozoa. Theriogenology 53: 1333-1344.
13. Lindsey AC, Schenk JL, Graham JK, Bruemmer JE, Squires EL (2002) Hysteroscopic insemination of low numbers of flow sorted fresh and frozen thawed stallion spermatozoa. Equine Vet J 34: 121-127.
14. Catt SL, Catt JW, Gomez MC, Maxwell WMC, Evans G (1996) The birth of a male lamb derived from an in vitro matured oocyte fertilized by intra-cytoplasmic injection of a single presumptive male sperm. Vet Rec 139: 494-495.
15. Hollinshead FK, O'Brien JK, Maxwell WMC, Evans G (2002) Production of lambs of pre-determined sex after the insemination of ewes with low numbers of X- or Y-chromosome bearing spermatozoa. Reprod Fertil Dev 14: 503-508.
16. Garner DL, Seidel GE Jr (2003) Past, present and future perspectives on sexing sperm. Can J Anim Sci 83: 375-384.
17. Maxwell WMC, Johnson LA (1997) Chlortetracycline analysis of boar spermatozoa after incubation, flow cytometric sorting, cooling or cryopreservation. Mol Reprod Dev 46: 408-418.
18. Hollinshead FK, O' Brien JK, Gillan L, Mayers M, Maxwell WMC, et al. (2004) Liquid storage of flow cytometrically sorted ram spermatozoa. Theriogenology 62: 587-605.
19. Catt SL, O'Brian JK, Maxwell WMC, Evans G (1997) Assessment of ram and boar spermatozoa during cell-sorting by flow cytometry. Reprod Domestic Anim 32: 251- 258.
20. Hollinshead FK, Gillan L, O'Brian JK, Evans G, Maxwell WMC (2003) In vitro and in vivo assessment of functional capacity of flow cytometrically sorted ram spermatozoa after freezing and thawing. Reprod Fertil Dev 15: 351-359.
21. De Graaf SP, Gillan L, Evans G, Maxwell WMC, O'Brian JK (2004). The effect of the sheath fluid on the quality of sex-sorted ram spermatozoa. Reprod Fertil

Dev 16: 284-285.

22. Perera GDRK, Pushpakumara PGA, De Silva LNA, Alexander B (2008) Production of genetically superior goats through embryo transfer technology in Sri Lanka. Tropi Agri Res 20: 177-184.

23. Bartlewski PM, Alexander BD, Rawlings NC, Barrett DMW, King WA (2008) Ovarian responses, hormonal profiles and embryo yields in anoestrus ewes superovulated with Folltropin-V after pretreatment with medroxyprogesterone acetate (MAP)- releasing vaginal sponges and a single dose of estradiol-17ß (E2-17ß). Reprod Domest Anim 43: 299 – 307.

24. Bartlewski PM, Alexander BD, King WA (2008a) Ovarian and endocrine determinants of superovulatory responses in anestrous ewes. Small Rumin Res 75: 210-216.

25. Gonzalez-Bulnes A, Baird DT, Campbell BK, Cocero MJ, Garcia-Garcia RM, et al. (2004) Multiple factors affecting the efficiency of multiple ovulation and embryo transfer in sheep and goats. Reprod Fertil Dev 16: 421-435.

26. Menchaca A, Vilarino M, Pinczak A, Kmaid S, Saldana JM (2009) Progesterone treatment, FSH plus eCG, GnRH administration, and Day 0 Protocol for MOET programs in sheep. Theriogenology 72: 477-483.

27. Alexander B, Coppola G, Mastromonaco GF, John ES, Reyes ER, et al. (2008) Early Pregnancy Diagnosis by Serum Progesterone and Ultrasound in Sheep Carrying Somatic Cell Nuclear Transfer - Derived Pregnancies. Repro Domest Anim 43: 207-211.

28. String fellow and Seidel (1998) Manual of the International Embryo Transfer Society.

29. Candappa IBR, Bainbridge HC, Price NT, Hourigan KR, Bartlewski PM (2009) A preliminary study on the suitability of Cervidil to induce cervical dilation for artifi cial insemination in ewes. Res Vet Sci 87: 204-206.

30. Garcia-Garcia R, Ward F, Fair S, O'Meara C, Wade M, et al. (2007) Development and quality of sheep embryos cultured in commercial G1.3/G2. 3 sequential media. Anim Reprod Sci 98: 233-240.

31. Cognie Y, Poulin N, Locatelli Y, Mermillod P (2004) State-of-the-art production, conservation and transfer of in-vitro-produced embryos in small ruminants. Reprod Fertil Dev 16: 437-445.

32. Wani NA (2002) In vitro maturation and in vitro fertilization of sheep oocytes. Small Rumin Res 44: 89-95.

33. Wang S, Liu Y, Holyoak GR, Evans RC, Bunch TD (1998) A protocol for in vitro maturation and fertilization of sheep oocytes. Small Rum Res 29: 83-88.

34. Alexander B, Coppola G, Di Berardino D, Rho GJ, St John E, et al. (2006) The effect of 6-dimethylaminopurine (6-DMAP) and cycloheximide (CHX) on the development and chromosomal complement of sheep parthenogenetic and nuclear transfer embryos. Mol Reprod Dev 73: 20-30.

35. Alberio R, Olivera J, Roche A, Alabart J, Folch J (2002) Performance of a

modified ovum pick-up system using three different FSH stimulation protocols in ewes. Small Rumin Res 46: 81-87.

36. Baldassarre H, Wang B, Kafi di N, Keefer C, Lazaris A, et al. (2002) Advances in the production and propagation of transgenic goats using laparoscopic ovum pickup and in vitro embryo production technologies. Theriogenology 57: 275-284.

37. Baldassarre H, Wang B, Kafi di N, Gauthier M, Neveu N, et al. (2003) Production of transgenic goats by pronuclear microinjection of in vitro produced zygotes derived from oocytes recovered by laparoscopy. Theriogenology 56: 83-839.

38. Wan PC, Haoa ZD, Zhoub P, Wua Y, Yanga L (2009) Effects of SOF and CR1 media on developmental competence and cell apoptosis of ovine in vitro fertilization embryos. Anim Reprod Sci 114: 279-288.

39. Baldassarre H, Furnus CC, De Matos DG, Pessi H (1996) In vitro production of sheep embryos using laparoscopic folliculocentesis: Alternative gonadotrophin treatments for stimulation of oocyte donors. Theriogenology 45: 707-717.

40. Gibbons A, Bonnet FP, Cueto MI, Catala M, Salamone DF, et al. (2007) Procedure for maximizing oocyte harvest for in vitro embryo production in small ruminants. Reprod Dom Anim 42: 423-440.

41. Cognie Y, Baril G, Poulin N, Mermillod P (2003) Current status of embryo technologies in sheep and goat. Theriogenology 59: 171-188.

42. Accardo C, Dattena M, Pilichi S, Mara L, Chessa B, et al. (2004) Effects of recombinant human FSH and LH on in vitro maturation of sheep oocytes: embryo development and viability. Anim Reprod Sci 81: 77-86.

43. Coppola G, Alexander B, Di Berardino D, St John E, Basrur PK, et al. (2007) Use of cross-species in-situ hybridization (ZOO-FISH) to assess chromosome abnormalities in day-6 in vivo- or in vitro-produced sheep embryos. Chromosome Res 15: 399-408.

44. Rho GJ, Hahnel AC, Betteridge KJ (2001) Comparison of oocyte maturation times and of three methods of sperm preparation for their effects on the production of goat embryos in vitro. Theriogenology 56: 503-516.

45. Comizzoli P, Mermillod P, Cognie Y, Chal N, Legendre X, et al. (2001) Successful in vitro production of embryos in red deer (Cervus elaphus) and the sika deer (Cervus nippon). Theriogenology 55: 649-659.

46. Berg DK, Thompson JG, Asher GW (2002) Development of In Vitro Embryo Production Systems for Red Deer (Cervus Elaphus). Part 3. In Vitro Fertilisation Using Sheep Serum as a Capacitating Agent and the Subsequent Birth of Calves. Anim Reprod Sci 70: 85-98.

47. O'Brien JK, Hollinshead FK, Maxwell WMC (2004) In vivo development capacity of in vitro produced embryos derived from sex-sorted and re-cryopreserved frozen-thawed ram-sperm. Reprod Fertil Dev 16: 286.

48. Hoshi H (2003) In vitro production of bovine embryos and their application for embryo transfer. Theriogenology 59: 675-685.

49. Holm P, Walker SK, Seamark RF (1996) Embryo viability, duration of gestation and birth weight in sheep after transfer of in vitro matured and in vitro fertilized zygotes cultured in vitro or in vivo. J Reprod Fertil 107: 175-181.

50. Sinclair KD, McEvoy TG, Maxfi eld EK, Maltin CA, Young LE, et al. (1999) Aberrant fetal growth and development after in vitro culture of sheep zygotes. J Reprod Fertil 116: 177-186.

51. Walker S, Hill J, Kleeman D, Nancarrow C (1996) Development of ovine embryos in synthetic oviductal fluid containing amino acids at oviductal fluid concentrations. Biol Reprod 55: 703-708.

52. Kelk DA (1997) Development characteristics of interspecific hybrid embryos. PhD Thesis. University of Guelph, Ontario, Canada.

53. Anderson GB (1988) Interspecific pregnancy: barriers and prospects. Biol Reprod 38: 1-15.

54. Pinheiro LEL, Guimaraes SEF, Almeida IL, Mikich AB (1989) The natural occurrence of sheep X goat hybrids. Theriogenology 32: 987 – 994.

55. Wilmut I, Schnieke AE, McWhir J, Kind AJ, Campbell KH (1997) Viable offspring derived from fetal and adult mammalian cells. Nature 385: 810-813.

56. Vajta G, Gjerris M (2006) Science and technology of farm animal cloning: state of the art. Anim Reprod Sci 92: 211-230.

57. Hill JR, Roussel AJ, Cibelli JB, Edwards JF, Hooper NL, et al. (1999) Clinical and pathological features of cloned transgenic calves and fetuses (13 case studies). Theriogenology 51: 1451-1465.

58. Rho GJ, Coppola G, Sosnowski J, Kasimanickam R, Johnson WH, et al. (2007) Use of somatic cell nuclear transfer to study meiosis in female cattle carrying a sex-dependent fertility-impairing X-chromosome abnormality. Cloning Stem Cells 9: 118-29.

59. Ma S, Lan G, Miao Y, Wang Z, Chang Z, et al. (2003) Hypoxanthine (HX) inhibition of in vitro meiotic resumption in goat oocytes. Mol Reprod Dev 66: 306-313.

60. Lan GC, Han D, Wu YG, Han ZB, Ma SF, et al. (2005) Effects of duration, concentration and timing of ionomycin and 6-dimethylaminopurine (6-DMAP) treatment on activation of goat oocytes. Mol Reprod Dev 71: 380-388.

61. Lan GC, Chang ZL, Luo MJ, Jiang YL, Han D, et al. (2006) Production of Cloned Goats by Nuclear Transfer of Cumulus Cells and Long-Term Cultured Fetal Fibroblast Cells Into Abattoir-Derived Oocytes. Mol Reprod Dev 73: 834-840.

62. Zou X, Wang Y, Cheng Y, Yang Y, Ju H, et al. (2002) Generation of cloned goats (Capra hircus) from transfected foetal fibroblast cells, the effect of donor cell cycle. Mol Reprod Dev 61: 164-172.

63. Behboodi E, Memili E, Melican DT, Destrempes MM, Overton SA, et al. (2004) Viable transgenic goats derived from skin cells. Transgenic Res 13: 215-224.

64. Keefer CL, Stone RK, Bhatia B, Lazaris A, Begin I, et al.(2000) Efficient production of viable goat offspring following nuclear transfer using adult somatic

cells. Biol Reprod 62: 192.

65. Keefer CL, Baldassarre H, Keyston R, Wang B, Bhatia B, et al. (2001) Generation of dwarf goat (Capra hircus) clones following nuclear transfer with transfected and nontransfected fetal fi broblasts and in vitro-matured oocytes. Biol Reprod 64: 849-856.

66. Xu J, Yang X (2003) Will cloned animals suffer premature aging - the story at the end of clone's chromosomes. Reprod Biol Endocrinol 1: 105-110.

67. Alexander B, Coppola G, Perrault SD, Peura TT, Betts DH, et al. (2007) Telomere length of sheep clones and their offspring. Mol Reprod Dev 74: 1525-1537.

68. Shiels PG, Jardine AG (2003) Dolly, no longer the exception: Telomeres and implications for transplantation. Cloning stem cells 5: 157-160.

69. Jiang L, Carter DB, Xu J, Yang X, Prather RS, et al. (2004) Telomere lengths in cloned transgenic pigs. Biol Reprod 70: 1589-1593.Betts DH, Bordingnon V, Hill JR, Winger Q, Westhusing ME, et al. (2001) Reprogramming of telomerase activity and rebuilding of telomere length in cloned cattle. Proc Natl Acad Sci USA 98: 1077-1082.

70. Betts DH, Bordingnon V, Hill JR, Winger Q, Westhusing ME, et al. (2001) Reprogramming of telomerase activity and rebuilding of telomere length in cloned cattle. Proc Natl Acad Sci USA 98: 1077-1082.

71. Tian XC, Xu J, Yang X (2000) Normal telomere lengths found in cloned cattle. Nat Genet 26: 272-273.

72. Lanza R, Cibelli JB, Blackwell C, Cristofalo VJ, Francis MK, et al. (2000) Extension of cell life- span and telomere length in animals cloned from senescent somatic cells. Science 288: 665-669.

73. Reik W, Dean W (2002) Epigenetic reprogramming: back to the beginning. Nature 420: 127.

74. Humpherys D, Eggan K, Akatsu H, Friedman A, Hoechedlinger K, et al. (2002) Abnormal gene expression in cloned mice derived from embryonic stem cell and cumulus cell nuclei. Proc Natl Acad Sci USA 99: 12889-12894.

75. Dean W, Santos F, Stojkovic M, Zakhartchenko V, Walter J, et al. (2001) Conservation of methylation reprogramming in mammalian development: aberrant reprogramming in cloned embryos. Proc Natl Acad Sci USA 98: 13734-13738.

76. Miyashita N, Shiga K, Yonai M, Keneyama K, Kobayashi S, et al. (2002) Remarkable differences in telomere lengths among cloned calves derived from different cell types. Biol Repro 66: 1649-1655.

77. Betts DH, Perrault SD, Petrik J, Lin L, Favetta LA, et al. (2005) Telomere length analysis in goat clones and their offspring. Mol Reprod Dev 72: 461-470.

78. Wheeler MB, Walters EM, Clark SG (2003) Transgenic animals in biomedicine and agriculture: outlook for the future. Anim Reprod Sci 79: 265-289.

79. Ebert KM, Selgrath JP, DiTullio P, Denman J, Smith TE, et al. (1991) Transgenic production of a variant of human tissue-type plasminogen activator in goat milk:

generation of transgenic goats and analysis of expression. Biotechnology 9: 835-838.

80. Baldassarre H, Wang B, Gauthier M (1999) Embryo transfer in a commercial transgenic production program using BELE® goat embryos. Theriogenology 51: 415.

81. Celebi C, Guillaudeux T, Auvray P, Vallet-Erdtmann V, Jegou B (2003) The making of "transgenic spermatozoa". Biol Reprod 68: 1477-1483.

82. Schnieke AE, Kind AJ, Ritchie WA, Mycock K, Scott AR, et al. (1997) Human Factor IX transgenic sheep produced by nuclear transfer of nuclei form transfected fetal fibroblasts. Science 278: 2130-2133.

83. Keefer CL, Lazaris A, Keyston R (2002) Cloning using somatic cells for the production and propagation of transgenic Nigerian dwarf and dwarf-crossed goats. Theriogenology 57: 422.

84. Houdebine LM (2000) Transgenic animal bioreactors Transgenic Res 9: 305-320.

85. Keefer CL (2004) Production of bioproducts through the use of transgenic animal models. Anim Reprod Sci 82: 5-12.

86. Baldassarre H, Wang B, Pierson J, Neveu N, Sneek L, et al. (2004) Prepubertal Propagation of Transgenic Cloned Goats by Laparoscopic Ovum Pick-Up and in Vitro Embryo Production. Cloning Stem Cells 6: 25-29.

87. Baldassarre H, Hockley DK, Dore M, Brochu E, Hakier B, et al. (2008a) Lactation performance of transgenic goats expressing recombinant human butyrylcholinesterase in the milk. Transgenic Res 17: 73-84.

88. Baldassarre H, Hockley DK, Olaniyan B, Brochu E, Zhao X, et al. (2008b) Milk composition studies in transgenic goats expressing recombinant human butyrylcholinesterase in the mammary gland. Transgenic Res 17: 863-872.

89. Niemann H, Halter R, Carnwath JW, Herrmann D, Lemme E, et al. (1999) Expression of blood clotting factor VIII in the mammary gland of transgenic sheep. Transgenic Res 8: 237-247.

90. Zhou Q, Kyazike J, Echelard Y, Meade HM, Higgins E, et al. (2005) Effect of genetic background on glycosylation heterogeneity in human antithrombin produced in the mammary gland of transgenic goats. J Biotechnol 117: 57-72.

91. Parker MH, Birck-Wilson E, Allard G, Masiello N, Day M, et al. (2004) Purification and characterization of a recombinant version of human alpha-fetoprotein expressed in the milk of transgenic goats. Protein Expr Purif 38: 177-183.

92. Young MW, Meade H, Curling JM, Ziomek CA, Harvey M (1998) Production of recombinant antibodies in the milk of transgenic animals. Res Immunol 149: 609- 610.

93. Pollock DP, Kutzko JP, Birck-Wilson E, Williams JL, Echerlard Y, et al. (1999) Transgenic milk as a method for the production of recombinant antibodies. J Immunol Methods 231: 147-157.

94. Behboodi E, Ayres SL, Memili E, O'Coin M, Chen LH, et al. (2005) Health and

reproductive profi les of malaria antigen-producing transgenic goats derived by somatic cell nuclear transfer. Cloning Stem Cells 7: 107-118.

95. Karatzas CN, Zhou FJ, Huang Y, Duguay F, Chretien N, et al. (1999) Production of recombinant spider silk (BioSteel®) in the milk of transgenic animals. Transgenic Res 8: 476-477.

96. Ko JH, Lee CS, Kim KH, Pang MG, Koo JS, et al. (2000) Production of biologically active human granulocyte colony stimulating factor in the milk of transgenic goat. Transgenic Res 9: 215-222.

97. Cerasoli DM, Griffi ths EM, Doctor BP, Saxena A, Fedorko JM, et al. (2005) In vitro and in vivo characterization of recombinant human butyrylcholinesterase (Protexia) as a potential nerve agent bioscavenger. Chem Biol Interact 158: 363-365.

98. Huang YJ, Huang Y, Baldassarre H, Wang B, Lazaris A, et al. (2007) Recombinant human butyrylcholinesterase from milk of transgenic animals to protect against organophosphate poisoning. Proc Natl Acad Sci USA 104: 13603-13608.

13
CHAPTER

ENVIRONMENTAL STRESS EFFECT ON ANIMAL REPRODUCTION

Córdova-Izquierdo Alejandro, Villa-Mancera Abel, Olivares Pérez Jaime and Sánchez-Aparicio Pedro

Keywords: Environmental stress, Animal reproduction

INTRODUCTION

At present, the need to produce more efficiently, the need to improve reproductive and productive parameters so that new technologies have been adopted which are becoming more sophisticated, all in order to increase the production in animals.

Despite the advances in technology are applied to animal production, few producers or farmers who have shown concern for improving the general environmental conditions of the animal production units, not realizing that these conditions are largely, are responsible for the health of animals, altering thereof and being significantly impacting on the presence of such problems reproductive and productive, significantly increasing production costs.

Stress is the result of confinement and an ambitious vision and uncaring by the man, who in the interest of improving production, has participated unconsciously, by manipulating animal production in its way, even trying to tame new species, impossible to adapt to our environment healthy.

At present, it has been concluded that stress is one of the environmental factors that affects handling and decreases the production of livestock, it is considered that the main factor which should be controlled in units Animal Production, since it is closely related to pathogens and infectious agents that may try to the health of animals.

The man plays a very important role in creating the stress, it can provide the means to alleviate remove or modify any potential stressor [1]. Some of the criteria that have been used to indicate the presence of stress in animals are the thyroid hormones such

Citation: Alejandro CI, Abel VM, Jaime OP and Pedro SA (2014) Environmental Stress Effect on Animal Reproduction. J Adv Dairy Res 2:114. doi: 10.4172/2329-888X.1000114.

as corticosteroids plasma or urinary catecholamine's and/or its metabolites, lactic acid, serum enzymes, blood glucose, alterations in the frequency respiratory or cardiac and somatic cell count in milk.

STRESS NEUROENDOCRINOLOGY

Mammals, respond to environmental changes by adaptation mechanisms, regulated by the release of hormones such as: Adrenocorticotropin (ACTH), Glucocorticoids (GH) and Catecholamine (CA), the amount of release depends on the type and intensity of the stress that comes along.

Under stress, there is a negative feedback effect of progesterone on luteinizing hormone (LH) increasing the first. Are elevated cortisol levels with subsequent depression, there are changes in the release of LH. The level of prostaglandin F2 alpha (PGF2α) and ACTH increase. The levels of epinephrine and norepinephrine, also increase [1]. All this adversely affects reproductive function in mammals [2].

TYPES OF STRESS

Caloric

Solar radiation, wind speed, air temperature and humidity are all factors which are concerned with the presentation of heat stress. An early reproductive response to heat stress is the decrease in the intensity of estrus and consequently low fertility [3]. It is known that mammalian females are more sensitive to heat stress, during the 12 days before estrus, increasing this sensitivity on day 2 before [4].

Studies by Coubrough 1985 [1] found that the absence of high temperatures cause libido, as well as affecting the quantity and quality of semen. They also reduce heat filing and when present, are less intense and of shorter duration. If the female is pregnant, the possibility of embryonic mortality is high, factors lead to a decrease in the fertility of domestic mammalian females [5].

Recent studies conducted by Roth et al. [6], mentioned for the first time, evidence of the delayed effect of heat stress on follicular steroidogenesis revealed that exposure of female domestic heat stress led to a decrease in steroidogenesis, 20 and 26 days later, in mid-sized preovulatory follicles. Furthermore, the theca cells more susceptible found that granulose cells and they expressed low estrogen production in medium-sized follicles and low viability in preovulatory follicles. In terms of production of steroids, theca cells appeared to be consistently susceptible to heat stress and expressed transport effects on androgen production in both types of follicles.

When the body temperature exceeds 40°C, the developed follicles can suffer severe damage and become non-viable [7], are secreted higher levels of progesterone, which inhibits the release of LH during estrus, interrupting the process of Ovulation [8].

A study by Roth et al. [9] has revealed that heat stress reduces dominant follicle size of the first and second follicular wave. In another study, Roth et al. [6], indicate that exposing domestic female mammals during heat stress is altered follicular dynamics and lowers the follicular dominance in the next cycle. In 1997, Wolfensohn et al. [10] argued that the preovulatory follicle is a key component in the reproductive system and the deterioration of their function during heat stress may trigger other reproductive effects, such as distortion in gonadotropin secretion, underdevelopment of the body luteum and embryo, resulting in a low fertility.

Studies by under in vitro conditions, suggesting that exposure of pregnant domestic females at high ambient temperatures and humidity, which often occur during the summer months, can disrupt the balance between the biochemical and endometrial factors responsible for the maintenance of pregnancy. Roth et al. [11] indicated that the lipid composition of membranes of oocytes is more deteriorated in summer than in winter. An important aspect to counteract the effect of heat stress on the quality of the oocyte and embryo development is the potential use of antioxidants to increase the strength of both the oocyte and embryo.

In 1994, Ealy et al. [12] found that the cooled for 2 to 3 to 5 to 6 days after the pregnancy, pregnancy rates improved as compared with females under heat stress. Moreover, Lee [8] found that thyroxine and triiodothyroxine can reduce heat stress. It has been shown that during periods of severe heat stress, only 10 to 20% of inseminations can result in normal pregnancies (Dave, 1999). In Israel, conception rates during the winter months reach more than 50%, while during the summer can lose up to 20% [13]. In cattle, it has been found that the Jersey breed is more tolerant to heat stress that Holstein.

Work performed in the United States; indicate that Holstein and Jersey cows set in hot, express signs of estrus for only 12 to 13 hours, which means a difference of 5-6 hours less than the normal duration of estrus animal's warm places [13].

Mazzucchelli and Tesouro [14], indicated that dairy cows with black layer absorb more heat from sunlight, cows of white layer. In this regard, it is known that the temperature from which affected reproductive efficiency in cattle, is in a range ranging from 21.1°C to 32.2°C, so that indicated [10], that the cows need to be cooled as much as possible throughout the summer, in order to raise fertility.

In other studies in Florida [15], showed that in summer with heat and humidity conditions, pregnancy rates are down to 10%.

The best way to control the effects of heat stress are: 1) Physical modification of the environment, 2) Breeding races less sensitive to heat stress and, 3) Good nutritional management [13].

Nutrition

In 1998, Zapiola said that in cattle, stress during transportation, has a detrimental effect on the physiology of the animal that the stress caused by lack of food and drink, for a period of equal length, which is because the blood flow is diverted from the internal organs, to the peripheral tissues in an attempt to reduce the body temperature by increasing heat losses. This mechanism leads to a reduction of blood flow, aimed at internal organs such as the uterus, fallopian tubes and ovaries; decreased blood supply to these organs also implies low nutrient availability and therefore low functional capacity [14].

Studies by Lee [8], reflected the corpus luteum of nutrition-stressed females, is lighter and also indicated by chemical constituents of allantoic fluid, the underfed female embryos, and degenerate.

Females consume less food during warm seasons, consequently, there may be insufficient nutrients are available, after the production of milk and therefore the ovaries do not function during the first 6 weeks of lactation [16].

Vitamin E may decrease moderately, the effects of heat stress during hot months, does not necessarily increase fertility [12]. A proactive factor could be the adequate food supply in quantity and quality, which have direct effects on reproduction and on the alterations caused by the indirect effects of nutritional stress [17].

By Handling

The interactions between employees and animals influence the response of cattle to stress management known as Human-animal interactions [17].

In 2000, Mazzucchelli and Tesouro, indicated that the way to handle livestock, animals may produce an unnecessary handling stress, which may affect the biological mechanisms of reproduction and the immune response. On the other hand, electrical bites, immobilization and other stress factors management weaken the female reproductive functioning.

There are many driving situations can cause stress seriously affecting the reproductive process of the species, such as excessive mobilization of inseminated females before, the use of aggressive or driving the separation of animals, mobilization of animals on the sleeve of management for different purposes, etc. [16].

The time when stress management, represents greater adverse effects on reproduction, it is just no time to estrus [14].

The temperament of cattle can be classified as follows: 1) Stand still while subject, 2) Is restless, 3) Shaken vigorously driving sleeve and 4) Shakes violently management sleeve and tries to escape. Overall, in cattle, the Bos indicus have a more aggressive and nervous temperament than Bos taurus [18].

CHARACTERISTICS OF STRESS IN REPRODUCTIVE PROCESS

Dobson and Smith [19] indicated that the process of reproduction is an important physiological system for the development of species, linked to stress, that Coubrough [1] has classified into two groups: environmental stress and handling. Environmental stress includes room temperature, cold and/or hot and cold, wind and moisture. The stress management includes animal density, to handling procedures, the flow of animals, to the interaction between animals of the same or different species and the existing social condition, as nonspecific psychological distress, noise, physical trauma, etc. The combination of both types of stress acts as stressors, which compromises the homeostasis of the animal.

Moreover, Dobson and Smith [2] mentioned the following types of stress: physical, which includes the transport of animals and physical damage, psychological, which provides for the isolation of animals, physiological, in which considered to hypoglycemia and changes in blood pressure, all aspects related to the reproductive process of mammals. Cattle, is one of the most adaptable and homeocinéticos have mechanisms to maintain body functions at the expense of critical changes in other physiological functions, such as reproduction, so that the reproductive function is determined largely by the environment [13].

Constant maintenance of a normal body temperature is known as homeotermia and is vital to the functioning of the brain tissue. In cattle, the normal diurnal variation in body temperature should be from 0.6 to 1.2°C. A further increase to 1.2°C is a sign of illness or poor adaptation to temperature elevation. This temperature range is known as a zone of comfort, varies according to the species and animal breeds. When animals have a body temperature below or above their level of comfort, decreases or increases metabolic rate than normal, appearing in imbalance in animal welfare. Heat or body temperature can be eliminated by the following mechanisms: radiation, conduction, convection and evaporation [13].

In the Male

Semen Quality

Environmental stress can cause low sperm quality, which is closely related to low fertility in females, probably due to a combination of low fertilization rates and increased embryonic mortality [1,20].

Direct exposure of the testis at high temperatures, causes changes in certain critical stages of spermatogenic cycle, which is also directly related to the quality of the ejaculate.

Recent studies conducted by Rutledge [21] suggested that the effect of stress on sperm quality can be improved by implementation of the seminal freezing technology; however, the uterus of female stress may represent heat to sperm.

In the Female

Sexual behavior and fertility rate are the main indicators of the mammalian female reproduction that are negatively affected by environmental stress. So that the programs undertaken in order to increase the fertility of domestic females, are less successful in hot weather than in temperate [20].

A temperature increase of 0.5°C uterine during hot days caused a decrease in the rate of fertilization. In cattle, heifers exposure to 32°C for 72 hours after insemination, inhibit embryonic development, however, it is known that 48% of the females held 21°C, can become pregnant without any problem. If heat stress is presented after 10 days after the service, fertility is not affected.

On Fertility

Fertility in lactating cows varies by season. In the winter, decreases about 50%, 20% in the summer and fall is lower than in the winter. A few years earlier, Gilad et al. [4] reported that conception rates in Israel fell from 52% in winter to 24% in the summer. In summer, 80% of estrus may be undetectable [15]. Also indicated, Ryan et al. [22] that when the rectal temperature of the animals increased from 38.5°C to 40°C in 72 hours after insemination service, pregnancy rates can decrease up to 50%.

Heifers and cows Studies have indicated that the decline in oocyte quality in the early postpartum period is associated with negative energy balance and low body condition of the animals, which is expressed in developing embryos increased and abnormal having embryos resulting in loss of the hottest months of the year [10].

On Embryos

Several studies [12,15,23] have indicated that in cattle, embryonic development is highly sensitive to high temperatures, in the top three to 11 days after service; acquiring more heat tolerance as the gestation period progresses. It is known that the embryos obtained by in vitro fertilization (IVF) are more susceptible to heat stress than those obtained under natural conditions. In this regard, Ambrose et al. [15] indicated that the greatest loss of bovine embryos from IVF, occur before 42 days, when females are under heat stress.

RECOMMENDATIONS FOR CONTROLLING ENVIRONMENTAL STRESS

For heat

- Provide a ventilation system that controls body temperature
- Implement water sprinklers
- Animal protection against solar radiation, directly and indirectly, through shadows or appropriate ceilings
- Provide shade in feeding and drinking, to increase feed intake in animals with heat stress
- Provide water sprays
- Provide animals, spray baths, in the hottest part of the day
- Try to have animals with white fur, as they are the most easily absorb heat and therefore are less sensitive to heat stress
- Develop genetically adapted in animals, as they may be less sensitive to heat stress
- Implementation of frozen embryos
- Insemination with frozen semen in less hot time

By handling

- Provide free areas in the Production Unit and shaded
- Provide the required area per animal, for comfort
- Bathing females before service and 3 to 5 days
- Breed or service in less hot periods
- Implement estrus synchronization programs, to schedule inseminations or services
- Do not isolate females long before artificial insemination or service

By feeding

- Balancing diets properly, providing the energy needed to offset declining intake presented
- Reduce your intake of fiber and protein and increase energy

REFERENCES

1. Coubrough RI (1985) Stress and fertility. A review. Onderstepoort J Vet Res 52: 153-156.
2. Dobson H, Smith RF (1995) Stress and reproduction in farm animals. J Reprod Fertil Suppl 49: 451-461.
3. Ullah G, Fuquay JW, Keawkhong T, Clark BL, Pogue DE, et al. (1996) Effect of gonadotropin-releasing hormone at estrus on subsequent luteal function and fertility in lactating Holsteins during heat stress. J Dairy Sci 79: 1950-1953.
4. Gilad E, Meidan R, Berman A, Graber Y, Wolfenson D (1993) Effect of heat stress on tonic and GnRH-induced gonadotrophin secretion in relation to concentration of oestradiol in plasma of cyclic cows. J Reprod Fertil 99: 315-321.
5. Rodríguez J, Baltasar (1997) How to minimize summer heat stress in dairy cows.
6. Roth Z, Meidan R, Shaham-Albalancy A, Braw-Tal R, Wolfenson D (2001) Delayed effect of heat stress on steroid production in medium-sized and preovulatory bovine follicles. Reproduction 121: 745-751.
7. Dave L (1999) Heat Stress Management.
8. Lee CN (1993) Environmental stress effects on bovine reproduction. Vet Clin North Am Food Anim Pract 9: 263-273.
9. Roth Z, Meidan R, Braw-Tal R, Wolfenson D (2000) Immediate and delayed effects of heat stress on follicular development and its association with plasma FSH and inhibin concentration in cows. J Reprod Fertil 120: 83-90.
10. Wolfenson D, Lew BJ, Thatcher WW, Graber Y, Meidan R (1997) Seasonal and acute heat stress effects on steroid production by dominant follicles in cows. Anim Reprod Sci 47: 9-19.

11. Roth Z, Arav A, Bor A, Zeron Y, Braw-Tal R, et al. (2001) Improvement of quality of oocytes collected in the autumn by enhanced removal of impaired follicles from previously heat-stressed cows. Reproduction 122: 737-744.
12. Ealy AD, Aréchiga CF, Bray DR, Risco CA, Hansen PJ (1994) Effectiveness of short-term cooling and vitamin E for alleviation of infertility induced by heat stress in dairy cows. J Dairy Sci 77: 3601-3607.
13. Yabuta O, Kunio A (2000) Heat stress in dairy cattle.
14. Mazzucchelli FY, Tesouro MA (2000) Influence of stress on the reproductive efficiency of dairy cattle.
15. Ambrose JD, Drost M, Monson RL, Rutledge JJ, Leibfried-Rutledge ML, et al. (1999) Efficacy of timed embryo transfer with fresh and frozen in vitro produced embryos to increase pregnancy rates in heat-stressed dairy cattle. J Dairy Sci 82: 2369-2376.
16. Varner MA (1998) Stress and reproduction.
17. Dobson H, Tebble JE, Smith RF, Ward WR (2001) Is stress really all that important? Theriogenology 55: 65-73.
18. Zapiola, Marcos (1998) Reducing stress management improves productivity and animal welfare. Prof Anim Sci 14.
19. Dobson H, Smith RF (2000) What is stress, and how does it affect reproduction? Anim Reprod Sci 60-61: 743-52.
20. Chemineau P (1993) Environment and animal breeding.
21. Rutledge JJ (2001) Use of embryo transfer and IVF to bypass effects of heat stress. Theriogenology 55: 105-111.
22. Ryan DP, Blakewood EG, Lynn JW, Munyakazi L, Godke RA (1992) Effect of heat-stress on bovine embryo development in vitro. J Anim Sci 70: 3490-3497.
23. Ealy AD, Drost M, Hansen PJ (1993) Developmental changes in embryonic resistance to adverse effects of maternal heat stress in cows. J Dairy Sci 76: 2899-2905.

14
CHAPTER

NUTRIGENOMICS APPROACHES TO FINE-TUNE METABOLISM AND MILK PRODUCTION

Massimo Bionaz

Ruminant nutrition is highly sophisticated. Immense research and application efforts have culminated in great successes. Today, we can satisfy with a relative high precision the nutritional needs of high producing dairy cows. Highly efficient genetic selection and improvement in overall management has resulted in a national average of ca. 10,000 kg of milk per year in USA. The 2012 USDA annual report indicated an average of 9,842 kg milk produced in one year (or ca. 32 kg/day), with a recent reported record of more than 32,000 kg/year (ca. 90 kg/day!). Dairy cows with a so sophisticated genetic makeup able to deliver extremely high milk yield require a more refined management, including feeding strategies that increase the precision in satisfying all requirements and account for the dynamic physiological adaptation of the animal to lactation and reproduction. This is evident by the increased importance of metabolic-related health problems and a continual decrease in fertility in high-yielding dairy cows. These problems can be addressed by improving overall nutritional management [1], including a strategy to use nutrients that can improve the metabolic adaptation of animals to lactation. As the genetic potential of dairy cows continues to improve, the advances in research for a more accurate feeding strategy are of utmost importance. In addition, there is an urgent need for research to improve efficiency of animal food production in an environmental sustainable way to feed the growing World human population. Not to mention, it is becoming increasingly evident that the quality of the food consumed by humans affects their health and consumers are playing important roles in World food markets.

With respect to research, we are now in a post-genomics era with the advent of systems biology. Today we have the opportunity to study in a holistic way the effect of any condition on the whole organism considering a large amount of molecular information (e.g., whole transcriptome, proteomics, metabolomics, and data from other omics-related technologies). We also have increasing availability of bioinformatics tools to biologically decipher such a large amount of data.

Citation: Bionaz M (2014) Nutrigenomics Approaches to Fine-Tune Metabolism and Milk Production. Adv Dairy Res 2: e107. doi:10.4172/2329-888X.1000e107.

Dietary nutrients alter gene expression either directly or indirectly; thus, affecting protein expression and metabolic and/or signaling status of cells, tissues, organs, and the entire organism. The concept that food components affects biological functions by interacting with transcriptome has revolutionized the field of nutrition in monogastrics [2], but has the potential to affect also the field of ruminant nutrition, and, in the case of dairy cows, to affect efficiency and quality of milk production.

All the above strongly prompt for a new approach to feed dairy animals by taking advantage of the capacity of nutritional molecules to affect the whole biology of the organism. In addition to satisfying the nutritional requirements of the animals, the diet can also modify their biology by altering gene expression. The ration formulation can then be used to "fine-tune" the transcriptome (and, as consequence, the whole biology) of the animal to improve its efficiency in production and fertility as well as its overall well-being. The fine-tuning approach can also be utilized to improve milk quality. To achieve this fine-tuning approach, research is needed to understand the molecular effects of nutritional molecules using systems biology approaches; i.e., systems nutrigenomics.

Nutrigenomics is defined as the study of "the genome-wide influences of nutrition" [3] and how this "affects the balance between health and disease by altering the expression and/or structure of an individual's genetic makeup" [4]. Nutrigenomics in dairy cows is a relatively new area of research [5]. Among all nutritional factors able to affect cell biology by changing gene expression, dietary energy, fatty acids, and amino acids have the strongest potential.

Dietary energy has a powerful and broad effect on the transcriptome [6]. Despite the potential benefit on the metabolism of dairy animals that can be achieved using dietary energy[7], the effect on the transcriptome is broad and is distributed in multiple tissues (thus on the organismal metabolism). Study of the transcriptomics effects of amino acids in ruminants is a new area of investigation. However, recent data seem to support a nutrigenomics effect of amino acids on milk protein synthesis [8]. Transcriptomics effects for fatty acids, particularly long chain fatty acids (LCFA), have been more widely studied. The effect of trans10,cis12-conjugated linoleic acid (t10,c12-CLA) on depressing milk fat synthesis (by decreasing the activity and expression of the nuclear receptor sterol regulatory element binding transcription factor 1, or SREBP1, that controls expression of key lipogenic genes) is the first and the most well-known nutrigenomics effect of LCFA on milk production [9]. However, even though this allows explaining milk fat depression in dairy cows, the practical application for milk synthesis fine tuning is quite limited. To date, the ability to increase milk fat synthesis through nutrigenomics still remains to be determined.

In monogastrics, it has been well established that LCFA can affect expression of genes mainly by their capacity to directly or indirectly (i.e., through their metabolites) bind and activate (or inhibit) transcriptional factors. Among all the transcriptional factors able to bind and be activated by LCFA the peroxisome proliferator-activated receptors (PPARs) are, so far, the more potent. This knowledge in monogastrics let few dairy researchers to start investigating the role of PPARs in ruminants [10]. Data in dairy ruminants support a role of PPARs in ruminant physiology, particularly lipid metabolism including regulation of milk fat synthesis[10]. In addition, there is a strong agonistic effect of LCFA, particularly the saturated fatty acids, in activating PPARs in ruminants. A model involving activation of PPAR isotypes by LCFA in several major tissues at precise physiological stages has been proposed [10]. In order to test such a complex nutrigenomics model, a systems physiology approach, together with sophisticated molecular biology techniques, must be carried out. The technologies to undertake nutrigenomics studies with a systems biology approach are costly and the main limitation for most of the research in these areas is the lack of adequate funding. Despite the high cost, investments in nutrigenomics studies using systems physiology approaches in dairy ruminants have the potential for a very high payoff in production, fertility, and animal well-being.

We are at the frontier of the nutrigenomics era in ruminants. The need to improve nutrition in dairy cows is great, especially considering the real urgency of increasing production by maximizing the limited resources but minimizing the environmental impact. Nutrigenomics can provide the means to effectively face these challenges. Thus, systems nutrigenomics can play a pivotal role in shaping the future of ruminant nutrition.

ACKNOWLEDGEMENTS

The author thanks Dr. Xenophon Markantonatos and Dr. Michelle Kutzler for critical revising the manuscript.

REFERENCES

1. Trevisi E, Bionaz M, Piccioli-Cappelli F, Bertoni G (2006) The management of intensive dairy farms can be improved for better welfare and milk yield. Livestock Science 103: 231-236.
2. Mutch DM, Wahli W, Williamson G (2005) Nutrigenomics and nutrigenetics: the emerging faces of nutrition. FASEB journal : official publication of the Federation of American Societies for Experimental Biology 19: 1602-1616.
3. Muller M, Kersten S (2003) Nutrigenomics: goals and strategies. Nature reviews Genetics 4: 315-322.
4. Kaput J, Rodriguez RL (2004) Nutritional genomics: the next frontier in the postgenomic era. Physiological genomics 16: 166-177
5. Loor JJ, Bionaz M, Drackley JK (2013) Systems Physiology in Dairy Cattle: Nutritional Genomics and Beyond. Annual Review of Animal Biosciences 1: 365-392.
6. Bionaz M, Loor JJ (2012) Ruminant metabolic systems biology: reconstruction and integration of transcriptome dynamics underlying functional responses of tissues to nutrition and physiological state. Gene Regul Syst Bio 6: 109-125.
7. Dann HM, Litherland NB, Underwood JP, Bionaz M, D'Angelo A, et al. (2006) Diets during far-off and close-up dry periods affect periparturient metabolism and lactation in multiparous cows. J Dairy Sci 89: 3563-3577.

8. Nan X, Bu D, Li X, Wang J, Wei H, et al. (2014) Ratio of Lysine to Methionine alters expression of genes involved in milk protein transcription and translation and mTOR phosphorylation in bovine mammary cells. Phisiol Genomics.

9. Bauman DE, Harvatine KJ, Lock AL (2011) Nutrigenomics, rumen-derived bioactive fatty acids, and the regulation of milk fat synthesis. Annu Rev Nutr 31: 299-319.

10. Bionaz M, Chen S, Khan MJ, Loor JJ (2013) Functional Role of PPARs in Ruminants: Potential Targets for Fine-Tuning Metabolism during Growth and Lactation. PPAR Res 2013: 684159.

15

CHAPTER

VALIDATION OF INDICATORS USED TO ASSESS UNCONSCIOUSNESS IN VEAL CALVES AT SLAUGHTER

M. T. W. Verhoeven, M. A. Gerritzen, L. J. Hellebrekers and B. Kemp

KEYWORDS: (un)consciousness, calves, electroencephalogram, reflexes, slaughter

IMPLICATIONS

Monitoring unconsciousness at slaughter is mandatory by European legislation and is often determined by the absence of behavioural indicators (i.e. loss of posture), physical signs (i.e. rhythmic breathing) and reflexes. Previous research in sheep has indicated that the absence of certain reflexes at slaughter is not a reliable indicator of unconsciousness. The present study showed that absence of the withdrawal- and threat reflex did not indicate unconsciousness in veal calves subjected to non-stunned slaughter and are therefore not recommended for assessment of unconsciousness. Absence of the eyelid- and corneal reflex, on the other hand, reflected unconsciousness well and are considered acceptable for assessment of unconsciousness during non-stunned slaughter.

INTRODUCTION

Animals are stunned before slaughter to minimize pain and distress as required by European law (Council Regulation (EC) No 1099/2009, 2009). Stunning, however, is not always compatible with specific methods of slaughter prescribed by religious rites and is legally determined not to be mandatory in animals subjected to this type of slaughter (Council Regulation (EC) No 1099/2009, 2009). Stunning after neck cutting is practiced in some European countries in preference to no stunning at all to avoid potential animal welfare problems (Farouk, 2013). Assessment of unconsciousness is a legal requirement during the slaughter process in Europe irrespective of the method used (Council Regulation (EC) No 1099/2009, 2009). During non-stunned slaughter, animals

Citation: M. T. W. Verhoeven, M. A. Gerritzen, L. J. Hellebrekers and B. Kemp. Validation of indicators used to assess unconsciousness in veal calves at slaughter. animal, available on CJO2016. doi:10.1017/S1751731116000422.

are restrained and bled by a transverse incision of the neck, severing the skin, muscles, trachea, oesophagus, carotid arteries, jugular veins and major nerves. The severe decrease in cerebral blood flow is intended to lead to a rapid onset of disorganized brain function and unconsciousness (Mellor *et al.*, 2009). In comparison to sheep, consciousness after the neck cut is prolonged in cattle, because the vertebral arteries, which are not severed by the neck cut, continue to supply blood to the circle of Willis and provide some blood to the brain (Baldwin and Bell,1963). There is considerable variation between cattle in time to loss of consciousness after the neck cut. Some studies suggest a rapid loss of consciousness (4 to 7 s) in cattle after the neck cut (Nangeroni and Kennett,1963; Bager *et al.*, 1992) whereas other studies report a spontaneous loss of brain activity 75±48 s after the neck cut (range 19 to 113 s) and the possibility of an intermittent return to consciousness for up to 123 to 323 s after slaughter (Newhook and Blackmore, 1982; Daly *et al.*, 1988). There is substantial debate about which indicators most adequately assess unconsciousness at slaughter and which merit further investigation (European Food and Safety Authority (EFSA), 2013). The use of recorded brain activity (as presented in an electroencephalogram (EEG)) is considered to be the most objective method available for assessing unconsciousness and is generally accepted as the current 'gold standard' (Erasmus *et al.*, 2010; EFSA, 2013; Verhoeven *et al.*, 2015a). Due to the complexity of collecting EEG data at slaughter, absence of behavioural indicators (i.e. loss of posture) physical signs (i.e. rhythmic breathing) and reflexes (i.e. righting-, threat-, withdrawal-, corneal- and eyelid reflex) is often used to assess unconsciousness. However, there is a lack of scientific publications reporting relationships between unconsciousness ascertained by EEG recordings with those assessed using behavioural indicators, physical signs or reflexes that could be used in slaughterhouse conditions in bovines. It has been formally agreed in the Netherlands that animals subjected to non-stunned slaughter should be unconscious within 40 s, based on absence of at least three of the following five indicators: (1) threat reflex; (2) withdrawal reflex; (3) corneal reflex; (4) (spontaneous) eyelid reflex; (5) rhythmic breathing (Staatscourant, 2012). Different studies showed that the corneal- and eyelid reflex and rhythmic breathing were lost long after animals were considered unconscious based on EEG recordings, making them distinctly conservative indicators when assessing unconsciousness in non-stunned slaughtered sheep (Verhoeven *et al.*, 2015b) and veal calves (Lambooij *et al.*, 2012). On the contrary, both threat- and withdrawal reflex were often lost before sheep were unconscious according to EEG recordings during non-stunned slaughter (Verhoeven *et al.*, 2015b). The objective of the current study was to assess absence and presence of the following indicators: threat-, withdrawal-, corneal- and eyelid reflex, and determine their relationship to (un)consciousness as identified by EEG recordings during stunned and non-stunned slaughter in veal calves.

MATERIAL AND METHODS

Approval

This study was approved (DEC 2014045.b) by the Ethical Committee of the Animal Sciences Group of Wageningen UR, The Netherlands and in compliance with Dutch legislation.

Experimental Set-up

On seven occasions, calves subjected to (non)stunned slaughter were observed at an abattoir in the Netherlands in the period from September to November 2014. Eighty-two cattle of mixed breeds (warm carcass weight 201±22 kg) were randomly selected from groups in lairage and held individually in a rotation box (Nawi, Borculo, the Netherlands) while equipped with non-invasive EEG electrodes. Calves were randomly assigned to one of the following four treatments: (1) Captive bolt stunning (Cash Magnum 9000 s) followed by neck cut within 30 s in an inverted position (180° rotation) (n=25); (2) Non-stunned slaughter in an upright position (n=7); (3) Non-stunned slaughter in an inverted position (180° rotation) (n=25); (4) Non-stunned slaughter in an upright position followed by captive bolt stunning 40 s after the neck cut (n=25). Treatment group 2 was limited to seven calves because of animal welfare concerns: that is delayed induction of unconsciousness due to insufficient bleeding. Calves were restrained with a head yoke and chin lift, but without a belly supporting plate. The neck cut was performed as a clean transverse incision of the neck, severing the skin, muscles, trachea, oesophagus, carotid arteries, jugular veins and major nerves. All stunning and slaughter procedures were performed by the same skilled halal slaughter man.

Electroencephalogram Recordings

After being positioned in the restrainer, the head of the animal was shaved with an electrical trimmer (Aesculap favorita II GT104, Braun Suhl GmbH, Germany) to enable placement of EEG electrodes. Thereafter, an elastic halter was placed around the calves head. A cross-shaped piece of cotton was attached to the halter containing Velcro parts at each end of the cross. A 'shooting-hole' was cut in the middle of the cross, through which the captive bolt could be administered. Five rubber sensor carriers were placed around the 'shooting-hole' (Figure 1a and b). Five Ag/Cl pellet electrodes (Twente Medical Systems International (TMSi), Oldenzaal, The Netherlands) were placed in the rubber sensor carriers to make contact with the shaved skin of the calf. Each electrode was wrapped in a small sponge soaked in saline solution, which served as interface with the skin. One electrode was placed over the frontal bone, on the sagittal midline on a line extending between the base of both ears. Two electrodes were placed 2 cm left and right from the sagittal midline and 3 cm frontal to the first electrode, the other two electrodes were placed 2 cm left and right from the sagittal midline and 6 cm frontal to the first electrode. An elastic band was wrapped once around the calves' head to secure the upper three electrodes, but leaving the ears and eyes free to move. All electrodes were connected via a 140 cm active shielded cable to the 32-channel Porti recording system (TMSi). Porti uses bipolar amplifier technology with high input impedance (>1 GΩ) that amplifies the potential difference between pairs of electrodes. The input amplifier is dimensioned as a multichannel instrumentation amplifier. Electrode impedance was <5 kΩ. The EEG was displayed with a band pass filter of 0.5 and 30 Hz, respectively, and raw, unfiltered, data were saved directly onto a computer. Sample rate was set at 0.5 kHz. Once the electrodes had been positioned correctly and a good live signal had been established, baseline EEG activity was recorded for at least 2 min. At the beginning of the slaughter procedure (T=0 s) calves were stunned (treatment 1) or neck cut in an upright position (treatment 2, 4) or inverted position (treatment 3). Calves of treatment 4 were captive bolt stunned 34±8 s after the neck cut. The EEG was recorded until a flat EEG (<10% baseline amplitude) was observed and all reflexes showed absence of response that was repeated sequentially three times. Hereafter, the equipment was carefully removed from the calf after which it was released from the restrainer to allow further processing of the carcass.

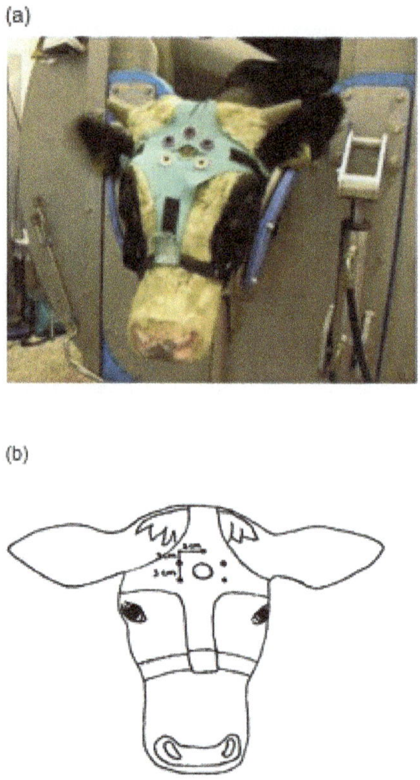

Figure 1: (a, b) Electroencephalogram (EEG) electrode placement on a veal calf.

Measurements around Stunning and Slaughter

The threat-, withdrawal-, corneal- and eyelid reflex were verified once before the slaughter treatment started. The threat reflex was assessed by checking the presence of a blinking reaction to an abrupt movement of the index finger towards the eye ball without touching the eye. The withdrawal reflex was studied by pinching the calves nose between two fingertips and determining whether or not it responded with a withdrawal of the nose or head. In case of non-stunned slaughter only withdrawal of the nose could be observed after the neck cut. The corneal reflex was assessed by a gentle touch of the cornea with the index finger, and the presence/absence of a blinking reaction was recorded. The eyelid reflex was assessed by a gentle touch of the eyelid, and the presence/absence of a blinking reaction was recorded. Reflexes were assessed every 20 s from T=15 s onwards for all treatments until all reflex tests showed a negative response that was repeated sequentially three times and a flat EEG was recorded. In addition, reflexes were assessed 5 s after captive bolt stunning in calves of treatment 1 and 4. Reflex tests were performed in a random order per calf.

Data Analyses

EEG data were displayed, stored and analysed using PolyBench software (TMSI). EEG activity (amplitude and frequency) of each calf was visually assessed to determine the start of the following stages: baseline, transitional, unconscious and minimal brain activity. Figure 2 shows an example of a representative series of 5 s of EEG registration during these different stages. The baseline stage consisted of a low amplitude, high frequency signal, indicating alert calves (Figure 2a). The transitional stage is characterized by low frequency, high amplitude (firing of neurons in a synchronized fashion) becoming more apparent compared with baseline, associated with reduced consciousness (Figure 2b, McKeegan et al., 2013). When low frequency, high amplitude dominated the EEG trace this was interpreted to indicate unconsciousness (Figure 2c, Baars et al., 2003). Minimal brain activity was reflected by a flat signal (<10% of baseline amplitude, Figure 2d). Time to loss of a reflex was defined as the first time at which a reflex showed a negative response that was repeated sequentially three times, since loss and return of a reflex was observed in some calves.

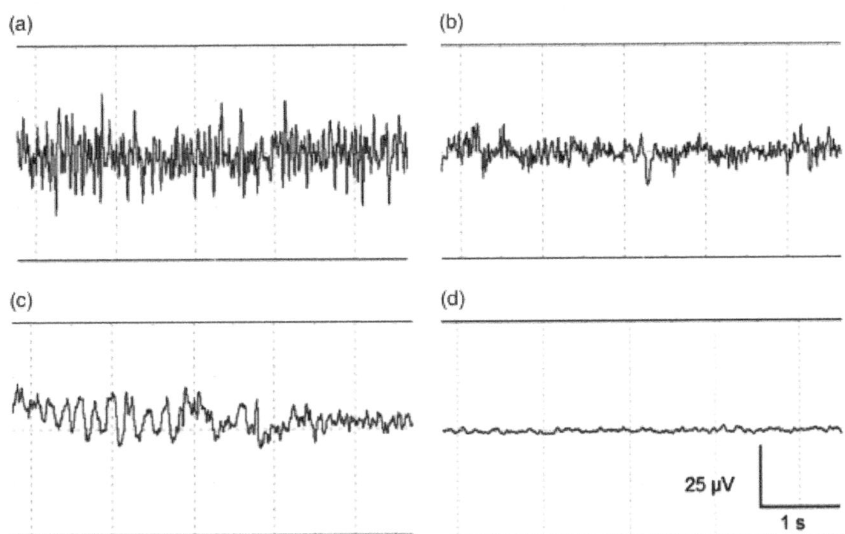

Statistical Analyses

To calculate sample size, a power analysis was performed using the Power Calculation version 2.03 programme, designed by van Wilgenburg and van Schaick Zillesen, 2003 version. Results indicated that a sample of at least 10 calves per treatment would be required to detect a difference of 20 s in loss of consciousness with 90% power using a t-test between means with α at 0.05. The sample size was increased to 25 calves per treatment to allow for possible loss of EEG recordings. Statistical analyses were performed using SAS (version 9.3; SAS Institute, 2004). Normality of variance was checked (PROC UNIVARIATE) for all the variables examined. Before the analyses, the variables 'time to lost eyelid reflex' and ' time to lost corneal reflex' were log transformed to normalize the variance within these variables. In order to determine whether or not non-stunned slaughter in an upright or inverted position had a significant effect on time to onset of different EEG stages and time to loss of reflexes, data were analysed using a GLM procedure according to the model:

$$Y_i = T_i + \varepsilon_i$$

where T_i is the treatment (2 and 3), and ε_{ij} the residual error term. Stunned slaughter (treatments 1 and 4) was not taken into account in this procedure, since unconsciousness was immediately induced after stunning in these animals. Differences were considered to be significant at the 5% probability level.

RESULTS

Animals

The total number of calves used in this study was 82. In one calf from treatment 4, the EEG signal was lost after stunning and consequently assessment of loss of consciousness was impossible. Data from this calf was excluded from further analyses.

Electroencephalogram Activity

A clear EEG signal was obtained from all calves during baseline recordings. Based on EEG recordings, calves stunned with a captive bolt (treatment 1) were unconscious 1 ± 0 s after stunning. No transitional stage was observed in these calves. One calf, however, lost consciousness 11 s after stunning based on EEG recordings. Minimal brain activity was observed in all calves from 22 ± 19 s after stunning. Calves of treatments 2 and 3 developed a transitional EEG based on EEG recordings 54 ± 13 and 27 ± 15 s, respectively, after the neck cut. Calves of treatments 2 and 3 were considered unconscious based on EEG recordings 109 ± 32 and 49 ± 25 s, respectively, after the neck cut. Calves of treatments 2 and 3 had minimal brain activity based on EEG recordings 157 ± 50 and 86 ± 34 s, respectively, after the neck cut.

Figure 3 represents the percentage of calves displaying different EEG stages over time subjected to non-stunned slaughter in an upright position (treatment 2) or an inverted position (treatment 3) at T = 0 s. The time span for 80% or 100% of the calves to reach a transitional EEG, unconsciousness or minimal brain activity is displayed in Table 1. Eighty per cent of the calves slaughtered in an upright position (treatment 2) lost consciousness on average 61 s later than calves slaughtered in an inverted position (treatment 3). All the calves slaughtered in an upright position (treatment 2) lost consciousness on average 31 s later than calves slaughtered in an inverted position (treatment 3). Of the treatment 2 calves, 20% had a transitional EEG, none were considered unconscious and none had minimal brain activity based on EEG recordings 40 s after the neck cut. Of the treatment 3 calves, 81% of the calves had a transitional EEG, 46% were considered unconscious and 8% had minimal brain activity based on EEG recordings 40 s after the neck cut. Treatment 4 calves (non-stunned slaughter in an upright position followed by captive bolt stunning) were stunned on average 34 ± 8 s after the neck cut. At stunning, 67% of the calves were conscious, 8% of the calves had a transitional EEG, 21% of the calves were unconscious and 4% of the calves displayed minimal brain activity. All calves that were conscious or had a transitional EEG before stunning were unconscious 1 ± 0 s after stunning based on EEG recordings.

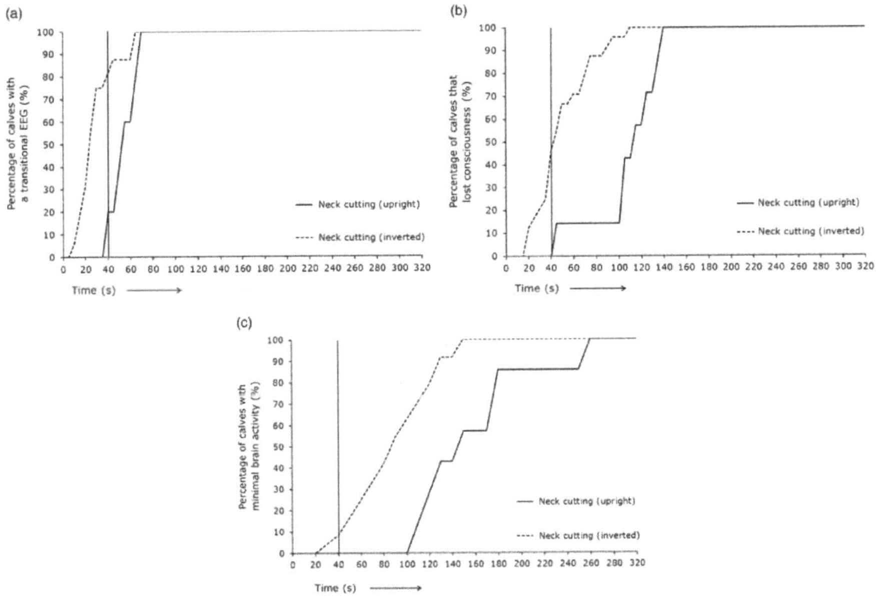

Loss of Reflexes for Unconsciousness

Reflexes were tested and present in all calves during baseline recordings. In calves of treatment 1, threat-, withdrawal-, corneal- and eyelid reflex were all permanently lost at the first testing moment (T = 5 s) after captive bolt stunning. Figure 4a presents the range of individual time points at which the different reflexes were lost in relation to onset of the different EEG stages (T = 0 min), based on visual assessment of EEG recordings during non-stunned slaughter of calves in an upright position (treatment 2). In all treatment 2 calves, the threat-, and withdrawal reflex were both lost before EEG-based loss of consciousness had been observed. The corneal-, and eyelid reflex were both lost in all calves of treatment 2 after EEG-based loss of consciousness was observed. Corneal- and eyelid reflex were lost in 71% and 86%, respectively, of calves of treatment 2 after onset of minimal brain activity. Figure 4b represents the range of individual time points at which the different reflexes were lost in relation to onset of the different EEG stages (T = 0 min), based on visual assessment of EEG recordings during non-stunned slaughter of calves in an inverted position (treatment 3). In calves of treatment 3, the threat- and withdrawal reflex were lost in 88% and 92% respectively, of the animals before EEG-based loss of consciousness was observed. Both corneal- and eyelid reflex were lost in 92% of calves of treatment 3 after EEG-based loss of consciousness had been observed. The threat- and withdrawal reflex were lost in 92% and 96%, respectively, of calves of treatment 3 before minimal brain activity was observed.

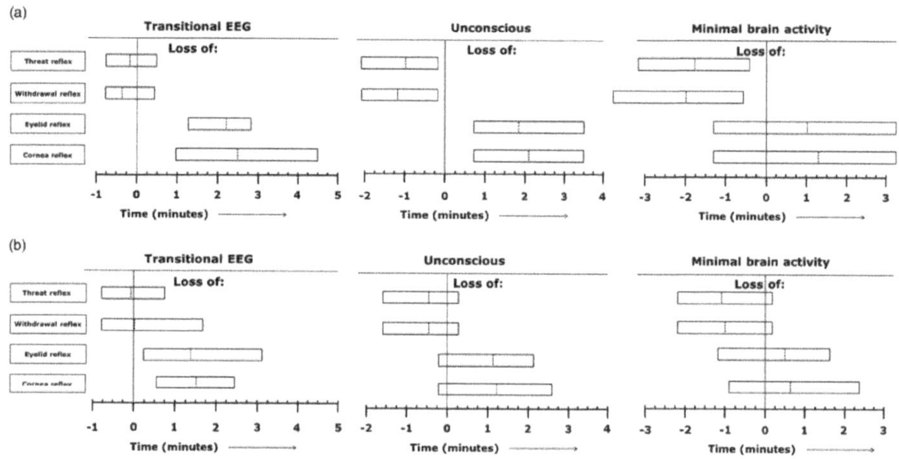

Figure 4: (a, b) Range of individual time points at which the different reflexes are lost in relation to start of the different electroencephalogram (EEG) stages (T = 0 min) based on visual assessment of EEG recordings during non-stunned slaughter of calves in an upright position (n = 7) (Figure 4a) and an inverted position (n = 25) (Figure 4b). Dotted lines indicate average values of all observations and left and right vertical lines of each block indicate minimum and maximum values for all observations.

The corneal- and eyelid reflex were lost in 88% and 67%, respectively, of calves of treatment 3 after minimal brain activity had been observed. Calves of treatments 2 and 3 lost the corneal reflex 235 ± 86 and 123 ± 34 s, respectively, after the neck cut. Calves of treatments 2 and 3 lost the eyelid reflex 218 ± 65 and 115 ± 36 s, respectively, after the neck cut. All calves of treatment 4, lost both the threat- or withdrawal reflex before EEG-based loss of consciousness. The corneal- or eyelid reflex were lost in 89% and 87%, respectively, of the calves after EEG-based loss of consciousness. The threat-, withdrawal-, corneal- and eyelid reflex were all permanently lost 5 s after the captive bolt stunning.

DISCUSSION

The main aim of this study was to provide insight into the validity of certain reflexes that are used to assess unconsciousness at slaughter in veal calves, that is the threat-, withdrawal-, corneal- and eyelid reflex. Brain activity, as presented in an electroencephalogram or EEG, served as a 'gold standard' for our assessment of unconsciousness. Consciousness involves the cerebral cortex and thalamus, together forming the thalamocortical complex, which is regulated by the brainstem. The central core of the brainstem is formed by the reticular formation, a large network of neural tissue located in the central region of the brain stem. Neuronal projections from the reticular formation to the cortex are called the ascending reticular activating system (ARAS). A good-functioning reticular formation and ARAS are essential for the maintenance of consciousness and damage to (one of) these regions causes rapid loss of consciousness. Unconsciousness induced by penetrating captive bolt stunning is caused by a combination of the concussive force generated by impact of the bolt on the cranium and irreversible physical damage produced by passage of the bolt through the brain. Generally, the pistol is placed on the forehead after which the bolt penetrates the skull to target the brain stem. Immediate loss of consciousness and reflexes should occur from the resulting trauma to the cerebral hemisphere and brain stem (Terlouw et al., accepted a). Depending on the method of stunning, the EEG shows a characteristic pattern of change when animals lose consciousness. Generally, an increase in low frequency activity is accompanied by an increase in amplitude. When neurons depolarize at the same time or frequency, they fire in a synchronized fashion creating slow high amplitude waves as seen in unconscious states suggesting a depression of the reticular formation (Lopes da Silva, 1983). EEG recordings in captive bolt stunned animals displayed high-amplitude, slow-frequency waves after stunning, indicating immediate unconsciousness, similar to findings of Lambooy and Spanjaard (1981) and Zulkifli et al. (2014). One calf in the current study displayed a period of baseline-like EEG activity, lasting for 11 s after captive-bolt stunning. This calf displayed a permanent loss of all reflexes from 5 s after stunning. Reflexes in the conscious calf before the beginning of the slaughter procedure were always present and it is therefore unlikely that these reflexes were lost, though the calf was conscious. The prolonged period of baseline-like EEG activity may have been caused by incorrect positioning of the stunner, a deflected shot, incorrect functioning of the stunning device itself or an ineffective EEG recording. Effectively shot cattle should collapse immediately due to damage to the reticular formation that also plays a role in maintaining posture (Laureys and Tononi, 2009). In our study, an immediate collapse was observed in some of the calves, but tight restraining of the calf often prevented collapse and made observation of this criterion difficult. Five seconds after captive bolt stunning all reflexes were permanently lost in all calves. Rhythmic breathing and the righting reflex are considered indicative of a (partial) return of consciousness and thus that the stunning has not been irreversible (Grandin, 2002). Both rhythmic breathing and the righting reflex could not be assessed because the animals were fully restrained in a rotatory box.

Non-stunned animals that are subjected to neck cutting lose consciousness gradually as a consequence of widespread anoxia in the brain that results in progressive dysfunction of cortical and subcortical structures (Terlouw et al., accepted b). During our experiment, calves in an inverted or upright position lost consciousness, based on EEG recordings, 49 ± 25 and 109 ± 32 s, respectively (mean ± SD), after the neck cut. Previous studies in cattle indicated loss of consciousness ranging from 10 s up to over 3 min after the neck cut (Daly et al., 1988; Gregory et al., 2010; Bourguet et al. 2011). Newhook and Blackmore (1982) suggested the possibility of an intermittent return to consciousness for up to 123 to 323 s after non-stunned slaughter in cattle. Detailed study of EEG recordings did not indicate signs of resurgence of consciousness in any of the calves in our study. The corneal- and eyelid reflex, however, were lost and regained in 9% and 16%, respectively, of the non-stunned calves (n = 32) during bleeding. Loss and return of the corneal reflex has been observed by Hoffman in an unspecified number of cattle (Hoffman, 1900 cited by Gregory et al. 2010). Bourguet et al. (2011) also observed a temporary return of the palpebral and corneal reflex in some of the non-stunned slaughtered cattle. A temporary return of blood pressure would be necessary for these reflexes to return and may possibly be explained by physiological changes, such as an increased heart rate, to alleviate the drop in blood pressure (Newhook & Blackmore, 1982; Vimini et al., 1983). There is no clear indication why reflexes were regained and lost again in some of the calves in the current experiment. Because this experiment took place at a commercial slaughter plant, the slaughter man was empowered to decide when an additional cut was necessary. No clear criteria were defined for this. In total 15 of 82 (18%) calves in our study were cut a second time and one of those 15 calves was cut a third time. The arteries of an animal in which poor bleeding is observed, are in practice often re-cut to allow adequate bleeding. Poor bleeding may be caused by occlusion, which is characterized by retraction and contraction of the elastic portion of the arterial wall and thrombus formation around the severed end of the vessel (Gregory et al., 2006). The prevalence of carotid arterial occlusion in 576 cattle slaughtered at abattoirs in the United Kingdom was found to be 16% and 25%, respectively, for adult cattle and bobby calves (Gregory et al., 2006). A study by Gregory et al. (2010) showed that 71% of cattle taking longer than 75 s to collapse, had a false aneurysm in the cardiac end of the carotid that was at least 3 cm in diameter. There was no further investigation into the background of this obstruction, but it may provide an explanation for the large variation in time to loss of consciousness between calves.

Duration to loss of consciousness was 50 s longer in non-stunned veal calves restrained upright than by those restrained in an inverted position. During this experiment, all calves were slaughtered in the same rotation box which normally is used for inverted slaughter. Hence, slaughter using a box designed for upright restraint may provide different results. Blood flow in calves slaughtered upright was often obstructed when the head fell limp against the head restrainer. Based on this observation and in consideration of the potential risk to animal welfare, the number of veal calves slaughtered non-stunned in an upright position was limited to seven calves. Inversion of cattle for slaughter in itself is widely debated and thought to cause distress because of a longer delay between entering the restrainer and full restraint. In addition there is discomfort from the inverted position resulting in hypoxemia and rumen pressure on the diaphragm (Dunn, 1990; Petty et al., 1994; Tagawa et al., 1994). Another potential welfare problem is aspiration of blood and refluxing gut content after neck cutting in nonstunned animals (EFSA, 2004). Though the latter problem is mainly associated with the inverted position, it has also been observed in non-stunned slaughtered calves in the upright position (Grandin and Regenstein, 1994; Gregory et al., 2009). A report by the Farm Animal Welfare Committee (2012) stated that there is little evidence of any welfare advantages of inversion in terms of speed and efficiency of the cut. Visits to 18 European cattle slaughter plants showed that the mean number of cuts was higher for cattle restrained in an upright position than those in other restraining systems (Dialrel, 2010). Although others claim, that the angle of the head in inverted cattle allows for a more effective downward cut compared to the upward cut in animals in an upright position (Slaughter of Animals (prevention of cruelty) Regulations 1958 – quoted by Dunn, 1990). The present study is, to our knowledge, the first to suggest that the neck cut may be more effective in terms of duration to loss of consciousness in non-stunned slaughtered animals in an inverted position.

Reflexes in non-stunned slaughtered animals were not lost immediately, as in captive bolt stunned animals, but disappeared gradually. Generally, the threat reflex and withdrawal reflex were lost first, followed by the cornealand eyelid reflex. Loss of reflexes in this order may be dependent on the difference in resistance to anoxia by different parts of the brain. Functions that rely on integration of cortical processes are more susceptible to anoxia than those originating from lower brain structures, such as the brain stem (Hansen, 1985). On average, the threat reflex was lost in non-stunned calves before EEG-based unconsciousness. The rapid loss of the threat reflex can possibly be explained by the sudden drop in blood flow after neck cutting causing hypoxia of the brain cells involved (Bourguet et al., 2011). A similar phenomenon was also observed for the withdrawal reflex that was lost on average before EEG-based loss of consciousness. When the neck is cut, the knife transects veins, arteries and other blood vessels, skin, muscle, trachea, oesophagus, sensory- and motor nerves, and connective tissue. Transecting these soft tissues will send a considerable amount of neural impulses to the brain (Gregory, 2004). The massive stimulation of all sensory nerves after the neck cut may overrule the neural input from other, milder sources of pain, such as pinching the nose. Opinions on pain perception during neck cutting are divided. Some (e.g. Grandin, 1994; Rosen, 2004) state that use of an exquisitely sharp knife, will prevent pain after the neck cut. Others (e.g. Gibson et al., 2007 and 2009; Mellor et al., 2009) state that the ventral neck cut is a noxious stimulus and that there is strong evidence that this cut would be perceived as painful from the time of the neck cut until loss of consciousness. Until now, neurophysiological methodology has not provided a final and definitive answer to this issue.

Both the corneal- and eyelid reflex require a functional afferent cranial nerve V (trigeminal) and efferent cranial nerve VII (facial) and the relevant eye muscles to function adequately (Adams and Sheridan, 2008). The connections between the spinal trigeminal- and facial nuclei pass through the reticular formation and failure of these reflexes often indicates a wider dysfunction of the brain, comprising part of the reticular formation, and thus a state of unconsciousness (Terlouw et al., accepted b). In the present study, the corneal reflex in non-stunned slaughtered calves was lost on average 85 ± 45 s after EEG-based loss of consciousness and on average 47 ± 63 s after minimal brain activity was observed. Lambooij et al. (2012) and Newhook and Blackmore (1982) observed that non-stunned slaughtered calves lost the corneal reflex up to 1 min after calves were unconscious (EEG-based) and this reflex was even observed in some calves after all EEG activity had been lost. Verhoeven et al. (2015b) observed that non-stunned slaughtered sheep lost the corneal reflex on average 00:59 ± 00:17 min (mean ± SD) after sheep were considered to be unconscious based on EEG recordings. The eyelid reflex in non-stunned slaughtered calves was lost on average 76 ± 50 s after EEG-based loss of consciousness and 37 ± 59 s after minimal brain activity was observed. Other studies report the presence of an eyelid reflex in unconscious sheep (Verhoeven et al., 2015b) and cattle (Newhook and Blackmore, 1982) and also during anaesthesia. The eyelid reflex is known to be present after an animal is considered unconscious (Dugdale, 2010). The order in which the eyelid- and corneal reflex cease may differ, providing uncertainty which reflex is more resistant to the effects of anoxia (Bourguet et al., 2011; Terlouw et al., accepted b). The results from this study support the hypothesis that a positive corneal- or eyelid reflex alone does not necessarily indicate consciousness, since positive brain stem reflexes might occur on the basis of residual brain stem activity and do not distinguish clearly between consciousness and unconsciousness (Anil, 1991). Consequently, a negative corneal- or eyelid reflex is considered a general indication of unconsciousness (Laureys, 2005; Terlouw et al., accepted b).

CONCLUSIONS

In both stunned and non-stunned calves, unconsciousness was induced and absence and presence of reflexes was successfully linked to EEG activity. The use of reflexes to assess unconsciousness can more easily be applied in daily practice, and EEG recordings can be used to validate the interpretation of such reflexes. After captive bolt stunning, absence of threat-, withdrawal-, corneal- and eyelid reflex indicated unconsciousness as determined by EEG recordings. On average, after non-stunned slaughter, the threat- and withdrawal reflex were lost before calves were considered unconscious based on EEG recordings. The eyelid- and corneal reflex were on average lost after calves were considered unconscious based on EEG recordings and appeared to be distinctly conservative indicators of unconsciousness in non-stunned slaughtered calves since they were observed up to 76 ± 50 and 85 ± 45 s (mean ± SD) respectively after EEG-based onset of unconsciousness.

ACKNOWLEDGEMENTS

The authors would like to thank Vincent Hindle and Samantha Surquin for their support while collecting the data. Jan Peuscher is acknowledged for his technical assistance while collecting the data and his help during data analyses. The authors also acknowledge the owner and employees of the slaughter plant for their assistance and hospitality during the study. Funding for this project was provided by the Dutch Ministry of Economic Affairs (KB-12-006.01-002).

REFERENCES

1. Adams, DB and Sheridan, AD 2008. Specifying the risks to animal welfare associated with livestock slaughter without induced insensibility. Retrieved on 19 November 2013 from http://www.australiananimalwelfare.com.au/app/webroot/files/upload/files/animal-welfare-livestock-slaughter.pdf
2. Anil, MH 1991. Studies on the return of physical reflexes in pigs following electrical stunning. Meat Science 30, 13–21.
3. Baars, BJ, Ramsøy, TZ and Laureys, S 2003. Brain, conscious experience and the observing self. Trends in Neurosciences 26, 671–675.
4. Bager, F, Braggins, T, Devine, C, Graafhuis, A, Mellor, D, Tavener, A and Upsdell, M 1992. Onset of insensibility at slaughter in calves: effects of electroplectic seizure and exsanguination on spontaneous electrocortical activity and indices of cerebral metabolism. Research in Veterinary Science 52, 162–173.
5. Baldwin, B and Bell, F 1963. The effect of temporary reduction in cephalic blood flow on the EEG of sheep and calf. Electroencephalography and Clinical Neurophysiology 15, 465–473.
6. Bourguet, C, Deiss, V, Tannugi, CC and Terlouw, E 2011. Behavioural and physiological reactions of cattle in a commercial abattoir: relationships with organisational aspects of the abattoir and animal characteristics. Meat Science 88, 158–168.
7. Council Regulation (EC) No 1099/2009 2009. Council Regulation No1099/2009 on the protection of animals at the time of killing. Official Journal of the European Union L303, 1–30.
8. Daly, C, Kallweit, E and Ellendorf, F 1988. Cortical function in cattle during slaughter: conventional captive bolt stunning followed by exsanguination compared with shechita slaughter. The Veterinary Record 122, 325–329.
9. Dialrel 2010. WP2.2. Religious slaughter: Evaluation of current practices. Retreived on 4 September 2015 from http://dialrel.eu/images/factsheet-practice.pdf
10. Dugdale, A 2010. Veterinary anaesthesia: principles to practice. Blackwell Publishing Ltd, Chichester, West-Sussex, UK.
11. Dunn, C 1990. Stress reactions of cattle undergoing ritual slaughter using two methods of restraint. Veterinary Record 126, 522–525.
12. Erasmus, MA, Turner, PV and Widowski, TM 2010. Measures of insensibility used to determine effective stunning and killing of poultry. Journal of Applied Poultry Research 19, 288–298.
13. European Food and Safety Authority (EFSA) 2004. Welfare aspects of stunning and killing methods. Scientific report of the scientific panel of animal health and welfare on a request from the commission. Question EFSA Q 2003-093. Adopted on the 15th June 2004. Bruxelles, Belgium.

14. European Food and Safety Authority (EFSA) 2013. Scientific opinion on monitoring procedures at slaughterhouses for bovines. EFSA Journal 11, 3460–3525.

15. Farm Animal Welfare Committee (FAWC) 2012. FAWC advice on cattle inversion for religious slaughter (independent report). Retrieved on 26 January 2015 from https://www.gov.uk/government/uploads/system/uploads/attachment_data/file/325002/FAWC_advice_on_cattle_inversion_for_religious_slaughter.pdf

16. Farouk, MM 2013. Advances in the industrial production of halal and kosher red meat. Meat Science 95, 805–820.

17. Gibson, TJ, Johnson, CB, Murrell, JC, Hulls, CM, Mitchinson, SL, Stafford, KJ, Johnstone, AC and Mellor, DJ 2009. Electroencephalographic responses of halothane anaesthetised calves to slaughter by ventral-neck incision without prior stunning. New Zealand Veterinary Journal 57, 77–83.

18. Gibson, TJ, Johnson, CB, Stafford, KJ, Mitchinson, SL and Mellor, DJ 2007. Validation of the acute electroencephalographic responses of calves to noxious stimulus with scoop dehorning. New Zealand Veterinary Journal 55, 152–157.

19. Grandin, T 1994. Euthanasia and slaughter of livestock. Journal of the American Veterinary Medical Association 204, 1354–1360.

20. Grandin, T 2002. Return-to-sensibility problems after penetrating captive bolt stunning of cattle in commercial beef slaughter plants. Journal of the American Veterinary Medical Association 221, 1258–1261.

21. Grandin, T and Regenstein, JM 1994. Religious slaughter and animal welfare: a discussion for meat scientists. Meat Focus International 3, 115–123.

22. Gregory, NG 2004. Physiology and behaviour of animal suffering (UFAW Animal Welfare Series). Blackwell Publishing, Oxford, UK.

23. Gregory, N, Fielding, H, Von Wenzlawowicz, M and Von Holleben, K 2010. Time to collapse following slaughter without stunning in cattle. Meat Science 85, 66–69.

24. Gregory, N, Shaw, F, Whitford, J and Patterson-Kane, J 2006. Prevalence of ballooning of the severed carotid arteries at slaughter in cattle, calves and sheep. Meat Science 74, 655–657.

25. Gregory, N, Wenzlawowicz, Mv and Von Holleben, K 2009. Blood in the respiratory tract during slaughter with and without stunning in cattle. Meat Science 82, 13–16.

26. Hansen, AJ 1985. Effect of anoxia on ion distribution in the brain. Physiological Reviews 65, 101–148.

27. Lambooij, E, van der Werf, J, Reimert, H and Hindle, V 2012. Restraining and neck cutting or stunning and neck cutting of veal calves. Meat Science 91, 22–28.

28. Lambooy, E and Spanjaard, W 1981. Effect of the shooting position on the stunning of calves by captive bolt. The Veterinary Record 109, 359–361.

29. Laureys, S 2005. Science and society: death, unconsciousness and the brain. Nature Reviews. Neuroscience 6, 899–909.

30. Laureys, S and Tononi, G 2009. The neurology of consciousness: cognitive neuroscience and neuropathology, 1st edition. Academic Press, London, UK.

31. Lopes da Silva, FH 1983. The assessment of unconsciousness: general principles and practical aspects. In Stunning of animals for slaughter (ed. G Eikelenboom), pp. 3–12. Martinus Nijhoff Publishers, The Hague, the Netherlands.

32. McKeegan, D, Reimert, H, Hindle, V, Boulcott, P, Sparrey, J, Wathes, C, Demmers, T and Gerritzen, M 2013. Physiological and behavioral responses of poultry exposed to gas-filled high expansion foam. Poultry Science 92, 1145–1154.

33. Mellor, D, Gibson, T and Johnson, C 2009. A re-evaluation of the need to stun calves prior to slaughter by ventral-neck incision: an introductory review. New Zealand Veterinary Journal 57, 74–76.

34. Nangeroni, L and Kennett, P 1963. An electroencephalographic study of the effect of shechita slaughter on cortical function in ruminants. Ithaca University, Ithaca.

35. Newhook, JC and Blackmore, DK 1982. Electroencephalographic studies of stunning and slaughter of sheep and calves: part 2 – the onset of permanent insensibility in calves during slaughter. Meat Science 6, 295–300.

36. Petty, D, Hattingh, J, Ganhao, M and Bezuidenhout, L 1994. Factors which affect blood variables of slaughtered cattle. Journal of the South African Veterinary Association 65, 41–45.

37. Rosen, S 2004. Physiological insights into Shechita. Veterinary Record 154, 759–765.

38. Staatscourant 2012. Convenant onbedwelmd slachten volgens religieuze riten. Retrieved on 13 January 2015 from http://www.rijksoverheid.nl/documenten-en-publicaties/convenanten/2012/06/05/convenant-onbedwelmd-slachten-volgens-religieuze-riten.html

39. Tagawa, M, Okano, S, Sako, T, Orima, H and Steffey, EP 1994. Effect of change in body position on cardiopulmonary function and plasma cortisol in cattle. The Journal of Veterinary Medical Science/the Japanese Society of Veterinary Science 56, 131–134.

40. Terlouw, EMC, Bourguet, C and Deiss, V (accepted a). Consciousness, unconsciousness and death in the context of slaughter. Part 1. Neurobiological mechanisms underlying stunning and killing. Meat Science.

41. Terlouw, EMC, Bourguet, C and Deiss, V (accepted b). Consciousness, unconsciousness and death in the context of slaughter. Part 2. Evaluation methods. Meat Science.

42. Verhoeven, M, Gerritzen, M, Hellebrekers, L and Kemp, B 2015a. Indicators used in livestock to assess unconsciousness after stunning: a review. Animal 9, 320–330. [CJO Abstract]

43. Verhoeven, M, Gerritzen, M, Kluivers-Poodt, M, Hellebrekers, L and Kemp, B 2015b. Validation of behavioural indicators used to assess unconsciousness in sheep. Research in Veterinary Science 101, 144–153.

44. Vimini, R, Field, R, Riley, M and Varnell, T 1983. Effect of delayed bleeding after captive bolt stunning on heart activity and blood removal in beef cattle. Journal of Animal Science 57, 628–631.

45. Zulkifli, I, Goh, Y, Norbaiyah, B, Sazili, A, Lotfi, M, Soleimani, A and Small, A 2014. Changes in blood parameters and electroencephalogram of cattle as affected by different stunning and slaughter methods in cattle. Animal Production Science 54, 187–193.

INDEX

E

F

G

H

I

J

L

M

N

O

P

Q

R

S

T

U

V

W